# ABOUT VECTORS

*As far as the laws of mathematics*
*refer to reality, they are not certain;*
*and as far as they are certain,*
*they do not refer to reality.*

ALBERT EINSTEIN

BANESH HOFFMANN
*Professor of Mathematics*
*Queens College of the City University of New York*

# ABOUT VECTORS

DOVER PUBLICATIONS, INC., NEW YORK

Books by Banesh Hoffmann

*The Strange Story of the Quantum*
*The Tyranny of Testing*
*About Vectors*
*Albert Einstein, Creator and Rebel*
(with Helen Dukas)



Published in Canada by General Publishing Company, Ltd., 30 Lesmill Road, Don Mills, Toronto, Ontario.
Published in the United Kingdom by Constable and Company, Ltd., 10 Orange Street, London WC 2.

This Dover edition, first published in 1975, is an unabridged, corrected republication of the work originally published by Prentice-Hall, Inc., New Jersey, in 1966.

*International Standard Book Number: 0-486-60489-6*
*Library of Congress Catalog Card Number: 75-11081*

Manufactured in the United States of America
Dover Publications, Inc.
180 Varick Street
New York, N.Y. 10014

# PREFACE

This book is written as much to disturb and annoy as to instruct. Indeed, it seeks to instruct primarily by being disturbing and annoying, and it is often deliberately provocative. If it should cause heated discussion and a re-examination of fundamentals in classroom and mathematics club it will have achieved one of its main purposes.

It is intended as a supplement and corrective to textbooks, and as collateral reading in all courses that deal with vectors. Because the exercises call for no great manipulative skill, and the book avoids using the calculus, it may at first sight seem to be elementary. But it is not. It has something for the beginner, to be sure. But it also has something for quite advanced students—and something, too, for their instructors.

I have tried to face awkward questions rather than achieve a spurious simplicity by sweeping them under the rug. To counteract the impression that axioms and definitions are easily come by and that mathematics is a thing of frozen beauty rather than something imperfect and growing, I have mixed pure and applied mathematics and have made the problem of defining vectors a developing, unresolved *leitmotif*. The book is unconventional, and to describe it further here would be to blunt its intended effect by giving away too much of the plot. A brief word of warning will not be amiss, however. There are no pat answers in this book. I often present ideas in conventional form only to show later that they need modification because of unexpected difficulties, my aim being to induce a healthy skepticism. But too much healthy skepticism can be decidedly unhealthy. The reader should therefore realize that the ideas could have been presented far more hearteningly as a sequence of ever-deepening insights and, thus, of successive mathematical triumphs rather than defeats. If he reads between the lines he will see that, in a significant sense, they are indeed so presented.

To my friends Professors Arthur B. Brown and Václav Hlavatý, who read the manuscript, go my warmest thanks. It is impossible to express the depth of my indebtedness to them for their penetrating comments, which have led to major improvements in the text. They should not be held accountable for the views expressed in the book: on some issues I resisted the urgent advice of one or the other of them. A ground-breaking book of this sort is unlikely to be free of debatable views and outright errors, and for all of these I bear the sole responsibility.

BANESH HOFFMANN

# CONTENTS

# 1

# INTRODUCING VECTORS

## 1. DEFINING A VECTOR

Making good definitions is not easy. The story goes that when the philosopher Plato defined *Man* as "a two-legged animal without feathers," Diogenes produced a plucked cock and said "Here is Plato's man." Because of this, the definition was patched up by adding the phrase "and having broad nails"; and there, unfortunately, the story ends. But what if Diogenes had countered by presenting Plato with the feathers he had plucked?

**Exercise 1.1** What? [Note that Plato would now have feathers.]

**Exercise 1.2** Under what circumstances could an elephant qualify as a man according to the above definition?

A *vector* is often defined as *an entity having both magnitude and direction.* But that is not a good definition. For example, an arrow-headed line segment like this

has both magnitude (its length) and direction, and it is often used as a drawing of a vector; yet it is not a vector. Nor is an archer's arrow a vector, though it, too, has both magnitude and direction.

To define a vector we have to add to the above definition something analogous to "and having broad nails," and even then we shall find ourselves not wholly satisfied with the definition. But it will let us start, and we can try patching up the definition further as we proceed—and we may even find ourselves replacing it by a quite different sort of definition later on. If, in the end, we have the uneasy feeling that we have still not found a completely satisfactory definition of a vector, we need not be dismayed, for it is the nature of definitions not to be completely satisfactory, and we shall have learned pretty well what a vector is anyway, just as we know, without being able to give a satisfactory definition, what a man is—well enough to be able to criticize Plato's definition.

**Exercise 1.3** Define a *door*.

**Exercise 1.4** Pick holes in your definition of a *door*.

**Exercise 1.5** According to your definition, is a movable partition between two rooms a door?

## 2. THE PARALLELOGRAM LAW

The main thing we have to add to the magnitude-and-direction definition of a vector is the following:

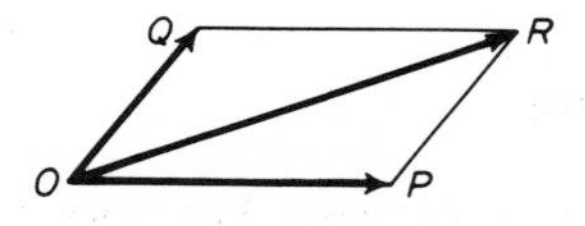


**Figure 2.1**

Let us think of vectors as having definite locations. And let the arrow-headed line segments $\overrightarrow{OP}$ and $\overrightarrow{OQ}$ in Figure 2.1 represent the magnitudes, directions, and locations of two vectors starting at a common point $O$. Complete the parallelogram formed by $\overrightarrow{OP}$ and $\overrightarrow{OQ}$, and draw the diagonal $OR$. Then, when taken together, the two vectors represented by $\overrightarrow{OP}$ and $\overrightarrow{OQ}$ are equivalent to a single vector represented by the arrow-headed line segment $\overrightarrow{OR}$. This vector is called the *resultant* of the vectors represented by $\overrightarrow{OP}$ and $\overrightarrow{OQ}$, and the above crucial property of vectors is called the *parallelogram law* of combination of vectors.

**Exercise 2.1** Find (*a*) by drawing and measurement, and (*b*) by calculation using Pythagoras' theorem, the magnitude and direction of the resultant of two vectors $\overrightarrow{OP}$ and $\overrightarrow{OQ}$ if each has magnitude 3, and $\overrightarrow{OP}$ points thus $\longrightarrow$ while $\overrightarrow{OQ}$ points perpendicularly, thus $\uparrow$. [*Ans.* The magnitude is $3\sqrt{2}$, or approximately 4.2, and the direction bisects the right angle between $\overrightarrow{OP}$ and $\overrightarrow{OQ}$.]

**Exercise 2.2** Show that the resultant of two vectors $\overrightarrow{OP}$ and $\overrightarrow{OQ}$ that point in the same direction is a vector pointing in the same direction and having a magnitude equal to the sum of the magnitudes of $\overrightarrow{OP}$ and $\overrightarrow{OQ}$. [Imagine the parallelogram in Figure 2.1 squashed flat into a line.]

**Exercise 2.3** Taking a hint from Exercise 2.2, describe the resultant of two vectors $\overrightarrow{OP}$ and $\overrightarrow{OQ}$ that point in opposite directions.

**Exercise 2.4** In Exercise 2.3, what would be the resultant if $\overrightarrow{OP}$ and $\overrightarrow{OQ}$ had equal magnitudes? [Do you notice anything queer when you compare this resultant vector with the definition of a vector?]

**Exercise 2.5** Observe that the resultant of $\overrightarrow{OP}$ and $\overrightarrow{OQ}$ is the same as the resultant of $\overrightarrow{OQ}$ and $\overrightarrow{OP}$. [This is trivially obvious, but keep it in mind nevertheless. We shall return to it later.]

In practice, all we need to draw is half the parallelogram in Figure 2.1—either triangle $OPR$ or triangle $OQR$. When we do this it looks as if we had combined two vectors $\overrightarrow{OP}$ and $\overrightarrow{PR}$ (or $\overrightarrow{OQ}$ and $\overrightarrow{QR}$) end-to-end like this, even

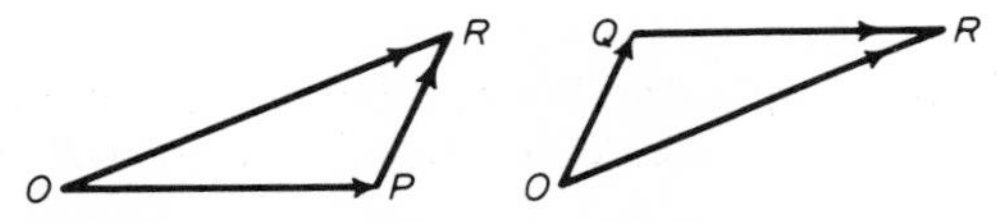


**Figure 2.2** (For clarity, the arrow heads meeting at R have been slightly displaced. We shall occasionally displace other arrow heads under similar circumstances.)

though they do not have the same starting point. Actually, though, we have merely combined $\overrightarrow{OP}$ and $\overrightarrow{OQ}$ by the parallelogram law.* But suppose we were dealing with what are called *free vectors*—vectors having the freedom to move from one location to another, so that $\overrightarrow{OP}$ and $\overrightarrow{QR}$ in Figure 2.2, for example, which have the same magnitude and the same direction, are officially counted not as distinct vectors but as the same free vector. Then we could indeed combine free vectors that were quite far apart by bringing them end-to-end, like $\overrightarrow{OP}$ and $\overrightarrow{PR}$ in Figure 2.2. But since we could also combine them according to the parallelogram law by moving them so that they have a common starting point, like $\overrightarrow{OP}$ and $\overrightarrow{OQ}$ in Figure 2.1, the parallelogram law is the basic one. Note that when we speak of the same direction we mean just that, and not opposite directions—north and south are not the same direction.

*Have you noticed that we have been careless in sometimes speaking of "the vector represented by $\overrightarrow{OP}$," at other times calling it simply "the vector $\overrightarrow{OP}$," and now calling it just "$\overrightarrow{OP}$"? This is deliberate—and standard practice among mathematicians. Using meticulous wording is sometimes too much of an effort once the crucial point has been made.

**Exercise 2.6** Find the resultant of the three vectors $\overrightarrow{OA}$, $\overrightarrow{OB}$, a $\overrightarrow{OC}$ in the diagram.

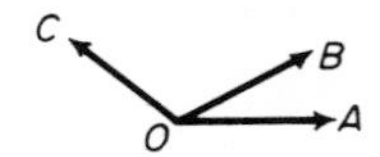


*Solution* We first form the resultant, $\overrightarrow{OR}$, of $\overrightarrow{OA}$ and $\overrightarrow{OB}$ like this:

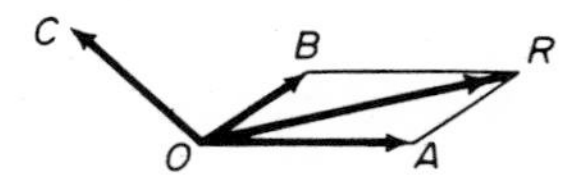


and then we form the resultant, $\overrightarrow{OS}$, of $\overrightarrow{OR}$ and $\overrightarrow{OC}$ like this:

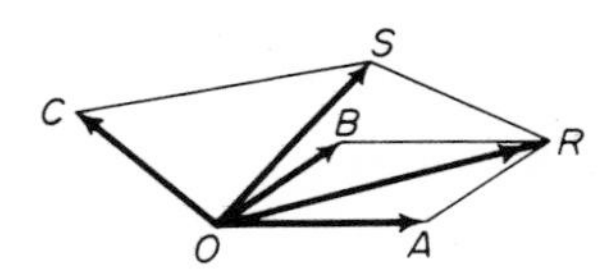


This figure looks complicated. We can simplify it by drawing only half of each parallelogram, and then even omitting the line $OR$, like this:

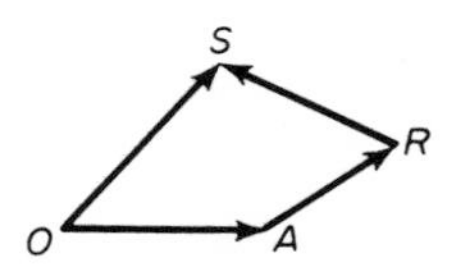


From this we see that the resultant $\overrightarrow{OS}$ can be found quickly by thinking of the vectors as free vectors and combining them by placing them end-to-end: $\overrightarrow{AR}$, which has the same magnitude and direction as $\overrightarrow{OB}$, starts where $\overrightarrow{OA}$ ends; and then $\overrightarrow{RS}$, which has the same magnitude and direction as $\overrightarrow{OC}$, starts where $\overrightarrow{AR}$ ends.

**Exercise 2.7** Find, by both methods, the resultant of the vectors in Exercise 2.6, but by combining $\overrightarrow{OB}$ and $\overrightarrow{OC}$ first, and then combining their resultant with $\overrightarrow{OA}$. Prove geometrically that the resultant is the same as before.

**Exercise 2.8**

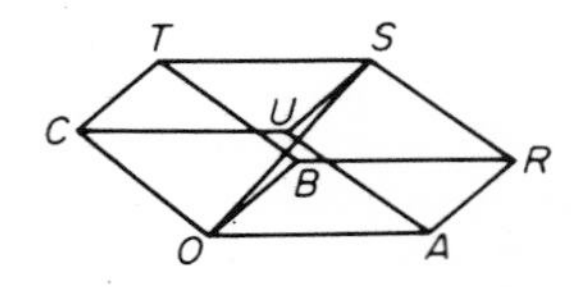


The above diagram looks like a drawing of a box. Show that if we drew only the lines $OA$, $AR$, $RS$, and $OS$ we would have essentially the last figure in Exercise 2.6; that if we drew only the lines $OB$, $BT$, $TS$, and $OS$ we would have a corresponding figure for Exercise 2.7; and that if we drew only $OA$, $AU$, $US$, and $OS$ we would have a figure corresponding to our having first combined $\overrightarrow{OA}$ with $\overrightarrow{OC}$ and then their resultant with $\overrightarrow{OB}$.

**Exercise 2.9** In Exercises 2.6, 2.7, and 2.8, is it essential that the three vectors $\overrightarrow{OA}$, $\overrightarrow{OB}$, and $\overrightarrow{OC}$ lie in a plane? Give a rule for finding the resultant of three noncoplanar vectors $\overrightarrow{OA}$, $\overrightarrow{OB}$, and $\overrightarrow{OC}$ that is analogous to the parallelogram law, and that might well be called the parallelepiped law. Prove that their resultant is the same regardless of the order in which one combines them.

**Exercise 2.10** Find the resultant of the three vectors $\overrightarrow{OA}$, $\overrightarrow{OB}$, and $\overrightarrow{OC}$ below by combining them in three different orders, given that vectors $\overrightarrow{OA}$ and $\overrightarrow{OC}$ have equal magnitudes and opposite directions. Draw both the end-to-end diagrams and the full parallelogram diagrams for each case.

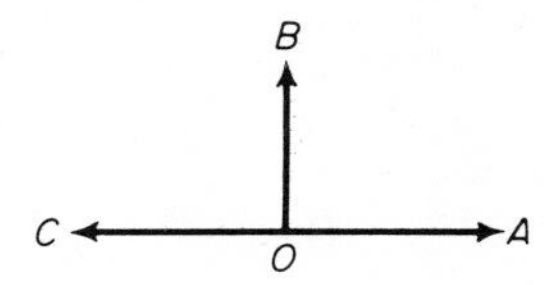


## 3. JOURNEYS ARE NOT VECTORS

It is all very well to start with a definition. But it is not very enlightening. Why should scientists and mathematicians be interested in objects that have magnitude and direction and combine according to the parallelogram law? Why did they even think of such objects? Indeed, do such objects exist at all —outside of the imaginations of mathematicians?

There are, of course, many objects that have both magnitude and direction. And there are, unfortunately, many books about vectors that give the reader the impression that such objects obviously and inevitably obey the parallelogram law. It is therefore worthwhile to explain carefully why most such objects do not obey this law, and then, by a process of abstraction, to find objects that do.

Suppose that I live at $A$ and my friend lives 10 miles away at $B$. I start from $A$ and walk steadily at 4 m.p.h. for $2\frac{1}{2}$ hours. Obviously, I walk 10 miles. But do I reach $B$?

You may say that this depends on the direction I take. But what reason is there to suppose that I keep to a fixed direction? The chances are overwhelming that I do not—unless I am preceded by a bulldozer or a heavy tank.

Most likely I walk in all sorts of directions; and almost certainly, I do not arrive at $B$. I may even end up at home.

**Exercise 3.1** Where are all the possible places at which I can end, under the circumstances?

Now suppose that I start again from $A$ and this time end up at $B$. I may take four or five hours, or I may go by bus or train and get there quickly. Never mind how I travel or how long I take. Never mind how many times I change my direction, or how tired I get, or how dirty my shoes get, or whether it rained. Ignore all such items, important though they be, and consider the abstraction that results when one concentrates solely on the fact that I start at $A$ and end at $B$. Let us give this abstraction a name. What shall we call it? Not a "journey." That word reminds us too much of everyday life—of rain, and umbrellas, and vexations, and lovers meeting, and all other such items that we are ignoring here; besides, we want to preserve the word "journey" for just such an everyday concept. For our abstraction we need a neutral, colorless word. Let us call it a *shift*.

Here are routes of four journeys from $A$ to $B$:

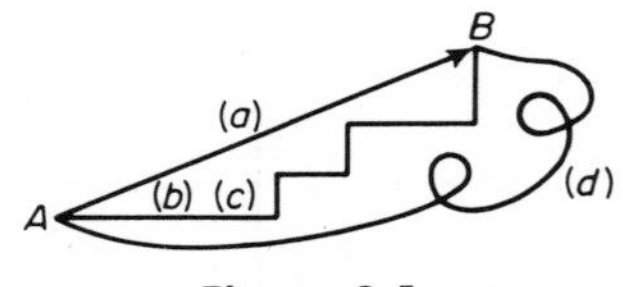


**Figure 3.1**

All four journeys are different—with the possible but highly improbable exception of ($b$) and ($c$).

**Exercise 3.2** Why "highly improbable"?

But though the four journeys are not all the same, they yield the same shift. We can represent this shift by the arrow-headed line segment $AB$. It has both magnitude and direction. Indeed, it seems to have little else. Is it a vector? Let us see.

Consider three places $A$, $B$, and $C$ as in Figure 3.2. If I walk in a straight

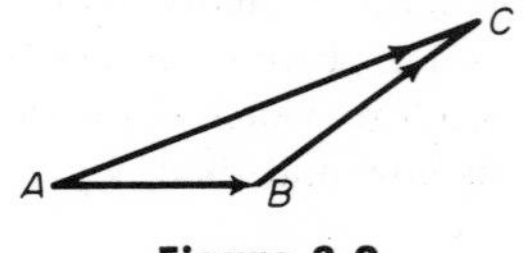


**Figure 3.2**

line from $A$ to $B$ and then in a straight line from $B$ to $C$, I make a journey from $A$ to $C$, but it is not the same as if I walked directly in a straight line from $A$ to $C$: the scenery is different, and so is the amount of shoe leather consumed, most likely, and we can easily think of several other differences.

**Exercise 3.3** Why "most likely"?

Thus, though we could say that the walks from $A$ to $B$ and from $B$ to $C$ combine to give a "resultant" journey from $A$ to $C$, it is not a journey in a straight line from $A$ to $C$: the walks do not combine in a way reminiscent of the way in which vectors combine; they combine more in the tautological sense that $2 + 1 = 2 + 1$ than $2 + 1 = 3$.

Journeys, then, are not vectors. But when we deal with shifts we ignore such things as the scenery and the amount of shoe leather consumed. A shift from $A$ to $B$ followed by a shift from $B$ to $C$ is indeed equivalent to a shift from $A$ to $C$. And this reminds us so strongly of the vectorial situation in Figure 2.2 that we are tempted to conclude that shifts are vectors. But there is a crucial difference between the two situations. We cannot combine the above shifts in the reverse order (compare Exercise 2.5). There is no single equivalent to the shift from $B$ to $C$ followed by the shift from $A$ to $B$. We can combine two shifts only when the second begins where the first ends. Indeed, in Figure 2.1, just as with journeys, we cannot combine a shift from $O$ to $P$ with one from $O$ to $Q$ in either order. Thus shifts are not vectors.

## 4. DISPLACEMENTS ARE VECTORS

Now that we have discovered why shifts are not vectors, we can easily see what further abstraction to make to obtain entities that are. From the already abstract idea of a shift, we remove the actual starting point and end point and retain only the relation between them: that $B$ lies such and such a distance from $A$ and in such and such a direction.* Shifts were things we invented in order to bring out certain distinctions. But this new abstraction is an accepted mathematical concept with a technical name: it is called a *displacement*. And it is a vector, as we shall now show.

In Figure 4.1, the arrow-headed line segments $AB$ and $LM$ are parallel and

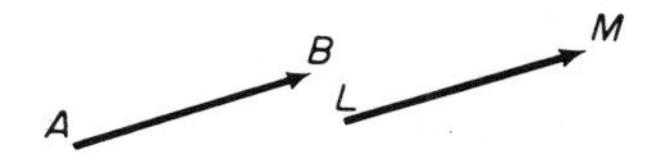


**Figure 4.1**

of equal length. Any journey from $A$ to $B$ is bound to be different from a journey from $L$ to $M$. Also, the shift from $A$ to $B$ is different from that from $L$ to $M$ because the starting points are different, as are the end points. But the two shifts, and thus also the various journeys, yield the same displacement: if, for example, $B$ is 5 miles north–northeast of $A$, so too is $M$ 5 miles north–northeast of $L$, and the displacement is one of 5 miles in the direction north–northeast.

*We retain, too, the recollection that we are still linked, however tenuously, with journeying, for we want to retain the idea that a movement has occurred, even though we do not care at all *how* or under what circumstances it occurred.

**Exercise 4.1** Starting from a point $A$, a man bicycles 10 miles due east to point $B$, stops for lunch, sells his bicycle, and then walks 10 miles due north to point $C$. Another man starts from $B$, walks 4 miles due north and 12 miles due east and then, feeling tired, and having brought along a surplus of travellers' checks, buys a car and drives 6 miles due north and 2 miles due west, ending at point $D$ in the pouring rain. What displacement does each man undergo? [*Ans.* $10\sqrt{2}$ miles to the northeast.]

Now look at Figure 2.1. The shift from $O$ to $P$ followed by the shift from $P$ to $R$ is equivalent to the shift from $O$ to $R$. The shift from $P$ to $R$ gives a displacement $\overrightarrow{PR}$ that is the same as the displacement $\overrightarrow{OQ}$. Therefore the displacement $\overrightarrow{OP}$ followed by the displacement $\overrightarrow{OQ}$ is equivalent to the displacement $\overrightarrow{OR}$.

**Exercise 4.2** Prove, similarly, that the displacement $\overrightarrow{OQ}$ followed by the displacement $\overrightarrow{OP}$ is also equivalent to the displacement $\overrightarrow{OR}$.

Thus, displacements have magnitude and direction and combine according to the parallelogram law. According to our definition, they are therefore vectors. Since displacements such as $\overrightarrow{AB}$ and $\overrightarrow{LM}$ in Figure 4.1 are counted as identical, displacements are free vectors, and thus are somewhat special. In general, vectors such as $\overrightarrow{AB}$ and $\overrightarrow{LM}$ are not counted as identical.

## 5. WHY VECTORS ARE IMPORTANT

From the idea of a journey we have at last come, by a process of successive abstraction, to a specimen of a vector. The question now is whether we have come to anything worthwhile. At first sight it would seem that we have come to so pale a ghost of a journey that it could have little mathematical significance. But we must not underestimate the potency of the mathematical process of abstraction. Vectors happen to be extremely important in science and mathematics. A surprising variety of things happen to have both magnitude and direction and to combine according to the parallelogram law; and many of them are not at all reminiscent of journeys.

This should not surprise us. The process of abstraction is a powerful one. It is, indeed, a basic tool of the mathematician. Take whole numbers, for instance. Like vectors, they are abstractions. We could say that whole numbers are what is left of the idea of *apples* when we ignore not only the apple trees, the wind and the rain, the profits of cider makers, and other such items that would appear in an encyclopedia article, *but also ignore even the apples themselves*, and concentrate solely on how many there are. After we have extracted from the idea of apples the idea of whole numbers, we find that whole numbers apply to all sorts of situations that have nothing to do with apples. Much the same is true of vectors. They are more complicated than whole numbers—so are fractions, for example—but they happen to embody an important type of mathematical behavior that is widely encountered in the world around us.

To give a single example here: forces behave like vectors. This is not something obvious. A force has both magnitude and direction, of course. But this does not mean that forces necessarily combine according to the parallelogram law. That they do combine in this way is inferred from experiment.

It is worthwhile to explain what is meant when we say that forces combine according to the parallelogram law. Forces are not something visible, though their effects may be visible. They are certainly not arrow-headed line segments, though after one has worked with them mathematically for a while, one almost comes to think they are. A force can be represented by an arrow-headed line segment $\overrightarrow{OP}$ that starts at the point of application $O$ of the force, points in the direction of the force, and has a length proportional to the magnitude of the force—for example, a length of $x$ inches might represent a magnitude of $x$ pounds. When a force is represented in this way, we usually avoid wordiness by talking of "the force $\overrightarrow{OP}$." But let us be more meticulous in our wording just here. To verify experimentally that forces combine according to the parallelogram law, we can make the following experiment. We arrange stationary weights and strings, and pulleys $A$ and $B$, as shown, the weight $W$ being the

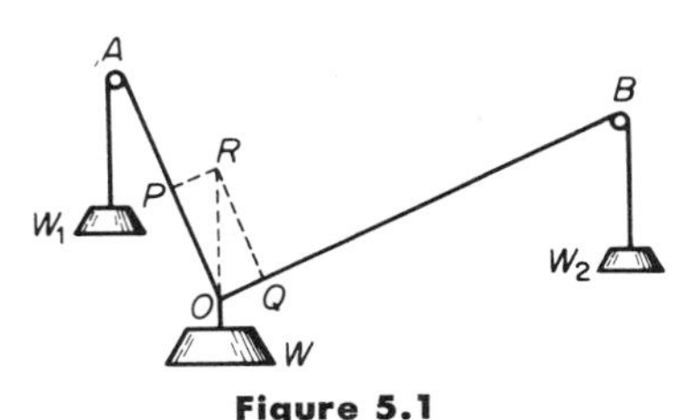


**Figure 5.1**

sum of the weights $W_1$ and $W_2$. Then along $OA$ we mark off a length $OP$ of $W_1$ inches, where $W_1$ is the number of pounds in the weight on the left and, thus, a measure of the force with which the string attached to it pulls on the point $O$ where the three pieces of string meet. Similarly, we mark off on $OB$ a length $OQ$ of $W_2$ inches. We then bring a vertical piece of paper up to the point $O$, and on it complete the parallelogram defined by $OP$ and $OQ$. We find that the diagonal $OR$ is vertical and that its length in inches is $W$, the number of pounds in the weight in the middle. We conclude that the resultant of the forces $W_1$ and $W_2$ in the strings would just balance the weight $W$. Since the forces $W_1$ and $W_2$ also just balance the weight $W$, we say that the resultant is equivalent to the two forces. We then do the experiment over again, with different weights, and reach a similar conclusion. After that, we do it yet again; and we keep at it till our lack of patience overcomes our skepticism, upon which we say that we have proved experimentally that forces combine according to the parallelogram law. And we bolster our assertion by pointing to other experiments, of the same and different types, that indicate the same thing.

We all know that it is much easier to get through a revolving door by pushing near the outer edge than by pushing near the central axis. The effect of a force depends on its location. Home runs are scarce when the bat fails to make contact with the ball. Thus forces do not behave like free vectors. Unlike displacements, vectors representing forces such as $\overrightarrow{AB}$ and $\overrightarrow{LM}$ in Figure 4.1, though they have the same magnitude and the same direction, are not counted as equivalent. Such vectors are called *bound* vectors.

Perhaps it worries us a little that there are different kinds of vectors. Yet we have all, in our time, survived similar complications. Take numbers, for example. There are whole numbers and there are fractions. Perhaps you feel that there is not much difference between the two. Yet if we listed the properties of whole numbers and the properties of fractions we would find considerable differences. For instance, if we divide fractions by fractions the results are always fractions, but this statement does not remain true if we replace the word "fractions" by "whole numbers." Worse, every whole number has a next higher one, but no fraction has a next higher fraction, for between any two fractions we can always slip infinitely many others. Even so, when trying to define *number* we might be inclined to insist that, given any two different numbers, one of them will always be the smaller and the other the larger. Yet when we become more sophisticated and expand our horizons to include complex numbers like $2 + 3\sqrt{-1}$, we have to give up even this property of being greater or smaller, which at first seemed an absolutely essential part of the idea of number. With vectors too, not only are there various types, but we shall learn that not every one of their attributes that seems at this stage to be essential is in fact so. One of the things that gives mathematics its power is the shedding of attributes that turn out not to be essential, for this, after all, is just the process of abstraction.

**Exercise 5.1** Find the resultants of the following displacements:

(a) 3 ft. due east and 3 ft. due north. [*Ans.* $3\sqrt{2}$ ft. to the northeast.]

(b) 5 ft. due north and 5 ft. due east.

(c) 9 cm. to the right and $9\sqrt{3}$ cm. vertically upwards. [*Ans.* 18 cm. in an upward direction making 60° with the horizontal towards the right.]

(d) 9 cm. to the left and $9\sqrt{3}$ cm. vertically downward.

(e) the resultants in parts (c) and (d).

(f) $x$ units positively along the $x$-axis and $y$ units positively along the $y$-axis. [*Ans.* $\sqrt{x^2 + y^2}$ units in the direction making an angle $\tan^{-1} y/x$ with the positive $x$-axis.]

**Exercise 5.2** Like Exercise 5.1 for the following:

(a) 8 km. to the left and 3 km. to the left.

(b) 5 fathoms vertically downward and 2 fathoms vertically upward.

(c) $\alpha$ units to the right and $\beta$ units to the left. [There are three different cases. What are they? Show how they can be summed up in one statement.]

(d) $h$ miles 60° north of east and $h$ miles 60° south of east.

**Exercise 5.3** What single force is equivalent to the following three horizontal forces acting on a particle at a point $O$? (1) magnitude 1 lb. pulling to the north; (2) magnitude 1 lb. pulling to the east; (3) magnitude $\sqrt{2}$ lb. pulling to the northwest. [*Ans.* 2 lbs. acting at point $O$ and pulling to the north.]

**Exercise 5.4** What force combined with a force at a point 0 of 1 lb. pulling to the east will yield a resultant force of 2 lbs. pulling in a direction 60° north of east?

**Exercise 5.5** Vector $\overrightarrow{OP}$ has magnitude $2a$ and points to the right in a direction 60° above the horizontal. What vector combined with it will yield a vertical resultant, $\overrightarrow{OR}$, of magnitude $2\sqrt{3}\,a$?

**Exercise 5.6** Find two forces at a point $O$, one vertical and one horizontal, that have a resultant of magnitude $h$, making 45° with the horizontal force. [*Ans.* The forces have magnitude $h/\sqrt{2}$.]

**Exercise 5.7** Find two forces at a point $O$, one vertical and one horizontal, that have a resultant of magnitude $h$ that makes an angle of 30° with the horizontal force.

**Exercise 5.8** Find two displacements, one parallel to the $x$-axis and the other to the $y$-axis, that yield a resultant displacement of magnitude $h$ ft. making a positive acute angle $\alpha$ with the positive $x$-direction.

**Exercise 5.9** What is the resultant of $n$ vectors, each starting at the point $O$, each having magnitude $h$, and each pointing to the pole star? [We could have shortened this by asking for the resultant of $n$ equal vectors. But we have not yet defined "equal" vectors—even though we have already spoken of the equality of free vectors! You may find it instructive to try to do so here; but be warned that it is not as easy as it seems, and that there is something lacking in the wording of the question.]

**Exercise 5.10** A particle is acted on by two forces, one of them to the west and of magnitude 1 dyne, and the other in the direction 60° north of east and of magnitude 2 dynes. What third force acting on the particle would keep it in equilibrium (i. e., what third force would make the resultant of all three forces have zero magnitude)? [*Ans.* Magnitude $\sqrt{3}$ dynes pointing due south.]

## 6. THE CURIOUS INCIDENT OF THE VECTORIAL TRIBE

It is rumored that there was once a tribe of Indians who believed that arrows are vectors. To shoot a deer due northeast, they did not aim an arrow in the northeasterly direction; they sent two arrows simultaneously, one due north and the other due east, relying on the powerful resultant of the two arrows to kill the deer.

Skeptical scientists have doubted the truth of this rumor, pointing out that not the slightest trace of the tribe has ever been found. But the complete disappearance of the tribe through starvation is precisely what one would expect under the circumstances; and since the theory that the tribe existed confirms two such diverse things as the NONVECTORIAL BEHAVIOR OF ARROWS and the DARWINIAN PRINCIPLE OF NATURAL SELECTION, it is surely not a theory to be dismissed lightly.

**Exercise 6.1** Arrow-headed line segments have magnitude and direction and are used to represent vectors. Why are they nevertheless not vectors?

**Exercise 6.2** Given the three vectors represented by $\overrightarrow{OP}$, $\overrightarrow{OQ}$, and $\overrightarrow{OR}$ in Figure 2.1, form three new entities having the same respective directions, but having magnitudes equal to five times the magnitudes of the respective vectors. Prove geometrically that these new entities are so related that the third is a diagonal of the parallelogram having the other two as adjacent sides.

**Exercise 6.3** If in Exercise 6.2 the new entities had the same respective directions as the vectors represented by $\overrightarrow{OP}$, $\overrightarrow{OQ}$, and $\overrightarrow{OR}$, but had magnitudes that were one unit greater than the magnitudes of the corresponding vectors, show that the new entities would not be such that the third was a diagonal of the parallelogram having the other two as adjacent sides.

**Exercise 6.4** Suppose we represented vectors by arrow-headed line segments that had the same starting points and directions as the vectors, but had lengths proportional to the squares of the magnitudes of the vectors, so that, for example, if a force of 1 lb. were represented by a segment of length 1 inch, then a force of 2 lbs. would be represented by one of 4 inches. Show that, in general, these representations of vectors would not obey the parallelogram law. Note that the statement of the parallelogram law in Section 2 therefore needs amending, and amend it accordingly. [If you think carefully, you will realize that this is a topsy-turvy question since, in proving the required result, you will assume that the vectors, when "properly" represented, obey the parallelogram law; and thus, in a sense, you will assume the very amendment you are seeking. But since you have probably been assuming the amendment all this while, you will be able to think your way through. The purpose of this exercise is to draw your attention to this rarely mentioned, usually assumed amendment.]

## 7. SOME AWKWARD QUESTIONS

When are two vectors equal? The answer depends on what we choose to mean by the word "equal"—we are the masters, not the word. But we do not want to use the word in an outrageous sense: for example, we would not want to say that two vectors are equal if they are mentioned in the same sentence.

Choosing a meaning for the word "equal" here is not as easy as one might imagine. For example, we could reasonably say that two vectors having the same magnitudes, identical directions, and a common starting point are equal vectors. And if one of the vectors were somehow pink and the other green, we would probably be inclined to ignore the colors and say that the vectors were still equal. But what if one of the vectors represented a force and the other a displacement? There would then be two difficulties.

The first difficulty is that the vector representing a displacement would be a free vector, but the one representing the force would not. If, in Figure 4.1, we counted free vectors represented by $\overrightarrow{AB}$ and $\overrightarrow{LM}$ as equal, we might find ourselves implying that forces represented by $\overrightarrow{AB}$ and $\overrightarrow{LM}$ were also equal, though actually they have different effects. [Even so, it is extremely convenient to say such things as "a force acts at $A$ and an equal force acts at $L$." We shall not do so in this book. But one can get by with saying such things once one has explained what is awkward about them, just as, in trigonometry, one gets by with writing $\sin^2 \theta$ after one has explained that this does not stand for $\sin(\sin \theta)$ but $(\sin \theta)^2$.]

As for the second difficulty about the idea of the equality of vectors, it takes us back to the definition of a vector. For if, in Figure 2.1, $\overrightarrow{OP}$ represents a force and $\overrightarrow{OQ}$ a displacement, the two vectors will not combine by the parallelogram law at all. We know this from experiments with forces. But we can appreciate the awkwardness of the situation by merely asking ourselves what the resultant would be if they did combine in this way. A "disforcement"?* [Compare Exercise 5.9.]

If two vectors are to be called equal, it seems reasonable to require that they be able to combine with each other. The situation is not the same as it is with numbers. Although 3 apples and 3 colors are different things, we can say that the numbers 3 are equal in the sense that, if we assign a pebble to each of the apples, these pebbles will exactly suffice for doing the same with the colors. And in this sense we can indeed combine 3 apples and 3 colors—not to yield 6 apples, or 6 colors, or 6 colored apples [it would surely be only 3 colored apples], but 6 *items*. There does not seem to be a corresponding sense in which we could reasonably combine a vector representing a force with one representing a displacement, quite apart from the question of bound versus free vectors: there does not seem to be a vectorial analogue of the numerical

*Actually, of course, lack of a name proves no more than that if the resultant exists, it has not hitherto been deemed important enough to warrant a name.

concept of a countable item.*

Though $\overrightarrow{OP}$ and $\overrightarrow{OQ}$ do not combine according to the parallelogram law if, for example, $\overrightarrow{OP}$ represents a force and $\overrightarrow{OQ}$ a displacement, they nevertheless represent vectors. Evidently our definition of a vector needs even further amendment. We might seek to avoid trouble by retreating to the definition of a vector as "an entity having both magnitude and direction," without mentioning the parallelogram law. But once we start retreating, where do we stop? Why not be content to define a vector as "an entity having direction," or as "an entity having magnitude," or, with Olympian simplicity, as just "an entity"? Alternatively, we could make the important distinction between the abstract mathematical concept of a vector and entities, such as forces, that behave like these abstract vectors and are called *vector quantities*. This helps, but it does not solve the present problem so much as sweep it under the rug. We might amend our definition of a vector by saying that vectors combine according to the parallelogram law only with vectors of the same kind: forces with forces, displacements with displacements, accelerations (which are vectors) with accelerations, and so on. But even that is tricky since, for example, in dynamics we learn that force equals mass times acceleration. So we would have to allow for the fact that though a force does not combine with an acceleration, it does combine with a vector of the type mass-times-acceleration in dynamics.

We shall return to this matter. (See Section 8 of Chapter 2.) But enough of such questions here. If we continue to fuss with the definition we shall never get started. Even if we succeeded in patching up the definition to meet this particular emegency, other emergencies would arise later. The best thing to do is to keep an open mind and learn to live with a flexible situation, and even to relish it as something akin to the true habitat of the best research.

*Even with numbers there are complications. For example, 3 ft. and 3 inches can be said to yield 6 items; yet in another sense they yield 39 inches, $3\frac{1}{4}$ ft., and so on—and each of these can also be regarded as a number of items, though the $3\frac{1}{4}$ involves a further subtlety. Consider also 3 ft. and 3 lbs., and then 2.38477 ft. and 2.38477 lbs.

# 2

# ALGEBRAIC NOTATION AND BASIC IDEAS

## 1. EQUALITY AND ADDITION

Instead of denoting vectors by symbols like $\overrightarrow{OA}$ and $\overrightarrow{PQ}$, it is often convenient to denote them by single letters printed in bold-face type, like this: **U**, **V**. (In written work, some people place semi-arrows above ordinary capital letters, as $\vec{U}$, $\vec{V}$. Others write double lines on the letters, like this: $\mathbb{U}$, $\mathbb{V}$. Yet others place lines underneath the letters: $\underline{U}$, $\underline{V}$.)

Let us agree to say that two vectors, **U** and **V**, are *equal* if they have equal magnitudes, identical directions, and a common starting point, and represent types of entities that can combine with each other according to the parallelogram law. If they are free vectors we need not insist that they have a common starting point.

We denote this equality by the familiar symbol =, writing, for example,

$$\overrightarrow{OA} = \overrightarrow{OB}, \qquad \mathbf{U} = \mathbf{V}. \tag{1.1}$$

We shall also use this symbol = in the broader sense of "equivalent to" as the phrase was used in the statement of the parallelogram law in Chapter 1. Also, of course, we shall use the symbol in its usual arithmetical and algebraic sense, as in $2 + 2 = 4$.

**Exercise 1.1** In the definition of equal vectors, what do you think we mean by "equal magnitudes"? Consider, for example, magnitudes of 6 ft., 6 lbs., and 2 yds.

The magnitude of a vector $\overrightarrow{OA}$ is usually denoted by $|OA|$ or just plain $OA$, and that of a vector **V** by $V$.

**Exercise 1.2** If **U** and **V** have the same starting point and *the same direction*, show that the magnitude of their resultant is $U + V$.

Note that we did not say in Exercise 1.2 that **U** and **V** were capable of having a resultant: **U** might have been a force and **V** a displacement, for example. Let us agree to assume that vectors in a given problem or discussion can combine unless the contrary is either obvious or else explicitly stated.

When vectors are related as in Exercise 1.2, their resultant is akin to a sum. For this reason, and for others that will appear later, it is appropriate to denote the combination of vectors according to the parallelogram law by the familiar symbol +. Thus, if $\overrightarrow{OR}$ is the resultant of $\overrightarrow{OP}$ and $\overrightarrow{OQ}$, as in Figure 2.1 of Chapter 1, we write:

$$\overrightarrow{OP} + \overrightarrow{OQ} = \overrightarrow{OR}, \tag{1.2}$$

and we even use arithmetical language, saying that we *add* the vectors $\overrightarrow{OP}$ and $\overrightarrow{OQ}$ and that their *sum* is $\overrightarrow{OR}$.

From Exercise 2.5 of Chapter 1, we see that

$$\mathbf{U} + \mathbf{V} = \mathbf{V} + \mathbf{U}. \tag{1.3}$$

Technically, we refer to this fact by saying that vectorial addition is *commutative*.

**Exercise 1.3** Show geometrically that vectorial addition is *associative*, that is,

$$(\mathbf{U} + \mathbf{V}) + \mathbf{W} = \mathbf{U} + (\mathbf{V} + \mathbf{W}). \tag{1.4}$$

Also write out what this relation says using the language of Chapter 1, e. g., resultant, combination, etc. [Note that we have not defined the symbol "("and the symbol")": their significance is easily guessed here.]

## 2. MULTIPLICATION BY NUMBERS

The resultant of **V** and **V** is a vector identical with **V** except that its magnitude is twice as great. We denote it by 2**V**, and this allows us to write the satisfactory equation:

$$\mathbf{V} + \mathbf{V} = 2\mathbf{V}.$$

Similarly, we denote the resultant of **V**, **V**, and **V** by 3**V**, and so on.

**Exercise 2.1** Show that $2\mathbf{V} + \mathbf{V} = 3\mathbf{V}$.

**Exercise 2.2** Describe $5\mathbf{V}$ in terms of $\mathbf{V}$.

If

$$3\mathbf{U} = 2\mathbf{V}$$

we write

$$\mathbf{U} = \frac{2}{3}\mathbf{V}, \quad \text{and} \quad \mathbf{V} = \frac{3}{2}\mathbf{U};$$

more generally, if $a$ and $b$ are positive whole numbers, and

$$a\mathbf{U} = b\mathbf{V}, \tag{2.1}$$

we write

$$\mathbf{U} = \frac{b}{a}\mathbf{V} \quad \text{and} \quad \mathbf{V} = \frac{a}{b}\mathbf{U}. \tag{2.2}$$

**Exercise 2.3** Show that $3(\frac{2}{3}\mathbf{V}) = 2\mathbf{V}$. [*Hint.* Start by expressing the left-hand side in terms of $\mathbf{U}$ as in the preceding paragraph.]

**Exercise 2.4** Show that if $a$ and $b$ are positive whole numbers,

$$a(\frac{b}{a}\mathbf{V}) = b\mathbf{V}.$$

**Exercise 2.5** Describe $\frac{2}{3}\mathbf{V}$ in terms of $\mathbf{V}$.

**Exercise 2.6** Show that if $a$ and $b$ are positive whole numbers, then $(b/a)\mathbf{V}$ is a vector identical with $\mathbf{V}$ except that its magnitude is $(b/a)V$.

If $h$ is a positive number, we write $h\mathbf{V}$ for the vector that is identical with $\mathbf{V}$ except that its magnitude is $h$ times that of $\mathbf{V}$. It may seem that this is merely a repetition of the result of Exercise 2.6, and contains no more than is contained in Equations (2.1) and (2.2); for we have only to express $h$ as a fraction $b/a$, where $a$ and $b$ are whole numbers, in order to link up with those equations. But not every number can be expressed as such a fraction. Those that can are called *rational* numbers, and the amount of rational numbers is negligible compared with the amount of numbers that cannot be so expressed. The latter are called *irrational* numbers, examples being $\sqrt{2}$ and $\pi$. A book about vectors is not the place for a discussion of irrational numbers. One can do arithmetic with them much as one does with rational numbers, and the general case $h\mathbf{V}$ can be regarded as a valid extension of the ideas embodied in Equations (2.1) and (2.2). If we feel so inclined, we can amend the definition of a vector to include the idea of $h\mathbf{V}$ where $h$ is irrational. We say that $h\mathbf{V}$ is the result of *multiplying* the vector $\mathbf{V}$ by the number $h$.

**Exercise 2.7** Let the vector $\mathbf{V}$ point along the $x$-axis. Form two new vectors, the first by rotating $\mathbf{V}$ about its starting point through 45° towards the positive $y$-direction, and the second by rotating $\mathbf{V}$ similarly through 45° towards the negative $y$-direction. Show that the resultant of these two vectors is $\sqrt{2}\,\mathbf{V}$.

**Exercise 2.8** What would be the resultant if we did Exercise 2.7 starting not with **V** but with $\sqrt{2}\,\mathbf{V}$?

**Exercise 2.9** Show by means of a diagram that multiplication of vectors by numbers obeys the *distributive* law:

$$h(\mathbf{U} + \mathbf{V}) = h\mathbf{U} + h\mathbf{V}. \tag{2.3}$$

[This can not be proved just by noting its similarity to a familiar algebraic equation. For a hint, compare Exercise 6.2 of Chapter 1, which is a special case of this theorem.]

**Exercise 2.10** If **U** and **V** both start at the origin and have equal magnitudes, **U** pointing along the positive $x$-axis and **V** along the positive $y$-axis, and if $a$ and $b$ are positive numbers, what is the tangent of the angle that the vector $a\mathbf{U} + b\mathbf{V}$ makes with the $x$-axis?

## 3. SUBTRACTION

Having managed to relate the symbol + to the parallelogram law without doing violence to its usual arithmetical meaning, we wonder whether we can be as successful with the symbol −. It turns out that we can. If **V** is a vector, we define its negative, which we write $-\mathbf{V}$, as a vector identical with it except that it points in the opposite direction. Thus in Figure 3.1, if **U** is **V** turned

**Figure 3.1**

through 180°, then $\mathbf{U} = -\mathbf{V}$. Note that **V** and $-\mathbf{V}$ have the same magnitude: the magnitude is always taken to be positive (when it is not zero, of course). To consolidate the link between the arithmetical and vectorial uses of the symbol −, we define $(-h)\mathbf{V}$ for positive $h$ as $h(-\mathbf{V})$.

**Exercise 3.1** If $\mathbf{U} = -\mathbf{V}$, show that $\mathbf{V} = -\mathbf{U}$. What is $-(-\mathbf{V})$?

**Exercise 3.2** What is the resultant of $2\mathbf{V}$ and $-\mathbf{V}$? [*Ans.* **V**.]

**Exercise 3.3** Prove that if $h$ is positive, $h(-\mathbf{V}) = -(h\mathbf{V})$.

**Exercise 3.4** What is the resultant of $a\mathbf{V}$ and $-b\mathbf{V}$?

We denote a vector of zero magnitude by the symbol 0, without bothering to use bold-face type. A zero vector, as it is called, does not quite fit our definition of a vector since it has no direction or, thought of in another way, it has all directions. We count it as a vector nevertheless, and we shall not pause to patch up the definition of a vector to accomodate it.

**Exercise 3.5** What is the resultant of **V** and $-\mathbf{V}$? What is the resultant in Exercise 3.4 if $a = b$?

**Exercise 3.6** Show that $\mathbf{V} + O = O + \mathbf{V} = \mathbf{V}$.

We now define $\mathbf{U} - \mathbf{V}$ as $\mathbf{U} + (-\mathbf{V})$, and say that to *subtract* $\mathbf{V}$ we add $-\mathbf{V}$. Since this subtraction is thus essentially a type of vectorial addition, it obeys the same rules as vectorial addition. For example,

$$(\mathbf{U} + \mathbf{V}) - \mathbf{W} = \mathbf{U} + (\mathbf{V} - \mathbf{W})$$

and

$$h(\mathbf{U} - \mathbf{V}) = h\mathbf{U} - h\mathbf{V}.$$

**Exercise 3.7** Consider the discussion of the equality of vectors in Section 7 of Chapter 1 in the light of the reasonable requirement that if $\mathbf{U} = \mathbf{V}$ then $\mathbf{U} - \mathbf{V} = 0$.

**Exercise 3.8** If $\mathbf{U} + \mathbf{V} = \mathbf{W}$, show that $\mathbf{U} = \mathbf{W} - \mathbf{V}$, first by means of a diagram involving two overlapping parallelograms, and second by the purely algebraic manipulation of adding $-\mathbf{V}$ to both sides of the equation and using the results of Exercises 3.5 and 3.6.

For displacements it is easy to see that

$$\overrightarrow{AB} = -\overrightarrow{BA}, \tag{3.1}$$

though if $\overrightarrow{AB}$ were not a free vector this would, in general, not be true, since $\overrightarrow{BA}$ starts at $B$ while $\overrightarrow{AB}$ starts at $A$.

Again, the resultant of the three displacements $\overrightarrow{AB}$, $\overrightarrow{BC}$ and $\overrightarrow{CD}$ is the dis-

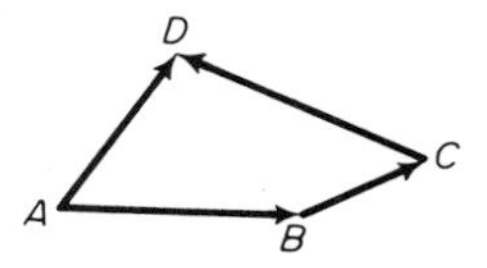


placement $\overrightarrow{AD}$. (Note that this is true even if $A$, $B$, $C$, and $D$ do not lie in a plane.) Let $D$ coincide with $A$. Then the resultant is zero. Therefore for any displacements $\overrightarrow{AB}$, $\overrightarrow{BC}$, $\overrightarrow{CA}$,

$$\overrightarrow{AB} + \overrightarrow{BC} + \overrightarrow{CA} = 0. \tag{3.2}$$

Suppose we replace the displacement $\overrightarrow{BC}$ by the algebraic expression $c - b$, $\overrightarrow{CA}$ by $a - c$, and $\overrightarrow{AB}$ by $b - a$. Then from Equations (3.1) and (3.2) we obtain the formally valid equations

$$b - a = -(a - b) \tag{3.3}$$

and

$$(c - b) + (a - c) + (b - a) = 0. \tag{3.4}$$

**Exercise 3.9** For displacements, show that $\overrightarrow{AB} + \overrightarrow{BC} + \overrightarrow{CD} + \overrightarrow{DE} = \overrightarrow{AE}$, write down the analogous algebraic equation, and verify that it is satisfied.

Later we shall see that this sort of relationship between displacement equations and algebraic equations is not accidental. We can begin to see why by considering vectors in a one-dimensional space.

**Exercise 3.10** If $A$ and $B$ are points on the $x$-axis with $x$-coordinates $a$ and $b$ respectively, and $b > a$, show that the magnitude of the displacement $\overrightarrow{AB}$ is $b - a$.

In one dimension, there are only two possible directions. Vectors confined to the $x$-axis, for example, can point to the right or to the left, but in no other direction. We could give these directions labels such as "right" and "left," or "first" and "second." But it is much more pleasing to label them "+" and "−." For example, in Exercise 3.10, the quantity $b - a$, being positive, can now be taken to describe not just the magnitude of the displacement $\overrightarrow{AB}$ but also its direction; while if we similarly relate the displacement $\overrightarrow{BA}$ to $a - b$, this quantity, being negative, gives both the magnitude and direction of $\overrightarrow{BA}$.

**Exercise 3.11** Show that $b - a$ in Exercise 3.10 represents both the magnitude and direction of the displacement $\overrightarrow{AB}$ whether $b > a$ or not.

**Exercise 3.12** In Exercise 3.11, if $O$ is the origin, show that $\overrightarrow{AB} = \overrightarrow{OB} - \overrightarrow{OA}$, *no matter what the relative positions of O, A, and B on the x-axis.* [Do this first by deducing it from Equation (3.2) with $O$ replacing $C$, and second, by considering the corresponding algebraic relation.]

No longer confining ourselves to one dimension, we have, from Equations (3.1) and (3.2),

$$\overrightarrow{AB} + \overrightarrow{CA} = -\overrightarrow{BC}$$

so that, by Equation (3.1),

$$\overrightarrow{AB} - \overrightarrow{AC} = \overrightarrow{CB}. \tag{3.5}$$

This gives us a convenient way to draw the difference of two displacements, $\overrightarrow{AB} - \overrightarrow{AC}$. We merely draw $\overrightarrow{AB}$ and $\overrightarrow{AC}$ and join their tips from $C$ to $B$ (not $B$ to $C$), as in Figure 3.2.

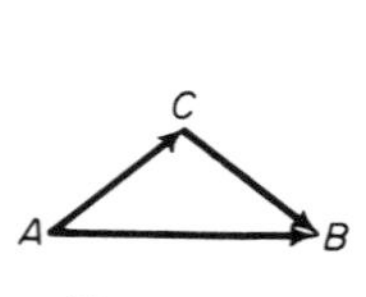


**Figure 3.2**

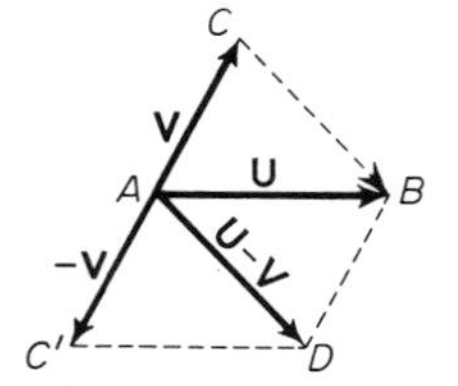


**Figure 3.3**

For vectors that are not free vectors but are tied to a particular starting point, the preceding still gives the magnitude and direction of their difference though not its location. Thus in Figure 3.3, if vectors **U**, **V** are represented by $\overrightarrow{AB}$, $\overrightarrow{AC}$ and we wish to find **U** − **V**, we should first construct − **V** by turning $\overrightarrow{AC}$ about $A$ through 180° to the position $AC'$. Then we should complete the parallelogram $AC'DB$ to obtain **U** − **V** as the vector represented by the diagonal $\overrightarrow{AD}$. But since triangles $AC'D$ and $CAB$ are congruent, and $CAC'$ is a straight line, it is easy to prove that $CB$ and $AD$ are parallel and of equal length. So we can find the magnitude and direction of **U** − **V** precisely as in Figure 3.2, and this is all we need since we already know it starts at $A$. [Compare Exercise 3.8.]

**Exercise 3.13** Prove that triangles $AC'D$ and $CAB$ are congruent, and that $CB$ and $AD$ are parallel and of equal length.

**Exercise 3.14** Find the magnitudes and directions of **U** − **V** in the following cases. [Use the method of Figure 3.2.]

(a) **U** has magnitude 5 and points northeast; **V** has magnitude 5 and points northwest. [*Ans.* magnitude $5\sqrt{2}$; pointing east.]

(b) **U** has magnitude $8\sqrt{3}$ and points east; **V** has magnitude 16 and points 30° north of east.

(c) **U** has magnitude 7 and points 20° north of east; **V** has magnitude $7\sqrt{2}$ and points 65° north of east.

**Exercise 3.15** A force of 8 lbs. acts along the positive direction of the $x$-axis. If it acts on a particle at the origin, what force combined with it would produce on the particle a resultant force of 8 lbs. pulling along the positive $y$-axis? [If the resultant force is **R**, the given force **X**, and the required force **Y**, then $\mathbf{X} + \mathbf{Y} = \mathbf{R}$; so $\mathbf{Y} = \mathbf{R} - \mathbf{X}$, whose magnitude and direction can be found by the method of Figure 3.2.]

**Exercise 3.16** Like Exercise 3.15 but with the resultant making 60° with the positive $x$-direction.

## 4. SPEED AND VELOCITY

In ordinary conversation we usually make no distinction between *speed* and *velocity*. In fact, most dictionaries treat the two words as synonymous. But in the physical sciences, speed and velocity have different meanings. Speed is the rate at which distance is covered per unit time, but velocity involves the direction also. Thus 5 m. p. h. is a speed, but 5 m. p. h. vertically upward is a velocity, and it is a different velocity from 5 m. p. h. in any other direction. A point moving with constant speed in a circle does not have an unchanging velocity: the velocity changes because its direction changes.

Figure 4.1 shows a platform with a line $AB$ marked on it. Suppose a point

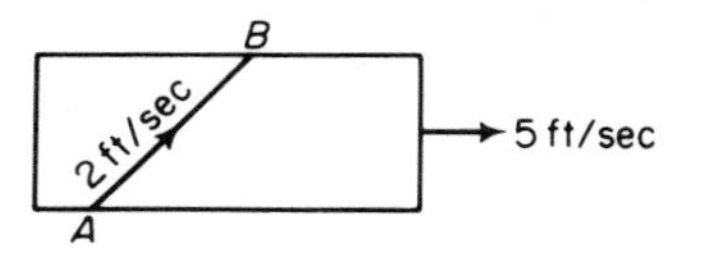


**Figure 4.1**

moves uniformly along this line from $A$ to $B$ at 2 ft./sec., and that the platform moves uniformly to the east at 5 ft./sec. relative to the ground. Then the point is being subjected to two different velocities at the same time. But it does not, therefore, move relative to the ground in two different directions at once. To say that it did would be to speak nonsense (except, perhaps, if we were talking of quantum phenomena, which we are not). It would remind us of the humorist Stephen Leacock's famous account of the angry man who "flung himself from the room, flung himself upon his horse and rode madly off in all directions."

How does the point actually move? We can find out by considering what happens in one second. For convenience, let us assume that $AB$ is of length 2 ft., so that in one second the point goes from $A$ to $B$. During this second, the platform moves 5 ft. to the east, and the ends of the line $AB$ drawn on it move to new positions, $A'$, $B'$, that are 5 ft. to the east of their original positions, as in Figure 4.2. The moving point travels from $A$ to $B'$ relative to the ground.

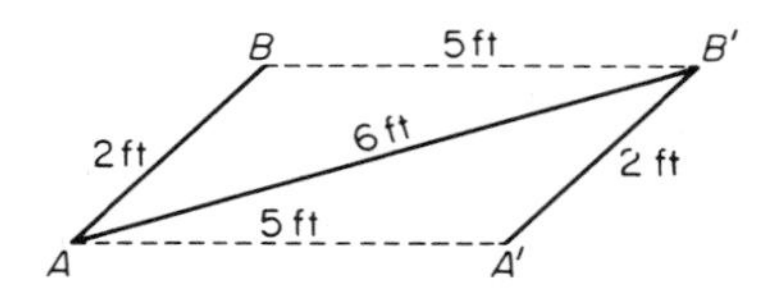


**Figure 4.2**

The distance $AB'$ depends on the angle $BAA'$; for simplicity, let us suppose that the distance comes to 6 ft. It is easy to see that the point travels from $A$ to $B'$ in a straight line, with uniform speed relative to the ground.

> **Exercise 4.1** Prove this. [To get the idea, consider first what happens in, say, $\frac{1}{2}$ sec., using similar triangles. Then consider what happens in $t$ secs.]

Figure 4.2 tells what happens in 1 sec. Let us, so to speak, divide it by 1 sec.—that is, let us replace ft. by ft./sec. wherever it occurs there. Then $\overrightarrow{AB}$ will represent the velocity of the point along $AB$ relative to the platform, $\overrightarrow{AA'}$ the velocity of the platform relative to the ground, and $\overrightarrow{AB'}$ that of the point relative to the ground. And we see that the first two combine by the parallelogram law to yield the third. In this sense, then, we may regard velocities as vector quantities.

**Exercise 4.2** Show that had Figure 4.2 been drawn to illustrate what happens in $t$ sec., the distances would be $2t$ ft., $5t$ ft., and $6t$ ft., but the ultimate relationship between the velocities would be the same as before.

**Exercise 4.3** A train is going at 50 m. p. h. due north, and a fly in one of the compartments flies at 20 m. p. h. northeast relative to the compartment. Find, by means of an accurate diagram, the actual velocity of the fly relative to the rails.

**Exercise 4.4** In Exercise 4.3, if the fly had been flying at 20 m. p. h. northeast relative to the rails, find, graphically, its velocity relative to the compartment.

**Exercise 4.5** A cyclist goes at 8 m. p. h. due east, and the wind is blowing at 10 m. p. h. due east. What is the velocity of the wind relative to the cyclist? What would it be if he turned around and travelled due west at 8 m. p. h.?

**Exercise 4.6** Like Exercise 4.5, but with the wind blowing due south at 8 m. p. h.

**Exercise 4.7** A cyclist goes at 6 m. p. h. due north and feels the wind coming against him with a relative velocity of 4 m. p. h. due west. What is the actual velocity of the wind?

**Exercise 4.8** The wind is blowing due south with a speed of 9 ft./sec. How fast must a car travel due east in order that the wind shall have, relative to the car, a speed of 15 ft./sec.?

**Exercise 4.9** A gun mounted on a train is pointed vertically upward. The train moves horizontally due east at 80 m. p. h. and the gun fires a bullet with a muzzle velocity of 80 m. p. h. What is the velocity of the bullet relative to the ground? [*Ans.* Since the bullet has the velocity of the gun, and thus of the train, as well as its vertical muzzle velocity relative to the gun, its velocity relative to the ground is $80\sqrt{2}$ m. p. h. toward the east at 45° to the horizontal.]

**Exercise 4.10** Rain is falling vertically with a speed of $u$ ft./sec. At what angle does it appear to be falling as viewed from a train moving horizontally at $v$ ft./sec.? [*Ans.* At $\tan^{-1}(u/v)$ to the horizontal.]

**Exercise 4.11** The wind is blowing steadily from the east at 5 m. p. h. A boat in still water is travelling in a circle, with a constant speed of 5 m. p. h. A pennant is flying at the top of the mast. Describe in words the way the direction of the pennant changes (a) relative to the deck of the boat, and (b) relative to the water. [It was a situation like this that led the astronomer Bradley to an understanding of the phenomenon of *aber-*

*ration*, a topic you may find well worth looking into on your own. Note that the velocities in parts (a) and (b) both change.]

**Exercise 4.12** In Exercise 4.10, if the rain were falling at an angle of $\theta$ to the horizontal, show that its apparent direction would make with its actual direction an angle $\sin^{-1}\{(u/v)\sin\theta\}$. [Compare this with the astronomical phenomenon of aberration which may be briefly described as follows: To catch the light from a star, a telescope must point in the apparent direction of the incoming light. Consequently, as the earth moves in its orbit the stars seem to move once a year in tiny ovals (actually ellipses) whose shapes depend on $\theta$, and thus on the positions of the stars.]

It is customary to define velocity as *the rate of change of displacement with respect to time*. Let us consider it from this point of view. If a point moves uniformly along a straight line from $P$ to $Q$, a distance of $s$ ft., in $t$ sec., it has a speed of $s$ ft./$t$ sec. or $\frac{s}{t}$ ft./sec. Its velocity is a quantity having the direction of the displacement $\overrightarrow{PQ}$ but a magnitude of $\frac{s}{t}$ ft./sec., which is $1/t$ times the magnitude of $\overrightarrow{PQ}$ and is, in fact, the speed. The displacement $\overrightarrow{PQ}$ combines with other displacements of the point according to the parallelogram law; and changing the magnitudes of these displacements by a factor $1/t$ will clearly yield entities that also combine according to the parallelogram law. So velocities, as here defined, are vector quantities.

The velocity vector above can be denoted by $(1/t)\,\overrightarrow{PQ}$, and this is reminiscent of the multiplication of a vector by a number. There is a difference, though. Vectors $5\mathbf{V}$ and $\mathbf{V}$ can combine with one another, but $\overrightarrow{PQ}$ and $(1/t)\,\overrightarrow{PQ}$ can not: the former is a displacement, and we have been measuring its magnitude here in ft.; but the latter is a velocity and its magnitude is not a length but a speed, here measured in ft./sec.

Suppose the motion is not uniform. Then when we divide the distance gone by the time taken we get the *average* speed. Correspondingly, we can define the *average* velocity as $(1/t)\,\overrightarrow{PQ}$. But this average velocity has an unexpected property. If I walk 10 miles in 2 hours my average speed is 5 m. p. h. But unless I walk in a straight line, my average velocity does not have a magnitude of 5 m. p. h. This is because my displacement is less than 10 miles if I change direction. If I walk a circuitous path and, having gone my 10 miles, end up where I started, I may end up exhausted from averaging 5 m. p. h. up hill and down dale for 2 solid hours, but my average velocity comes to zero.

I think we simply have to accept the fact that the vectorial concept of average velocity has unpleasant aspects when applied to motion not in a straight line. When motion is uniform, the speed is the same at every instant. When the motion is not uniform, the concept of *instantaneous* speed is a rather sophisticated one. It involves ideas belonging to the calculus into which we can not go here; but roughly speaking, the instantaneous speed is the value that the average speed approaches as the time interval gets smaller and smaller.

When these ideas are applied to the average velocity, they yield an *instantaneous* velocity that, except in highly artificial situations, behaves much as one would expect it to behave, and this despite the peculiar characteristics of the average velocity from which it is obtained. The instantaneous velocity is a vector quantity, and its magnitude is the instantaneous speed. At every instant, it has the direction of the line tangent to the path of the moving point at the instantaneous position of the point.

There is a subtle difference between velocity considered in terms of the moving platform and velocity defined as the rate of change of displacement with respect to time. Look back at the discussion of the former, and at the exercises based on it, and you will see that in each case the three velocities that are involved in the parallelogram law are not all measured with respect to the same standard of reference. In the platform situation, for example, the 5 ft./sec. and the 6 ft./sec. are velocities relative to the ground, but the 2 ft./sec. is a velocity relative to the moving platform. With the definition in terms of displacements, the three velocities in a parallelogram-law relationship are really velocities with respect to the same standard of reference.* This may well seem like an unimportant distinction, especially since the two approaches to the concept of velocity both lead to the theorem that velocity is a vector quantity, and both yield the same resultant for two given velocities. But actually the distinction is a basic one physically. The resultants are the same only because of the nature of Newtonian space and time. In Einstein's special theory of relativity, the resultants are different.

When defined in terms of displacements, velocities, like displacements, are free vectors. Yet in physical situations it seems essential to give velocity a specific location. For example, the velocity of a particle should surely be located at the particle and thus be represented by an arrow-headed line segment starting at the particle. In a storm, the velocity of the air is different at different places. To regard velocity as a free vector under the circumstances would be inappropriate. If a thrown object spins as it flies through the air, different points of it have different velocities; one can not free these velocities and transfer them from point to point and still describe the motion correctly.

**Exercise 4.13** What is the resultant of a velocity of 2 ft./sec. due east and a velocity of $2\sqrt{2}$ ft./sec. northwest? [Note that the wording of this exercise lets it pertain to the definition of velocity as the time rate of change of displacement.]

**Exercise 4.14** A certain boat of negligible length has a maximum speed of 4 yds./sec. in still water. The water in a straight canal, 100 yds. wide, flows at 3 yds./sec. If the boat goes at maximum speed while pointing perpendicularly to the bank, what is its speed relative to the bank? How long does it take to cross the canal? [The answer is neither $33\frac{1}{3}$ secs.

*The unpleasant aspects of average velocity are not exorcized by the moving platform concept.

nor 20 secs.] In what direction should the boat be pointed in order to cross the canal perpendicular to the banks, and how long would it take to cross? [It would take $100/\sqrt{7}$ secs.]

## 5. ACCELERATION

Acceleration is the rate of change of velocity with respect to time. The awkwardness associated with a changing velocity is compounded when we consider acceleration; but, just as in the case of velocity, when ideas belonging to the calculus are brought in, these awkward aspects no longer obtrude. Suffice it to say here that acceleration is a vector quantity. The *average* acceleration of a particle may be said to be the change in velocity divided by the time taken, the velocity being regarded as a free vector. Thus if a point has a velocity of $15\sqrt{3}$ ft./sec. due east at noon and a velocity of 30 ft./sec. 30° north of east at 5 secs. past noon, the change in its velocity is a velocity of 15 ft./sec. due north, as is easily seen from a figure. Since this change of velocity took place in 5 secs., the average acceleration is 3 ft./sec.$^2$ due north; we obtain this by dividing the 15 ft./sec. by 5 secs.

**Exercise 5.1** A point changes velocity from 144 ft./sec. due east to 144 ft./sec. due north in 8 secs. What is the average acceleration? [*Ans.* $18\sqrt{2}$ ft./sec.$^2$ to the northwest.]

**Exercise 5.2** A man has a velocity of 5 ft./sec. due east and an acceleration of 5 ft./sec.$^2$ due north. Describe the resultant. [One word suffices.]

**Exercise 5.3** A point moves in a circle of radius $r$ with constant speed $v$. Find its average acceleration for (a) one revolution, (b) half of a revolution, and (c) a quarter of a revolution. [Remember that the instantaneous velocity has the direction of the tangent to the circle at the position of the particle. Remember also that though the speed is constant, the velocity changes because its direction changes. In case (c) the average acceleration comes out to have a magnitude of $2\sqrt{2}\,v^2/\pi r$.]

**Exercise 5.4** A point moves in a circle with constant speed. Prove geometrically that the average acceleration between two positions, $A$ and $B$, of the point has a direction parallel to the bisector of the angle between the radii to $A$ and $B$.

From Exercise 5.4 we see that, in finding the average acceleration of a point moving in a circle with constant speed, the smaller we take the interval of time, the closer together will be the two points, $A$ and $B$, and the closer will the direction of the average acceleration be to the directions of the radii to $A$ and $B$. When the time interval is so small that these radii are close to being coincident, the direction of the average acceleration will also be close to being coincident with them. It will come as no great surprise, then, that when one applies ideas belonging to the calculus one finds that the *instantaneous* acceleration at the instant when the point is at $A$ lies along the radius to $A$ and points towards the center of the circle. If the speed were not constant, the instantaneous acceleration would, in general, not lie along the radius.

## 6. ELEMENTARY STATICS IN TWO DIMENSIONS

If a body placed at rest* remains at rest, it is said to be in *equilibrium*, and the forces acting on it are also said to be in equilibrium. In *statics* we study the conditions under which bodies and forces are in equilibrium.

If a body is so small that we can neglect its size and obtain satisfactory results by treating it mathematically as a point having mass, we call it a particle. (In some investigations quite large bodies—stars, for example—are treated as particles.) According to Newtonian dynamics, a particle of mass $m$ acted on by forces having a resultant $\mathbf{F}$ will undergo an acceleration $\mathbf{a}$ where

$$\mathbf{F} = m\mathbf{a}. \tag{6.1}$$

For a particle to be in equilibrium, it must be unaccelerated, and therefore, by Equation (6.1), the resultant of the forces acting on it must be zero.

If a particle is in equilibrium under the influence of three forces, the forces must be coplanar. This is easily proved. Denote the forces by $\mathbf{F}_1$, $\mathbf{F}_2$, $\mathbf{F}_3$, and let $\mathbf{R}$ be the resultant of $\mathbf{F}_2$ and $\mathbf{F}_3$. Then the three forces are equivalent to $\mathbf{F}_1$ and $\mathbf{R}$. For equilibrium, these two forces must have zero resultant. So $\mathbf{R} = -\mathbf{F}_1$. But the parallelogram construction ensures that $\mathbf{R}$, $\mathbf{F}_2$, and $\mathbf{F}_3$ are coplanar. Therefore $\mathbf{F}_1$, $\mathbf{F}_2$, and $\mathbf{F}_3$ must be too.

> **Exercise 6.1** A particle is acted on by three forces. Prove that if arrow-headed line segments representing, in the usual way, their magnitudes and directions (but not necessarily their locations) can be appropriately arranged to form a triangle, the particle is in equilibrium. Also prove the converse.

Consider now a rigid body that is too large to be treated as a particle. If

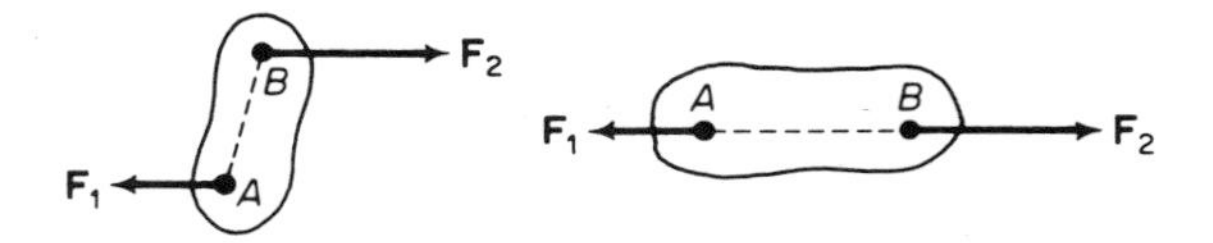


**Figure 6.1**

it is acted on at points $A$ and $B$ by two parallel forces $\mathbf{F}_1$ and $\mathbf{F}_2$ pointing in opposite directions as shown on the left in Figure 6.1, experiments show that it will be apt to turn and, if friction is present (but not in excessive amounts),

*Let us not ask awkward questions here about the word "rest."

it will take up the final position shown on the right,* with the two forces acting along a common line; and if $\mathbf{F}_1$ and $\mathbf{F}_2$ have equal magnitudes the forces will then be in equilibrium.

From such experiments we infer two things. The first is that a rigid body can transmit the effect of a force unchanged from its point of application to any other point in the body that lies on the line along which the force acts. Because force is a vector quantity that has a definite point of application, it is often referred to as a *bound* vector (a term we have already been using) to distinguish it from a free vector. But when a force acts on a rigid body in equilibrium, its point of application can, in a certain sense, be thought of as free to move along the line of action of the force to another position within the body. Therefore, it is sometimes referred to as a *sliding* vector. Note, though, that if on the left in Figure 6.1 we moved the point of application of the force $\mathbf{F}_1$ from $A$ to another point $A'$ in the body lying on the line of action of $A$, the body would end up on the right in a position having $A'B$ instead of $AB$ lined up with $\mathbf{F}_1$ and $\mathbf{F}_2$; this position would not be the same as the one illustrated. There is danger, therefore, in thinking of forces as other than bound vectors, though the presence of danger need not prevent us from thinking of them as sliding vectors in appropriate circumstances, such as when a rigid body is held immovable.

The second thing we infer from these experiments is that the conditions for the equilibrium of a rigid body involve the locations of the forces in a significant way, since even if $\mathbf{F}_1$ and $\mathbf{F}_2$ had equal magnitudes, the rigid body would not be in equilibrium when it was in the position on the left, though it would be when in the position on the right.

Two conditions must be fulfilled by the forces acting on a rigid body to produce equilibrium. They are consequences of Equation (6.1), but we shall not give their derivation here. *One condition of equilibrium is that if the forces acting on the body are treated as free vectors their resultant must be zero*—just as if they were all acting on a particle. This condition ensures that the body has no overall translational acceleration. But it must be supplemented by another condition since the forces may still tend to twist the body.

Consider the simple case of a seesaw of negligible mass, as shown, with a boy of weight 75 lbs. at $B$, $x$ ft. from $O$, and a man of weight 150 lbs. at $M$

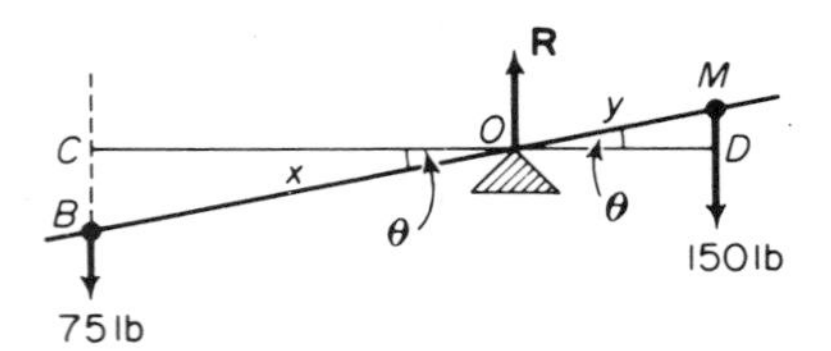


**Figure 6.2**

*You may think friction is unnecessary. But the situation is not the same as it would be if we were pulling on strings attached to $A$ and $B$. Can you see why not? The point is a subtle one.

$y$ ft. from $O$. The turning effect, or *moment*, of the boy's weight is counterclockwise, and it is measured by the product of his weight and the perpendicular distance $OC$ from $O$ to the line of action of the weight. It is therefore $75x \cos \theta$, where $\theta$ is as shown. Similarly the turning effect of the man's weight is $150y \cos \theta$, the effect being clockwise. For equilibrium, the clockwise and counterclockwise turning effects must balance. So

$$75x \cos \theta = 150y \cos \theta, \tag{6.2}$$

from which we see that $x = 2y$. Thus, the boy balances the man if he sits twice as far as the man does from the point of support; and the balance then holds for all values of $\theta$.

By regarding counterclockwise moments as positive and clockwise moments as negative, we can rewrite Equation (6.2) in the form:

$$75x \cos \theta + (-150y \cos \theta) = 0. \tag{6.3}$$

It then exemplifies *the second condition for the equilibrium of a rigid body: the total moment of the forces acting on the body must be zero.* [All moments must be about the same point, but this point can be anywhere.] For simplicity we are confining ourselves here to situations in which all the forces lie in a plane. The two conditions of equilibrium, asserted here without proof, hold for the general case; we shall discuss this case when we have developed the necessary vectorial mathematics.

Note that the weights of the boy and the man add up to 225 lbs. so that, at first sight, it seems as if the first condition of equilibrium is not satisfied. But there is a third force acting on the seesaw, namely the reaction on it of the fulcrum at $O$. (We are assuming that the mass of the seesaw is negligible.) Denote the reaction by $\mathbf{R}$, and its magnitude by $R$. Then the first condition tells us that $\mathbf{R}$ is vertical, and

$$R - 75 - 150 = 0$$

so that $R = 225$ lbs.

**Exercise 6.2** Show that if we took moments about $B$ instead of $O$ the total moment would still be zero. [Note that $\mathbf{R}$ will now have a nonzero moment. Its moment about $O$ is zero, so it would not have made any difference to Equation (6.2) or (6.3), even if we had thought of it at the time.]

**Exercise 6.3** Like Exercise 6.2, but taking moments about $M$.

**Exercise 6.4** Prove that if a rigid body is in equilibrium under the influence of three nonparallel coplanar forces, the lines of action of these three forces must meet in a point. [Take moments about the point of intersection of two of these lines of action.]

**Exercise 6.5** *ABCD* is a rigid square of side $a$. Force $\mathbf{F}_1$ acts at $B$ in the direction from $A$ to $B$, and force $\mathbf{F}_2$, of equal magnitude, acts at $C$ in the direction from $B$ to $C$. Find the magnitude, direction, and location of the single additional force that would maintain the square in equilib-

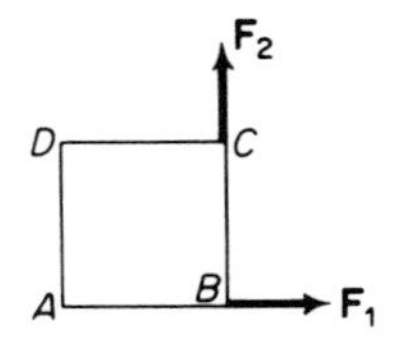


rium. [Use Exercise 6.4.] Verify that the total moment of the three forces about $D$ is zero.

**Exercise 6.6** If $n$ parallel coplanar forces acting on a rigid body have zero total moment about a point $O$ in their plane, prove that they do not have zero total moment about every other point in their plane unless they are in equilibrium. [Let the forces have magnitudes $\mathbf{F}_1, \mathbf{F}_2, \ldots, \mathbf{F}_n$ and be at distances $x_1, x_2, \ldots, x_n$ from $O$. Then $x_1 F_1 + x_2 F_2 + \ldots + x_n F_n = 0$. But $(x_1 + a)F_1 + (x_2 + a)F_2 + \ldots + (x_n + a)F_n$ will not, in general, be zero.]

## 7. COUPLES

Consider two parallel forces $\mathbf{F}_1$ and $\mathbf{F}_2$ acting on a rigid body at points $A$ and $B$ as in Figure 7.1, *the forces having equal magnitudes, F, and opposite directions.* Let their distances apart be $a$. Because the magnitudes of the forces

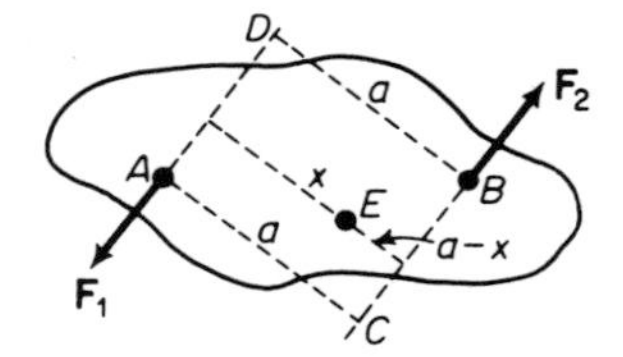


**Figure 7.1**

are equal, the first condition of equilibrium is satisfied since, when treated as free vectors, $\mathbf{F}_1$ and $\mathbf{F}_2$ have zero resultant. The second condition is not satisfied, though. The forces do not have zero total moment. They tend to turn the body. If we take moments about $A$, we obtain the values 0 and $aF$ for a total of $aF$. If we take moments about $B$ we obtain $aF$ and 0, also yielding a total of $aF$. If we take moments about a point $E$, in the plane containing the forces, that is distant $x$ from the line of action of $F_1$ and thus $a - x$ from that of $F_2$, we obtain $xF + (a - x)F$ which gives the same total, $aF$, as before.

A pair of parallel forces of equal magnitudes and opposite directions constitutes a *couple*. The magnitude of the moment of a couple is measured by the product of the magnitude of either force and the perpendicular distance between their lines of action. Thus the two forces in Figure 7.1 constitute a couple of moment $aF$ acting counterclockwise. Usually the total moment of a system of forces depends on the position of the point about which the moments are taken. With a couple, though, as we have seen, the moment does not change when the point is changed. Therefore there is no need to specify a point when talking about the moment of a couple.

Suppose a force $\mathbf{F}_1$ of magnitude $F$ acts on a rigid body at point $A$, as in Figure 7.2. At point $B$, let us introduce two self-cancelling forces $\mathbf{F}_2$ and $\mathbf{F}_3$

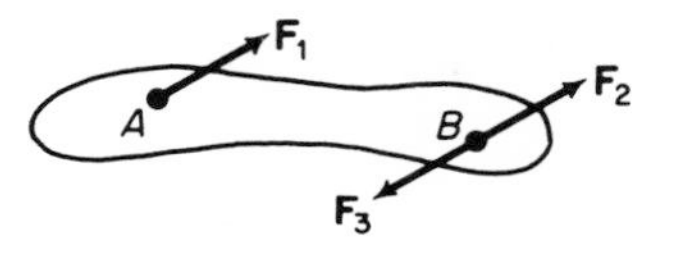


**Figure 7.2**

parallel to $\mathbf{F}_1$ and having the same magnitude $F$. Now let us regroup these three forces in our minds: let us think of $\mathbf{F}_1$ and $\mathbf{F}_3$ as forming a couple, and $\mathbf{F}_2$ as standing apart. Then we see that the single force $\mathbf{F}_1$ at $A$ is equivalent to the combination of the single parallel force $\mathbf{F}_2$ of equal magnitude acting at $B$ and the couple formed by $\mathbf{F}_1$ and $\mathbf{F}_3$.

In effect, then, when we move a force to a different line of action (parallel to the original one), we automatically introduce a couple.

**Exersise 7.1** In Figure 7.2, if the forces make an angle $\theta$ with the line $AB$, and the distance $AB$ is $a$, what is the moment of the couple? [*Ans.* $aF \sin \theta$.]

**Exercise 7.2** $A$, $B$, $C$, and $D$ are the vertices of a square of side $a$. A force of magnitude 10 lbs. acts at $A$ and points in the direction $\overrightarrow{AB}$. What couple would be introduced if the force were moved parallel to itself so as to act (a) at $B$, (b) at $C$, and (c) at $D$? [*Ans.* 0, $-10a$, $-10a$.]

**Exercise 7.3** A force of magnitude $F$ lbs. pointing in the positive $x$-direction acts at the origin. A couple in the $xy$-plane has a moment of $G$ ft. lbs. Where in the $xy$-plane could the point of application of the force be moved to in order to yield a total couple of $H$ ft. lbs.?

## 8. THE PROBLEM OF LOCATION. VECTOR FIELDS

The problem of the location of vectors is a vexing one for which there seems to be no wholly satisfactory elementary solution. *The easy way out is to ignore the matter, or to say that such an entity as a force is more than a vector, being a vector equipped with the additional, nonvectorial attribute of location. The purely vectorial aspect of such an entity—its vector, so to speak —is then regarded as a free vector, all vectors being taken to be free vectors.*

There is something highly attractive about this approach. It makes for a neat mathematical theory of vectors. But we really do not understand vectors until we grapple with the problem of location, and ultimately realize that we do not understand it. In this book, therefore, we shall continue to look into the awkward aspects of this problem. The reader who feels that he is uninterested in the problem of location is invited, at any stage in this book, to make the ultimate (or seemingly ultimate) abstraction of denying vectors location and then to ignore all further mentions of the problem. If he does this he will be happier than those who try to face the issue. But he will understand vectors less intimately in the end.

Let us recall the discussion of velocity as the rate of change of displacement with respect to time. Since displacement is a free vector, we at first thought that velocity must be too. But then we thought of the velocity of a particle and felt that this should be located at the particle. And after that, we thought of the velocity of the wind in a storm and realized that there was a different velocity vector at each point.

There are many situations analogous to that of the wind in a storm. The force exerted by a given magnet on another magnet, for instance, is different at different locations of the other magnet. The force exerted by an electric charge on another charge stationary with respect to it also depends on the location. When we have such a situation—an enormous number of vectors, one at each place—we say we have a *vector field.* Mathematically, we would say that there were infinitely many vectors, one at each point. Evidently, in a nonuniform field, we can not switch vectors around to different locations without altering the field. In this sense, the vectors of a vector field have to be considered bound vectors.

By a process of abstraction similar to that that led us from shifts to displacements, we can imagine free vectors associated with the vectors of a field. But then we wonder how much it will profit us to do so. For in a field of force, for example, the force at $P$ does not combine physically by the parallelogram law with the force at a different point $Q$.

Nevertheless we can ask what is the difference of these two forces in magnitude and direction. And in answering we could go off into a margin and work the problem out, treating the two forces as free vectors and moving them to a common starting point. In working in the margin in this way, we would be working with what the mathematicians call a *linear vector space.* When dealing with forces, we can work in one linear vector space. When dealing with, say, displacements, we can work in another. By judiciously using different linear vector spaces for different situations, we can make the structures of all the linear vector spaces the same, and say that in a *given* linear vector space not only do all vectors have a common starting point, but *every* vector combines with *every other* vector therein according to the parallelogram law.

Pure mathematicians often prefer to start by *defining* vectors as abstract elements in a linear vector space that obey certain rules laid down as axioms. These axioms, indeed, simultaneously define the vectors and the vector space that they inhabit. Often pure mathematicians think of vectors in a vector space as "ordered sets of numbers," by which they mean unpictorial entities like $(1, 9, 1, \frac{1}{2})$ that are counted as different from, say, $(9, 1, 1, \frac{1}{2})$, the amount of numbers in each ordered set depending on the dimensionality of the vector space. By taking this abstract stance, they gain enormously in precision and elegance and power—as generally happens when good mathematicians make abstractions. But the person who studies vectors only as a branch of abstract algebra, regarding them solely as elements in a vector space, obtains the precision and elegance and power at a price. For they come to him too patly for him to realize their worth. He gains no inkling of the motives that led to the choice of just these axioms and abstractions rather than others, and he is apt not to realize how ugly are the seams that join such neat algebraic abstractions to their applications—even to applications to other branches of mathematics such as geometry.

Note, for example, that in asking about the difference of the two forces at $P$ and $Q$ we were not being as reasonable as we thought at the time. We were assuming not only that such a comparison could be made, but also that it could be made in a unique way. That sounds innocent enough. But in effect we were assuming the validity of Euclid's axiom of parallels (among other things) and these days we all know that this axiom can be dispensed with. To pursue this particular aspect of the matter further would be outside the scope of this book. We have pushed it this far in order to show that the problem is not as simple as it sometimes seems.

# 3

# VECTOR ALGEBRA

## 1. COMPONENTS

Let **V** be a vector starting at a point $O$. Take any plane containing **V**, and in it draw any two noncoincident lines, $A'A$ and $B'B$, passing through $O$. Then we can always find two vectors, **X** and **Y**, lying respectively along these

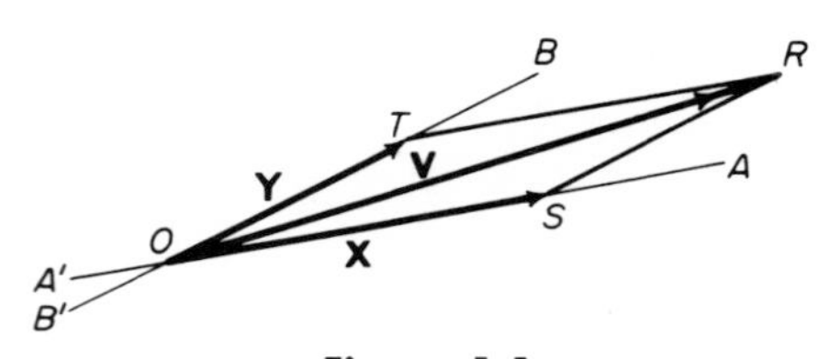


**Figure 1.1**

two lines, such that **X** and **Y** are together equivalent to **V** —— that is, such that $\mathbf{X} + \mathbf{Y} = \mathbf{V}$. Moreover, the vectors **X** and **Y** are uniquely determined by **V** and the two lines. For, through the end $R$ of **V** we can draw lines $RT$ and $RS$ respectively parallel to $A'A$ and $B'B$, intersecting those lines at $S$ and $T$ respectively. Then $OSRT$ is a parallelogram. Therefore, the vectors represented by $\overrightarrow{OS}$ and $\overrightarrow{OT}$ are uniquely the **X** and **Y** we seek.

**Exercise 1.1** Prove the uniqueness of **X** and **Y**.

Let $\mathbf{e}_x$ and $\mathbf{e}_y$ be nonzero vectors starting at $O$ and lying respectively along the lines of action of **X** and **Y**. Denote the ratio of the magnitude of **X** to that of $\mathbf{e}_x$ by $V_x$, taking $V_x$ to be positive if **X** and $\mathbf{e}_x$ have the same direction, and negative if they have opposite directions. Similarly, denote the ratio of the magnitude of **Y** to that of $\mathbf{e}_y$ by $V_y$, these quantities, $V_x$ and $V_y$, being, of course, not vectors. Then we can write:

$$\mathbf{X} = V_x\mathbf{e}_x\,, \qquad \mathbf{Y} = V_y\mathbf{e}_y\,. \tag{1.1}$$

$V_x$ and $V_y$ are called the *components* of the vector **V** relative to (or with respect to, or in) the *frame of reference* defined by $\mathbf{e}_x$ and $\mathbf{e}_y$. The point $O$ is called the *origin* of the frame of reference, and $\mathbf{e}_x$, $\mathbf{e}_y$ its *base vectors.* Some books use the word components for the vectors **X** and **Y**, and then fail to make a sharp verbal distinction between **X**, **Y** and $V_x$, $V_y$. To avoid confusion, we shall refer to **X** and **Y** as *component vectors* of **V**.

**Exercise 1.2** A vector of magnitude 144 makes 45° with each of two perpendicular lines. What are its component vectors along these lines? [There are four possible answers to the question as worded.]

**Exercise 1.3** In Exercise 1.2, if $\mathbf{e}_x$ and $\mathbf{e}_y$ lie along the perpendicular lines, and $\mathbf{e}_x$ has a magnitude of 4 and $\mathbf{e}_y$ a magnitude of 6, what are the components of the vector relative to the reference frame defined by the base vectors $\mathbf{e}_x$ and $\mathbf{e}_y$? [*Ans.* $\pm 18\sqrt{2}$, $\pm 12\sqrt{2}$.]

**Exercise 1.4** If $\mathbf{e}_x$ and $\mathbf{e}_y$ have unit magnitude and are perpendicular, what are the components of a vector of magnitude $h$ pointing along the positive direction of $\mathbf{e}_x$? Would the components be different if $\mathrm{e}_y$ were not perpendicular to $\mathbf{e}_x$? [In situations of this sort we shall assume that the vectors have the same starting point unless they are free vectors, in which case the assumption is unnecessary.]

**Exercise 1.5** If $\mathbf{e}_x$ has a fixed direction but $\mathbf{e}_y$ a varying one, is $V_x$ constant or not?

**Exercise 1.6** Find the component vectors of a vector of magnitude 10 pointing 30° north of east relative to (a) the east and north directions, (b) the east and northwest directions, and (c) the east direction and the direction 30° north of east.

**Exercise 1.7** In Exercise 1.6, find the component of the given vector relative to the reference frame $\mathbf{e}_x$, $\mathbf{e}_y$, if $\mathbf{e}_x$ has magnitude 2 and $\mathbf{e}_y$ magnitude 1, and if, in each part, $\mathbf{e}_x$ points to the east while $\mathbf{e}_y$ points in the second direction mentioned. [*Ans.* (b) $5(\sqrt{3}+1)/2$, $5\sqrt{2}$.]

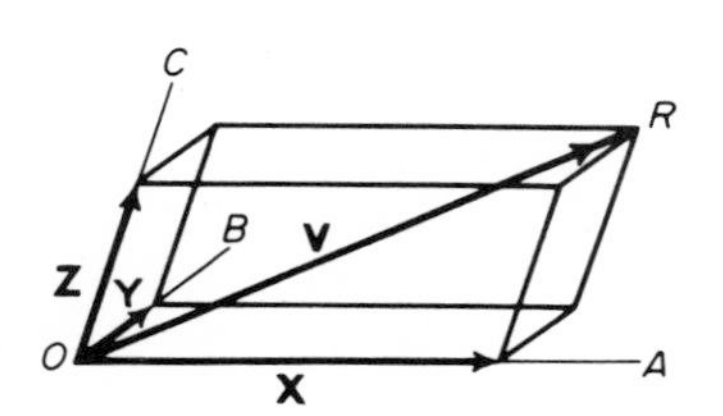


**Figure 1.2**

If, in Figure 1.1, the vector $V$ did not lie in the plane of the lines $A'A$ and $B'B$, we could not have constructed the parallelogram. But if we take *three* lines $OA$, $OB$, and $OC$ that do not lie in a plane, we can always* find, along the respective lines, vectors $\mathbf{X}, \mathbf{Y}$, and $\mathbf{Z}$ that are together equivalent to $\mathbf{V}$; these three vectors are uniquely determined by $\mathbf{V}$ and the lines. The method is to construct a parallelepiped with edges along and parallel to the three chosen lines $OA$, $OB$, and $OC$, as in Figure 1.2.

**Exercise 1.8** Explain how to construct the parallelepiped in Figure 1.2 by passing appropriate planes through $R$.

**Exercise 1.9** Prove that $\mathbf{X} + \mathbf{Y} + \mathbf{Z} = \mathbf{V}$ in Figure 1.2 (a) by first combining $\mathbf{X}$ and $\mathbf{Y}$ and combining their resultant with $\mathbf{Z}$; (b) by first combining $\mathbf{Y}$ and $\mathbf{Z}$; (c) by first combining $\mathbf{Z}$ and $\mathbf{X}$; and (d) by using the end-to-end method of finding the resultant of several vectors.

If we introduce three nonzero base vectors $\mathbf{e}_x$, $\mathbf{e}_y$, $\mathbf{e}_z$ along $OA$, $OB$, and $OC$, respectively, we can write:

$$\mathbf{X} = V_x\mathbf{e}_x, \qquad \mathbf{Y} = V_y\mathbf{e}_y, \qquad \mathbf{Z} = V_z\mathbf{e}_z, \tag{1.2}$$

the quantities $V_x$, $V_y$, and $V_z$ being the components of $\mathbf{V}$ with respect to the reference frame defined by $\mathbf{e}_x$, $\mathbf{e}_y$, and $\mathbf{e}_z$. Then, since $\mathbf{V} = \mathbf{X} + \mathbf{Y} + \mathbf{Z}$, we have:

$$\mathbf{V} = V_x\mathbf{e}_x + V_y\mathbf{e}_y + V_z\mathbf{e}_z. \tag{1.3}$$

**Exercise 1.10** In two dimensions, if $\mathbf{U} = \mathbf{U}_x\mathbf{e}_x + \mathbf{U}_y\mathbf{e}_y$ and $\mathbf{V} = V_x\mathbf{e}_x + V_y\mathbf{e}_y$, show by means of a diagram that $\mathbf{U} + \mathbf{V} = (U_x + V_x)\mathbf{e}_x + (U_y + V_y)\mathbf{e}_y$ and $\mathbf{U} - \mathbf{V} = (U_x - V_x)\mathbf{e}_x + (U_y - V_y)\mathbf{e}_y$. [Note how neatly the symbols, + and −, applied to vectors turn out to be related to the algebraic symbols, + and −, applied to their components.]

**Exercise 1.11** Prove the relations in Exercise 1.10 algebraically by noting that $\mathbf{U} + \mathbf{V}$ is really the sum of four vectors $U_x\mathbf{e}_x$, $U_y\mathbf{e}_y$, $V_x\mathbf{e}_x$, and $V_y\mathbf{e}_y$, adding the first and third, then the second and fourth, and combining the results.

**Exercise 1.12** Relative to $\mathbf{e}_x$ and $\mathbf{e}_y$, let $\mathbf{U}$ have components $U_x$ and $U_y$, let $\mathbf{V}$ have components $V_x$ and $V_y$, and let $\mathbf{W}$ have components $W_x$ and $W_y$. What are the components of $\mathbf{U} + \mathbf{V} - \mathbf{W}$?

**Exercise 1.13** In Exercise 1.12, what must be true of the components of $\mathbf{U}$, $\mathbf{V}$, and $\mathbf{W}$ if $\mathbf{U} + \mathbf{V} - \mathbf{W}$ (a) points in the $\mathbf{e}_x$ direction, (b) points in the negative $\mathbf{e}_y$ direction (i. e., in the direction opposite to that in which $\mathbf{e}_y$ points), (c) lies along the direction of the resultant of $\mathbf{e}_x$ and $\mathbf{e}_y$?

**Exercise 1.14** In the situation epitomized in Equation (1.3), what will happen to the components of $\mathbf{V}$ if the magnitudes of the base vectors

*Beware of the word "always." It is almost "always" wrong. What if we were discussing three lines in four dimensions, for example?

$\mathbf{e}_x$, $\mathbf{e}_y$, and $\mathbf{e}_z$ are increased to 5 times their present values? [Note that the components become smaller, not larger.]

**Exercise 1.15** In Equation (1.3) what new magnitudes for $\mathbf{e}_x$, $\mathbf{e}_y$, and $\mathbf{e}_z$ would make the components of $\mathbf{V}$ all equal to 1? Are there exceptional cases?

**Exercise 1.16** What are the components of $\mathbf{e}_x$, $\mathbf{e}_y$, and $\mathbf{e}_z$ respectively in the reference frame defined by $\mathbf{e}_x$, $\mathbf{e}_y$, and $\mathbf{e}_z$?

**Exercise 1.17** If $\mathbf{U} = U_x\mathbf{e}_x + U_y\mathbf{e}_y + U_z\mathbf{e}_z$ and $\mathbf{V} = V_x\mathbf{e}_x + V_y\mathbf{e}_y + V_z\mathbf{e}_z$, show that the components of $a\mathbf{U} \pm b\mathbf{V}$ are $aU_x \pm bV_x$, etc., $a$ and $b$ being numbers.

**Exercise 1.18** If, relative to $\mathbf{e}_x$, $\mathbf{e}_y$, and $\mathbf{e}_z$, $\mathbf{U}$ has components (6, 2, − 4), what can you say about the components of $\mathbf{V}$ (a) if $\mathbf{U} + 2\mathbf{V}$ lies along the $\mathbf{e}_x$ direction? (b) if $3\mathbf{U} - \mathbf{V}$ lies in the plane defined by $\mathbf{e}_y$ and $\mathbf{e}_z$? (c) if $\mathbf{U} + \mathbf{V}$ is the same as $6\mathbf{e}_x + 2\mathbf{e}_y - 4\mathbf{e}_z$? (d) if $\mathbf{e}_x + \mathbf{U} = 3\mathbf{e}_y + \mathbf{V}$? [*Ans.* (a) Since $\mathbf{e}_x$ has components (1, 0, 0), we must have $6 + 2V_x =$ some constant, $2 + 2V_y = 0$, $-4 + 2V_z = 0$. So we can say that $V_y = -1$, $V_z = 2$.]

**Exercise 1.19** Relative to $\mathbf{e}_x$, $\mathbf{e}_y$, and $\mathbf{e}_z$, the velocity of an airplane has components (100, 150, 20) and the wind velocity has components (30, − 20, 0). What are the components of the velocity with which the wind passes the plane? [*Ans.* (30 − 100, − 20 − 150, 0 − 20), or (− 70, − 170, − 20), just like that! Note how much easier it is here to work with components than with arrow-headed line segments and three-dimensional diagrams.]

**Exercise 1.20** A particle in equilibrium is acted on by three forces, two of which have components (6, 5, − 2) and (3, − 10, 8) respectively, relative to a particular reference frame. What are the components of the third force? [*Ans.* (− 9, 5, − 6).]

## 2. UNIT ORTHOGONAL TRIADS

The formidable title of this section has to do with something that is simpler than what was studied in the preceding section, being a special case of it.

If we know the components of a vector with respect to a reference frame, we know the vector, for it is given by Equation (1.3). Theoretically, then, we ought to be able to calculate its magnitude once we know its components in a given reference frame. And indeed we can. But when the reference frame is a general one of the type considered in the preceding section, the formula is rather complicated. Consider, for example, the two-dimensional case, with the reference vectors $\mathbf{e}_x$ and $\mathbf{e}_y$ making an angle $\theta$ with each other. If the components of $\mathbf{V}$ are $V_x$ and $V_y$, and the magnitudes of $\mathbf{e}_x$ and $\mathbf{e}_y$ are $\mathrm{e}_x$ and $\mathrm{e}_y$, then the lines $OS$ and $OT$ in Figure 2.1 are respectively of length $V_x e_x$ and $V_y e_y$. In triangle $OSR$, $SR$ has the length of $OT$, namely $V_y e_y$, and angle $OSR$ is $180° - \theta$. So, by the law of cosines, since $\cos(180° - \theta) = -\cos\theta$, we have:

$$V^2 = V_x^2 e_x^2 + V_y^2 e_y^2 + 2V_x e_x V_y e_y \cos\theta, \tag{2.1}$$

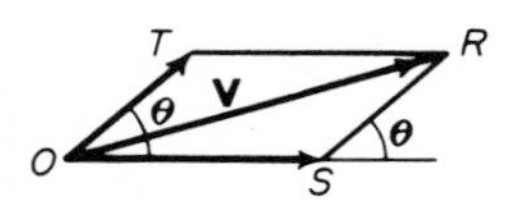

**Figure 2.1**

a none-too-pleasant formula even without our taking the square roots of both sides. And this is just for the two-dimensional case.

In many situations, of which the preceding is one, the work is significantly simplified if we use a special type of reference frame, though the simplification is obtained at a price: loss of generality and of depth of understanding.

The idea is to use reference frames in which $\mathbf{e}_x$, $\mathbf{e}_y$, and $\mathbf{e}_z$ are mutually perpendicular and of unit magnitude; hence the title of this section. It is customary to denote the vectors $\mathbf{e}_x$, $\mathbf{e}_y$, and $\mathbf{e}_z$ of such a unit orthogonal reference triad by the special symbols **i**, **j**, and **k**. Note that **i** does not stand for $\sqrt{-1}$ here. For one thing, it is printed in boldface type.

The parallelepiped in Figure 1.2 now becomes rectangular, as in Figure 2.2. [Note the change in letters.] And instead of Equation (1.3) we have:

$$\mathbf{V} = V_x\mathbf{i} + V_y\mathbf{j} + V_z\mathbf{k}. \tag{2.2}$$

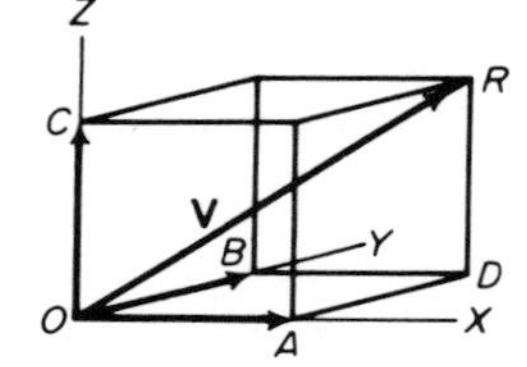

**Figure 2.2**

What is the magnitude of **V** in terms of $V_x$, $V_y$, and $V_z$? If **V** is represented by $\overrightarrow{OR}$ in Figure 2.2, Equation (2.2) tells us that

$$OA = V_x, \qquad OB = V_y, \qquad OC = V_z, \tag{2.3}$$

since the magnitudes of **i**, **j**, and **k** are all unity. [But we shall have to return to this in the next chapter.] Applying Pythagoras' theorem to the right triangle $OAD$, we have $OD^2 = OA^2 + AD^2$. And applying it to the right triangle $ODR$, we have $OR^2 = OD^2 + DR^2$. So $OR^2 = OA^2 + AD^2 + DR^2 = OA^2 + OB^2 + OC^2$. But $OR$ is the magnitude of **V**. Therefore, by Equation (2.3),

$$V^2 = V_x^2 + V_y^2 + V_z^2, \tag{2.4}$$

from which we obtain

$$V = \sqrt{V_x^2 + V_y^2 + V_z^2}. \tag{2.5}$$

Since the magnitude is always counted as non-negative, we do not use a $\pm$ sign in Equation (2.5).

**Exercise 2.1** A vector has components (1, 1, 1) relative to **i**, **j**, and **k**. What is its magnitude? Describe as best you can its direction in relation to **i**, **j**, and **k**.

**Exercise 2.2** Like Exercise 2.1 for a vector having components (1, 1, $-$ 1).

**Exercise 2.3** What are the magnitudes of the vectors having the following components relative to **i**, **j**, and **k**? (a) (1, 2, 2); (b) (– 4, – 4, 7); (c) ($\sqrt{3}$, $\sqrt{7}$, – $\sqrt{6}$); (d) (10, – 20, – 30); (e) ($V_x$, $V_y$, 0).

**Exercise 2.4** Relative to a unit orthogonal triad, a vector **U** has components (5, 12, 5) and a vector **V** has components (4, – 3, $z$). What can you say about $z$ if **U** and **V** have equal magnitudes?

**Exercise 2.5** If **U** and **V** have components ($U_x$, $U_y$, $U_z$) and ($V_x$, $V_y$, $V_z$) relative to **i**, **j**, and **k**, what are the magnitudes of **U** + **V** and **U** – **V**?

**Exercise 2.6** A particle is acted on by two forces **F** and **F**′ having components ($F_x$, $F_y$, $F_z$) and ($F'_x$, $F'_y$, $F'_z$) relative to **i**, **j**, and **k**. What are the components and magnitude of the force that will keep the particle in equilibrium?

**Exercise 2.7** What is the locus of the tip of a vector of fixed magnitude $r$ and fixed initial point $O$? If the components of the vector relative to **i**, **j**, and **k** are ($x$, $y$, $z$) what equation must these quantities satisfy?

For given **i** and **j**, we can take **k** in either of the two opposite directions perpendicular to the plane of **i** and **j**, as shown in Figure 2.3. The two configurations (a) and (b) are distinct. We can not maneuver one of them as a rigid body so as to make it coincide with the other. We therefore give the configurations different names; we refer to (a) as a *right-handed system* and to (b) as a *left-handed system.*

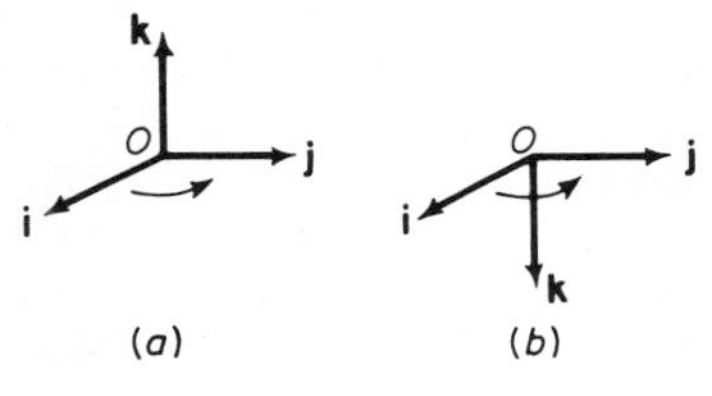

**Figure 2.3**

Imagine a right-handed corkscrew pointing along the **k**-direction in the right-handed system in (a). If we twist the handle as if we were trying to make **i** take up the position now occupied by **j** (via the 90° route, not the 270° one!) the corkscrew will move in the direction in which **k** points. In the left-handed system (b) it will move in the direction opposite to that in which **k** points.

**Exercise 2.8** Verify that the preceding paragraph will remain valid if we replace **i**, **j**, and **k** by **j**, **k**, and **i** respectively, or by **k**, **i**, and **j** respectively.

**Exercise 2.9** In Figure 2.2, do the lines $OA$, $OB$, and $OC$ form a right-handed or a left-handed system? What if we took them in the order $OB$, $OC$, and $OA$? What if we took them in the order $OB$, $OA$, and $OC$? [*Ans.* right-handed; right-handed; left-handed.]

**Exercise 2.10** Starting with a right-handed system, what does it become if we reverse the direction of one axis? Of two? Of three?

**Exercise 2.11** Give a definition of right-handedness that would apply to a general reference frame defined by $\mathbf{e}_x$, $\mathbf{e}_y$, and $\mathbf{e}_z$ in that order [Note the importance of the words *in that order*. We usually take them for granted without mentioning them. The definition is a simple generalization of that for the orthogonal case; but one must be careful to specify that, for example, the rotation of $\mathbf{e}_x$ towards $\mathbf{e}_y$ is via the shorter route. What is the maximum angle one could permit in this connection?]

## 3. POSITION VECTORS

Let the period at the end of this sentence represent a point *P*.

Where is the point *P*? We could say it is right there, at the end of the sentence. But where is the sentence? On page 40 of a book? Then where is the book? On a table in a certain room on the earth? Then where is the earth?

After pushing this silent dialogue with ourselves a little further, we begin to realize that the concept of "where" is fundamentally a puzzling one.

But suppose we erect a frame of reference by means of three base vectors $\mathbf{e}_x$, $\mathbf{e}_y$, and $\mathbf{e}_z$. Then, without knowing *where* this frame of reference is in any deep sense of the word, we can nevertheless ask where a point is in relation to the reference frame; and this relative sort of *where* is a far more tractable concept.

To specify the location of a point relative to a frame of reference, it suffices to give the components of the vector that starts at *O* and ends at the point. For example, in Figure 1.2 the location of the point *R* relative to the reference frame is given by the components of the vector **V** relative to the frame.

**Exercise 3.1** Point *R* is the midpoint of $\mathbf{e}_x$. Denote the vector $\overrightarrow{OR}$ by **V**. What are the components of **V** relative to $\mathbf{e}_x$, $\mathbf{e}_y$, and $\mathbf{e}_z$?

**Exercise 3.2** What would the components of the vector **V** in Exercise 3.1 become if, keeping the same origin, we rotated to a new reference frame with base vectors $\mathbf{e}'_x$, $\mathbf{e}'_y$, and $\mathbf{e}'_z$ such that $\mathbf{e}'_x = \mathbf{e}_y$, $\mathbf{e}'_y = \mathbf{e}_z$, and $\mathbf{e}'_z = \mathbf{e}_x$? [*Ans.* $(0, 0, \frac{1}{2})$.]

We could dispense with a reference frame, retaining only its origin, *O*, and specifying the point *R* by means of the vector **V** itself rather than by means of its components relative to some reference frame. Or so it would seem. Indeed, the idea becomes quite attractive when we recall that if we change from one reference frame to another having the same origin, the components of **V** change but the vector itself does not.

But what do we mean when we say that the vector **V** does not change? Change relative to what? How do we know it isn't whirling around *O* at an alarming rate? You say you know because you are holding your book steady? Then hurl it into the air, and now try answering the question.

It is clear that when we think we are dispensing with a reference frame we really are not. In doing Exercise 3.2 above, did we so much as pause to ask ourselves what was meant by "the vector **V** in Exercise 3.1"? Did we not assume that the meaning was obvious? The meaning is indeed obvious. But let us face what it implies. In thinking of **V** as being the same in Exercises 3.1 and 3.2, we have in mind some master reference frame relative to which **V** is indeed unchanging. It could be the frame $\mathbf{e}_x$, $\mathbf{e}_y$, and $\mathbf{e}_z$, but it need not be; and the chances are that we do not think of it as being that reference frame. However, we do here think of $\mathbf{e}_x$, $\mathbf{e}_y$, and $\mathbf{e}_z$ as being unchanging vectors relative to this master frame, and we think of $\mathbf{e}'_x$, $\mathbf{e}'_y$, and $\mathbf{e}'_z$ as being so too. And if we should have occasion to speak of a *moving* reference frame, we would really mean that it was moving relative to some other frame, whether we mentioned this other frame or not.

Having looked into the situation, we may, for the most part, go on speaking about fixed vectors, fixed points, and the like as we have been doing all along, realizing now that we are really doing so in relation to some mentioned or unmentioned master reference frame.

Given a fixed reference point $O$, then, we can specify the position of a point $P$ by means of a vector having magnitude and direction represented by the arrow-headed line segment $\overrightarrow{OP}$. This vector is called the *position vector* of $P$ relative to $O$.

In what sense can we locate a point $P$ relative to a point $O$ by means of a *vector?* In thinking of $\overrightarrow{OP}$ not just as an arrow-headed line segment but as a vector, we commit ourselves to seemingly nonsensical things about the positions of points; for example, that if $OPRQ$ is a parallelogram, the position of $P$ relative to $O$ and the position of $Q$ relative to $O$ are together somehow equivalent to the position of $R$ relative to $O$. Stated this way, it is indeed nonsense. It makes sense if we regard these positions as displacements, even though they were all conceived as positions relative to the same point $O$; but it does not make the sort of sense we had in mind when we thought we were specifying the locations of points relative to $O$. Even so, the idea of representing these positions by means of vectors that we then regard as displacements is a powerful one, as the following exercises will indicate.

**Exercise 3.3** If points $A$ and $B$ have position vectors **a** and **b** relative to $O$, what is the position vector of the mid-point $M$ of $AB$?

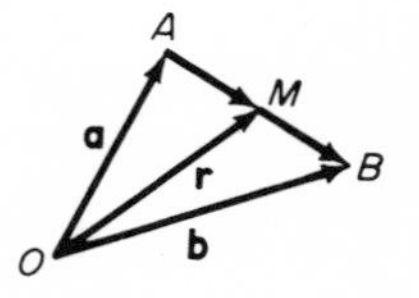


**Figure 3.1**

*Solution* Denote the position vector of $M$ by **r**. Then, regarding **a** and **b** as displacements, which are free vectors, we have:

$$\overrightarrow{AM} = \overrightarrow{OM} - \overrightarrow{OA} = \mathbf{r} - \mathbf{a}$$

and

$$\overrightarrow{MB} = \overrightarrow{OB} - \overrightarrow{OM} = \mathbf{b} - \mathbf{r}.$$

But the displacements $\overrightarrow{AM}$ and $\overrightarrow{MB}$ are equal. Therefore,

$$\mathbf{r} - \mathbf{a} = \mathbf{b} - \mathbf{r},$$

from which we find that

$$\mathbf{r} = \frac{1}{2}(\mathbf{a} + \mathbf{b}). \tag{3.1}$$

This relation is called the MID-POINT FORMULA.

**Exercise 3.4** Points $A_1$ and $A_2$ have position vectors $\mathbf{a}_1$ and $\mathbf{a}_2$ relative to $O$. A point $P$ on $A_1A_2$ is such that $A_1P/PA_2 = m_1/m_2$, where $m_1$ and $m_2$ are numbers. Show that the position vector, $\mathbf{r}$, of the point $P$ is given by:

$$\mathbf{r} = \frac{m_2\mathbf{a}_1 + m_1\mathbf{a}_2}{m_1 + m_2}. \tag{3.2}$$

[This relation is called the POINT-OF-DIVISION FORMULA.]

**Exercise 3.5** Using vectors, prove that the medians of a triangle meet at a point that divides each median in the ratio 2:1 starting from its vertex.

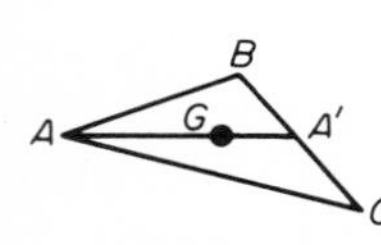


**Figure 3.2**

*Solution* Denote the position vectors of the vertices $A$, $B$, and $C$ of the triangle by $\mathbf{a}$, $\mathbf{b}$, and $\mathbf{c}$ respectively. Then, by the mid-point formula, the position vector of the mid-point $A'$ of side $BC$ is $\frac{1}{2}(\mathbf{b} + \mathbf{c})$. Let $G$ be the point of $AA'$ such that $AG = 2GA'$, so that $AG{:}GA' = 2{:}1$. Then, by the point-of-division formula, the position vector of $G$ is:

$$\frac{1\mathbf{a} + 2\left[\frac{1}{2}(\mathbf{b} + \mathbf{c})\right]}{3},$$

which is just $\frac{1}{3}(\mathbf{a} + \mathbf{b} + \mathbf{c})$. This quantity is symmetrical in $\mathbf{a}$, $\mathbf{b}$, and $\mathbf{c}$: though we found it by going two-thirds of the way down the median through $A$, it contains the vector $\mathbf{a}$ on an equal footing with $\mathbf{b}$ and $\mathbf{c}$. Consequently, if we did the corresponding calculations for the other medians they would have to give essentially the same result — if we made no mistakes, or all our mistakes cancelled. Therefore, the point $G$ is common to all three medians.

**Exercise 3.6** Do Exercise 3.5 in terms of the median through $B$.

**Exercise 3.7** Using vectors, prove that the line segments joining the mid-points of opposite sides of any quadrilateral, whether plane or skew, bisect each other.

A position vector is a curious thing. It is akin to a displacement, but instead of being free it is tied to a particular origin $O$. If the position vector of a point with respect to an origin $O$ has the same magnitude and direction as that of $P'$ relative to $O'$, the points $P$ and $P'$ will not coincide unless $O$ and $O'$ do. With $O$ and $O'$ distinct, the position vectors $\overrightarrow{OP}$ and $\overrightarrow{O'P'}$ do not define the same position, though they would be counted as equal vectors if they were displacements. Yet if we think of a position vector as a bound vector starting at $O$ we have to wonder in what sense, for example, the position vectors $\overrightarrow{OA}$ and $\overrightarrow{OM}$ in Figure 3.1 can combine to give a vector $\overrightarrow{AM}$ that not only does not start at $O$ but does not even pass through $O$. And we may conclude that $\overrightarrow{OM} - \overrightarrow{OA}$ does not yield $\overrightarrow{AM}$ but a vector starting at $O$ having the same magnitude and direction as $\overrightarrow{AM}$. This is awkward, but Exercise 3.3 and others like it can, in fact, be worked out in terms of vectors all of which start at $O$.

**Exercise 3.8** Work out Exercise 3.3 in terms of such vectors. (Note that a crucial step is to require that the vectors defined by $\overrightarrow{OM} - \overrightarrow{OA}$ and $\overrightarrow{OB} - \overrightarrow{OM}$ be identical. Draw the two parallelograms involved.)

Perhaps we find it distasteful to work in this way here in terms of bound vectors all starting at $O$. But before we decide to reject them, let us consider Figure 3.3. What does it represent? Three guesses: A design with match sticks? A lopsided equation telling that $X = 1$? A preliminary study for a modern work of art? All wrong. It is Figure 3.1 done in terms of free vectors (with letters and arrowheads omitted).

**Figure 3.3**

When one is dealing with position vectors, there seems to be no way of avoiding awkwardness, except by ignoring it. But that is not a bad method.

**Exercise 3.9** It is instructive to try to do Exercise 3.3 solely in terms of shifts, using only relations of the type $OA + AM = OM$. Try this, and note why the attempt does not succeed [For example, $AM$ and $MB$ are not identical shifts, and Equation (3.1) is not applicable to shifts, though it is to displacements corresponding to shifts. One sees from this how powerful is the concept of a vector as compared with that of a shift, even though the shift seems to contain more data of importance.]

Suppose, in Exercise 3.3, we changed from the origin $O$ to a new origin $O'$. Then all the position vectors would be changed. Yet (relative to our master reference frame) the points $A$, $M$, and $B$ would not move, and $M$ would remain the mid-point of the line segment $AB$. Clearly, the mid-point formula (3.1) will have to hold for the new position vectors as well as for the old. If we are dubious (though the foregoing is really a sufficient proof), we can check very easily. Let the position vector of $O'$ relative to $O$ be $\mathbf{h}$. Then, from Figure 3.4, we see that the old position vector $\mathbf{a}$ of point $A$ is related to its new position vector $\mathbf{a}'$ by

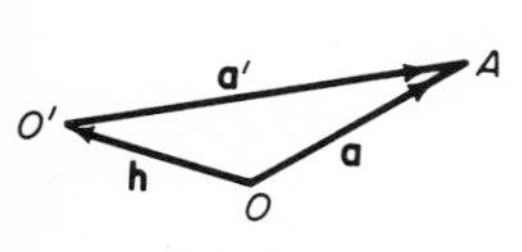


**Figure 3.4**

$$\mathbf{a} = \mathbf{a}' + \mathbf{h}. \tag{3.3}$$

Similar relations hold for all other position vectors. Let us introduce new

position vectors for old ones in Equation (3.1), namely $\mathbf{r} = \frac{1}{2}(\mathbf{a} + \mathbf{b})$. Then we have

$$\mathbf{r}' + \mathbf{h} = \tfrac{1}{2}\,[(\mathbf{a}' + \mathbf{h}) + (\mathbf{b}' + \mathbf{h})],$$

which reduces to

$$\mathbf{r}' = \tfrac{1}{2}\,(\mathbf{a}' + \mathbf{b}'),$$

as we foretold that it must.

**Exercise 3.10** Show by the two methods above that the point-of-division formula (3.2) also holds for new position vectors when the origin is changed.

**Exercise 3.11** Like Exercise 3.10 for the formulas for the position vectors of the ultimate points of intersection in Exercises 3.5 and 3.7.

There is a simple theorem that tells us what sort of linear relation between position vectors preserves its form when the origin is changed. Let the relation be:

$$\alpha_1\mathbf{a}_1 + \alpha_2\mathbf{a}_2 + \ldots + \alpha_n\mathbf{a}_n = 0, \tag{3.4}$$

where the **a**'s are position vectors relative to $O$, and the $\alpha$'s are numbers. Make the change to a new origin $O'$ by means of equations of which Equation (3.3) is a specimen. Then Equation (3.4) becomes:

$$\alpha_1(\mathbf{a}_1' + \mathbf{h}) + \alpha_2(\mathbf{a}_2' + \mathbf{h}) + \ldots + \alpha_n(\mathbf{a}_n' + \mathbf{h}) = 0$$

or

$$\alpha_1\mathbf{a}_1' + \alpha_2\mathbf{a}_2' + \ldots + \alpha_n\mathbf{a}_n' + (\alpha_1 + \alpha_2 + \ldots + \alpha_n)\mathbf{h} = 0,$$

and this will reduce to Equation (3.4), with **a**′'s replacing **a**'s, if and only if

$$\alpha_1 + \alpha_2 + \ldots + \alpha_n = 0. \tag{3.5}$$

**Exercise 3.12** Verify that the condition (3.5) is satisfied by Equation (3.1). [Write Equation (3.1) as $\mathbf{r} - \frac{1}{2}\mathbf{a} - \frac{1}{2}\mathbf{b} = 0$ and note that the coefficients $1, -\frac{1}{2}$, and $-\frac{1}{2}$ add up to zero.]

**Exercise 3.13** Like Exercise 3.12 for the results of Exercises 3.4, 3.5, and 3.7. [Note that in Figure 3.2 we did not even bother to draw the origin.]

If, without drawing the origin, we can draw a diagram that nevertheless contains all the essential data, the expression of these data in terms of position vectors will retain the same form when the origin is changed. And this will be true even if the relations between the vectors are nonlinear. The reason is so simple as to be almost primitive. The position vectors depend on the origin. If they are used in order to represent a configuration that does not even involve an origin, then they must do so in a way that does not give one particular origin preference over any other. Therefore, the relation between the position vectors must retain the same form when the origin is changed, since otherwise all origins would not enjoy equal status—or perhaps we should say here, lack of status.

Examples of such situations are easy to find. For instance, we can draw a diagram of the fact that $M$ is the mid-point of the line segment $AB$ without needing to draw in an origin. Other such situations have been encountered in Exercises 3.4, 3.5, and 3.7. Yet others are that four given points lie in a plane, or that they lie at the vertices of a regular tetrahedron, or that one of the points is the intersection of the angle bisectors of the triangle formed by the other three. [The vectorial formula for this last would probably be encountered as a formula for the position vector of the point of intersection of the angle bisectors in terms of the position vectors of the vertices of the triangle.]

**Exercise 3.14** Show by means of Equation (3.3) that $\overrightarrow{OB} - \overrightarrow{OA} = \overrightarrow{O'B} - \overrightarrow{O'A}$, and therefore, that any vector relationship built up solely of combinations of the form $\overrightarrow{OB} - \overrightarrow{OA}$ will retain its form when the origin is changed. [Note how this is related to the preceding discussion of diagrams in view of the fact that $\overrightarrow{OB} - \overrightarrow{OA} = \overrightarrow{AB}$.]

**Exercise 3.15** Show that if $\mathbf{a}_1$, $\mathbf{a}_2$, and $\mathbf{a}_3$ are position vectors and $\beta_{23}$, $\beta_{31}$, and $\beta_{12}$ are numbers, the linear relation among differences of $\mathbf{a}$'s,

$$\beta_{23}(\mathbf{a}_2 - \mathbf{a}_3) + \beta_{31}(\mathbf{a}_3 - \mathbf{a}_1) + \beta_{12}(\mathbf{a}_1 - \mathbf{a}_2) = 0,$$

satisfies the condition (3.5) for being independent of the origin.

**Exercise 3.16** Extend Exercise 3.15 to the case of $n$ position vectors $\mathbf{a}_1, \mathbf{a}_2, \ldots, \mathbf{a}_n$.

**Exercise 3.17** Show that if $\mathbf{a}_1$, $\mathbf{a}_2$, and $\mathbf{a}_3$ are position vectors and the linear relation

$$\alpha_1\mathbf{a}_1 + \alpha_2\mathbf{a}_2 + \alpha_3\mathbf{a}_3 = 0$$

is independent of the origin so that $\alpha_1 + \alpha_2 + \alpha_3 = 0$, then the linear relation can be written as a linear relation among differences of position vectors. [Thus we can rearrange the left-hand side by writing it as $\alpha_1(\mathbf{a}_1 - \mathbf{a}_2) + (\alpha_1 + \alpha_2)\mathbf{a}_2 + \alpha_3\mathbf{a}_3$, which is equal to $\alpha_1(\mathbf{a}_1 - \mathbf{a}_2) + (\alpha_1 + \alpha_2)(\mathbf{a}_2 - \mathbf{a}_3) + (\alpha_1 + \alpha_2 + \alpha_3)\mathbf{a}_3$, and the last term vanishes because of the conditions on the $\alpha$'s. Now work the problem starting with $\mathbf{a}_3$ instead of $\mathbf{a}_1$, and note that the resulting relation has a different form.]

**Exercise 3.18** At first sight, it may seem surprising that there should be various ways of expressing a linear relation among position vectors in terms of differences of such vectors, in view of the fact that these differences directly represent line segments in a diagram that does not contain $O$. To understand what is involved, show that the mid-point formula in Exercise 3.3 can be rewritten in the following three forms, and state their geometric interpretation in terms of line segments involving the points $A$, $M$, and $B$: $(\mathbf{r} - \mathbf{a}) + (\mathbf{r} - \mathbf{b}) = 0$, $(\mathbf{r} - \mathbf{a}) - \frac{1}{2}(\mathbf{b} - \mathbf{a}) = 0$; $(\mathbf{r} - \mathbf{b}) - \frac{1}{2}(\mathbf{a} - \mathbf{b}) = 0$.

**Exercise 3.19** Denoting the position vector of the point $G$ in Exercise 3.5 by $\mathbf{g}$, express the relation $\mathbf{g} = \frac{1}{3}(\mathbf{a} + \mathbf{b} + \mathbf{c})$ in a form involving only differences of position vectors, and interpret the result geometrically. [One such form is $(\mathbf{g} - \mathbf{a}) + (\mathbf{g} - \mathbf{b}) + (\mathbf{g} - \mathbf{c}) = 0$. This tells us that the vectors represented by $\overrightarrow{GA}$, $\overrightarrow{GB}$, and $\overrightarrow{GC}$ have zero resultant. So if, for example, we complete the parallelogram $CGBK$, we see that $GK$ is an extension of $AG$ and of the same length as $AG$. Thus, by merely rearranging the original relation between position vectors in terms of differences of these vectors (a rearrangement that is only possible if the original relation is independent of the origin), we come out with a geometrical theorem.]

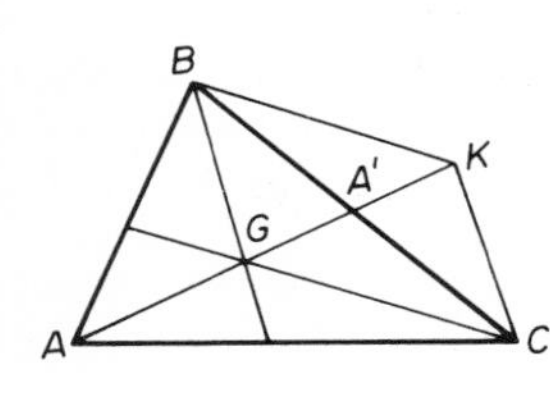


**Exercise 3.20** In Exercise 3.19, one could rewrite the relation in the form $3(\mathbf{g} - \mathbf{a}) + 2(\mathbf{a} - \mathbf{b}) + (\mathbf{b} - \mathbf{c}) = 0$. Interpret this geometrically. [It is easier to see what is involved if one rewrites this as $\frac{3}{2}(\mathbf{g} - \mathbf{a}) + (\mathbf{a} - \mathbf{b}) + \frac{1}{2}(\mathbf{b} - \mathbf{c}) = 0$.]

**Exercise 3.21** $A$, $B$, $C$, and $D$ are four points not necessarily in the same plane. Using position vectors, prove that if the line segments $AC$ and $BD$ bisect each other, $ABCD$ is a parallelogram. [*Hint:* rearrange $(\mathbf{a} + \mathbf{c})/2 = (\mathbf{b} + \mathbf{d})/2$ to obtain $\mathbf{a} - \mathbf{b} = \mathbf{c} - \mathbf{d}$.]

**Exercise 3.22** Extend Exercise 3.17 to the case of $n$ position vectors $\mathbf{a}_1, \mathbf{a}_2, \ldots, \mathbf{a}_n$.

Let $A$ be a fixed point and $\mathbf{c}$ a given vector. How can we represent the line through $A$ parallel to $\mathbf{c}$? If $P$ is any point on this line, and the position vectors of $A$ and $P$ are $\mathbf{a}$ and $\mathbf{r}$, respectively, then, since $AP$ is parallel to $\mathbf{c}$,

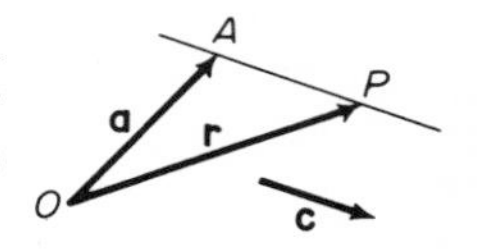


$$\mathbf{r} - \mathbf{a} = \mathbf{c}t, \tag{3.6}$$

where $t$ is a parameter. We may rewrite this as

$$\mathbf{r} = \mathbf{a} + \mathbf{c}t, \tag{3.7}$$

and this represents the line in the sense that it gives the position vector of every point on the line in terms of the parameter $t$.

Consider now the analogous problem of representing the straight line through two points $A$ and $B$ with position vectors $\mathbf{a}$ and $\mathbf{b}$. Here we have merely to replace $\mathbf{c}$ in Equation (3.6) or (3.7) by $\mathbf{b} - \mathbf{a}$, since the line is certainly parallel to the vector $\mathbf{b} - \mathbf{a}$ inasmuch as it actually lies along $AB$. So we have

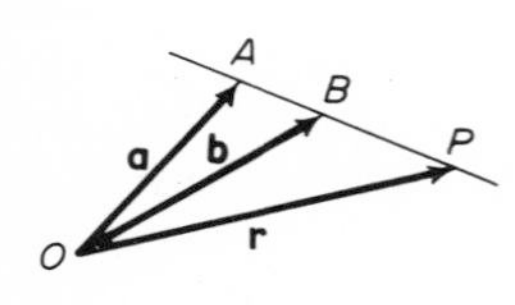


$$\mathbf{r} - \mathbf{a} = (\mathbf{b} - \mathbf{a})t \tag{3.8}$$

or

$$\mathbf{r} = \mathbf{a} + (\mathbf{b} - \mathbf{a})t. \tag{3.9}$$

Now a seeming puzzle arises. If we write Equation (3.7) in the form:

$$\mathbf{r} - \mathbf{a} - t\mathbf{c} = 0$$

and Equation (3.9) in the form:

$$\mathbf{r} + (t - 1)\mathbf{a} - t\mathbf{b} = 0,$$

we see that the sum of the coefficients in the latter is $1 + (t - 1) - t$, which is zero, whereas the corresponding sum in the former is $1 - 1 - t$, which is not zero. It appears, then, that Equation (3.9) is independent of the origin but (3.7) is not. This is not true, though, or rather, it is true that this *appears* to be the case, but it is actually not the case. Both equations are independent of the origin. This is clear from the fact that (3.6) involves only $(\mathbf{r} - \mathbf{a})$ and $\mathbf{c}$, each of which is independent of the origin. Where is the fallacy? [*Hint:* Why is $\mathbf{c}$ independent of the origin? Is it a position vector? What is the nature of the vectors $\mathbf{a}_1, \ldots, \mathbf{a}_n$ in Equation (3.5)? Does Equation (3.7) fall within the category of (3.5), or is it more general?]

All this preoccupation with vectorial relations that are independent of the origin can be misleading. It is a standard preoccupation, and it is forced on us because we are using those hybrid things, position vectors. But it obscures a crucial point. With what we may refer to as nonposition vectors, i. e., vectors of the usual type, relations such as

$$\overrightarrow{AM} = \overrightarrow{MB} \quad \text{and} \quad \mathbf{U} + 2\mathbf{V} = 5\mathbf{W}$$

are automatically independent of any choice of origin. They are, in fact, objective. And the objectivity of such vectorial relations is one of the most valuable attributes of vectors. Bear it in mind. It will play a crucial role later.

## 4. COORDINATES

Let $\mathbf{r}$ be the position vector of a point $P$ relative to a reference frame defined by vectors $\mathbf{e}_x$, $\mathbf{e}_y$, and $\mathbf{e}_z$ starting at an origin $O$. The components of $\mathbf{r}$ relative to this frame are called the *coordinates* of $P$ relative to the frame, and they are usually denoted by $(x, y, z)$. [We are here considering the three-dimensional case, of course.] Thus,

$$\mathbf{r} = x\mathbf{e}_x + y\mathbf{e}_y + z\mathbf{e}_z. \tag{4.1}$$

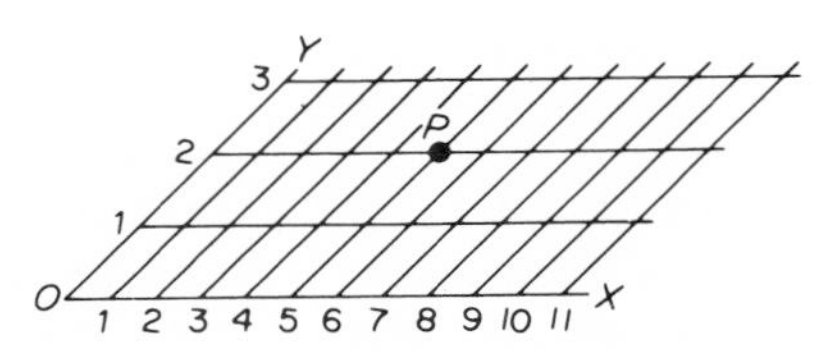


**Figure 4.1**

Note that while all vectors have components, only the components of *position* vectors are called coordinates. The coordinates here defined are of a restricted sort called *Cartesian coordinates.* Polar coordinates, and other curvilinear or nonuniform types of coordinates are not of this sort.

If we use Cartesian coordinates in two dimensions we are, in effect, using *graph paper* of the type shown in Figure 4.1, the point $P$ having the coordinates (5, 2). Note that the scales along the two axes, though uniform, are not necessarily the same.

**Exercise 4.1** In Figure 4.1, which vector has the greater magnitude, $\mathbf{e}_x$ or $\mathbf{e}_y$?

It is not so easy to draw a corresponding diagram of a three-dimensional Cartesian coordinate mesh: the diagram tends to become confusing.

**Exercise 4.2** Draw such a diagram and see for yourself.

**Exercise 4.3** Two points $A$ and $B$ have coordinates $(x_1, y_1, z_1)$, $(x_2, y_2, z_2)$ relative to a given reference frame. Prove that the coordinates of the mid-point, $M$, of $AB$ are: $\left(\frac{x_1 + x_2}{2}, \frac{y_1 + y_2}{2}, \frac{z_1 + z_2}{2}\right)$. [*Hint*: Merely consider the coordinates of the position vectors in the vectorial form of the mid-point formula in Equation (3.1).]

**Exercise 4.4** In Exercise 4.3, if $P$ is a point on $AB$ such that $AP/PB = m_1/m_2$, show that the coordinates of $P$ are: $\left(\frac{m_2 x_1 + m_1 x_2}{m_1 + m_2}, \frac{m_2 y_1 + m_1 y_2}{m_1 + m_2}, \frac{m_2 z_1 + m_1 z_2}{m_1 + m_2}\right)$.

**Exercise 4.5** Work out Exercises 3.5 and 3.7 using coordinates throughout. Note that, in effect, one does the same thing three times, once for the $x$-coordinates, once for the $y$, and once for the $z$.

**Exercise 4.6** Prove, first in terms of vectors and then in terms of coordinates, that the lines joining the vertices of any tetrahedron to the points of intersection of the medians of the opposite faces intersect at a point that divides each of the lines in the ratio 3: 1.

**Exercise 4.7** Show that, as $t$ changes, the point with coordinates $(a_x + tc_x, a_y + tc_y, a_z + tc_z)$ moves along a straight line, the $a$'s and $c$'s being constants. [Use Equation (3.7).]

When the reference frame is defined by a unit orthogonal triad of vectors, the corresponding coordinates are called *rectangular Cartesian coodinates.* [I suppose that, strictly speaking, this term should apply also to the case in which $\mathbf{e}_x$, $\mathbf{e}_y$, $\mathbf{e}_z$ are mutually perpendicular but do not have unit magnitudes or even equal magnitudes. But we shall use it only for the above coordinates.] In two dimensions, the corresponding graph paper is of the familiar sort.

**Exercise 4.8** Show that, in rectangular Cartesian coordinates, the distance from the origin of a point with coordinates $(x, y, z)$ is $\sqrt{x^2 + y^2 + z^2}$.

**Exercise 4.9** Points $A$ and $B$ have rectangular Cartesian coordinates $(x_1, y_1, z_1)$, $(x_2, y_2, z_2)$. Show that the vector represented by $\overrightarrow{AB}$ has components $(x_2 - x_1, y_2 - y_1, z_2 - z_1)$ and thus that the distance $AB$ is $\sqrt{(x_2 - x_1)^2 + (y_2 - y_1)^2 + (z_2 - z_1)^2}$.

## 5 DIRECTION COSINES

In Figure 5.1 the lines $OX$, $OY$, and $OZ$ are mutually perpendicular and lie along the positive directions of the vectors (not shown) **i**, **j**, and **k** of a unit orthogonal triad. These three lines are called *coordinate axes.* [They extend indefinitely on both sides of $O$, of course. The diagram shows only a

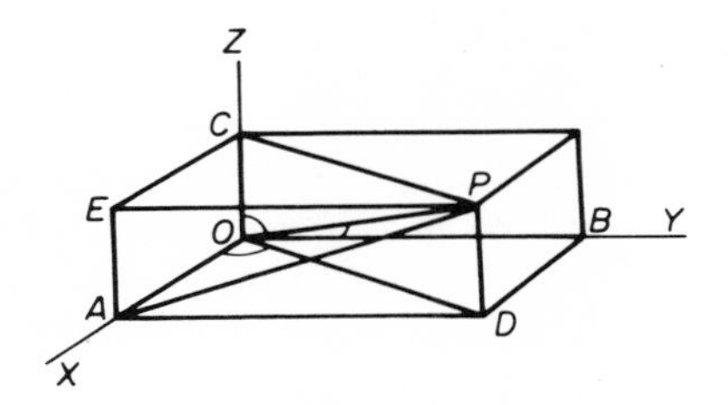


**Figure 5.1**

portion of each coordinate axis.] The line segment $OP$ is a diagonal of the rectangular parallelepiped shown. There are many right angles in the diagram that are not readily recognized as such.

**Exercise 5.1** Which pairs of lines meeting at the point $D$ are at right angles? [There are four such pairs.] Also, which pairs of lines meeting at the point $A$ are at right angles? [There are four—and that is without counting $AX$ and $AO$ as distinct lines; otherwise the count would be seven.]

We wish to specify in a convenient way the direction of $OP$ relative to **i**, **j**, and **k**. There are various possible ways. For example, we could state the angle through which the plane $OPC$ has swung about $OC$ from the plane $AOC$, and the angle in the plane $OPC$ that the line $OP$ makes with the line $OC$; that is, we could state the two angles $AOD$ and $COP$. Alternatively, we could use other appropriate pairs of angles. In certain cases describing the direction in terms of two angles is both convenient and useful. Indeed, the two angles $AOD$ and $COP$ are those used in what are called spherical polar coordinates.

There is a different method of specifying the direction of $OP$. It has the advantage of being symmetrical. But it uses *three* angles, and these angles, when first encountered, seem about as awkward a trio as one could imagine.

They are the three angles $AOP$, $BOP$, and $COP$, that $OP$ makes with the three axes $OX$, $OY$, and $OZ$. These angles, denoted respectively by $\alpha$, $\beta$, and $\gamma$, do not lie in one plane. Moreover, since any two of them suffice to determine the direction of $OP$, the three can not be independent of one another and, therefore, there must be a relation between them.

**Exercise 5.2** Is it strictly correct to say that any two of these angles suffice to determine the direction of $OP$? [For a given $\alpha$, $OP$ will be on a certain cone. For given $\beta$, it will be on another cone. Consider how these cones intersect.]

We may by now be wondering what merits the angles $\alpha$, $\beta$, and $\gamma$ can have that will outweigh the above disadvantages. But this is because we have, in a sense, told the story backwards. Mathematicians did not arbitrarily pick three queer angles to do the work of two seemingly more sensible ones. Let us look at the situation from a different point of view.

Consider $OP$ as a vector $\mathbf{V}$ with components $(V_x, V_y, V_z)$ relative to $\mathbf{i}$, $\mathbf{j}$, and $\mathbf{k}$. Then if we know the components, we know the vector, and therefore, in particular, its direction. But if $h$ is any number (greater than zero) the vector $h\mathbf{V}$ has the same direction as $\mathbf{V}$. Let us specify the direction, then, by means of the components of a vector of unit magnitude lying along $\mathbf{V}$. This unit vector is $(1/V)\mathbf{V}$, where $V$ is the magnitude of $\mathbf{V}$:

$$V = \sqrt{V_x^2 + V_y^2 + V_z^2}. \tag{5.1}$$

The components of the unit vector, denoted by $l$, $m$, and $n$, are given by

$$l = \frac{V_x}{V}, \qquad m = \frac{V_y}{V}, \qquad n = \frac{V_z}{V}; \tag{5.2}$$

and we easily see that

$$l^2 + m^2 + n^2 = 1, \tag{5.3}$$

so that any two of the quantities $l$, $m$, and $n$ automatically determine the third, to within a sign.

Having thus come upon these quantities $l$, $m$, and $n$, we now ask what they look like on a diagram. Let us look at triangle $POA$ in Figure 5.1. Since $OA$ is perpendicular to the plane $PDA$, it is perpendicular to every line in that plane, and therefore to the line $AP$. So despite appearances to the contrary, angle $OAP$ is a right angle. In the right triangle $OAP$, the length of $OA$ is $V_x$ and the length of the hypotenuse $OP$ is $V$. Therefore $V_x/V$, which is just $l$, is the cosine of the angle $AOP$ that we have called $\alpha$.

The coordinate planes $OYZ$, $OZX$, and $OXY$ (called respectively the *yz*-plane, the *zx*-plane, and the *xy*-plane) separate the whole three-dimensional space into eight regions called *octants*. We have here discussed only the case in which $OP$ points into what is called the first octant. When $OP$ points into other octants, some or all of the angles $AOP$, $BOP$, and $COP$ are obtuse and the corresponding cosines are negative.

**Exercise 5.3** Prove that angle $OBP$ is a right angle, and that $m$, which is $V_y/V$, equals $\cos\beta$. Prove also that $n = \cos\gamma$.

We see, then, that

$$l = \cos\alpha, \qquad m = \cos\beta, \qquad n = \cos\gamma; \tag{5.4}$$

and from Equation (5.3) we see that

$$\cos^2\alpha + \cos^2\beta + \cos^2\gamma = 1. \tag{5.5}$$

The angles $\alpha$, $\beta$, and $\gamma$ that the line $OP$ makes with the coordinate axes are called the *direction angles* of $OP$ with respect to the reference frame; and the cosines of these angles are called its *direction cosines.* In practice, one works much more with the direction cosines than directly with the direction angles.

Though the line segments $OP$ and $PO$ are the same, the directions $OP$ and $PO$ are opposite, and if the direction cosines of the direction $OP$ are $l$, $m$, and $n$, those of the direction $PO$ are $-l$, $-m$, and $-n$. When we wish to stress that we are thinking of a given line (or line segment) as pointing in one rather than the other of the two directions associated with it, we call it a *directed* line (or line segment). Thus the $x$-axis, in its role as a coordinate axis, is a directed line rather than just a line, though we often think of it as just a line—as when we say that a point is 5 units away from it.

If a directed line or line segment does not pass through $O$, its direction can still be given in terms of direction cosines, since the direction is the same as that of a parallel directed line that does pass through $O$. Let points $A$ and $B$ have position vectors $\mathbf{r}_a$ and $\mathbf{r}_b$ relative to $\mathbf{i}$, $\mathbf{j}$, and $\mathbf{k}$, and let the coordinates of these points be $(x_a, y_a, z_a)$, $(x_b, y_b, z_b)$. Then the displacement $\overrightarrow{AB}$ has components $(x_b - x_a, y_b - y_a, z_b - z_a)$, and therefore its direction cosines are

$$\frac{x_b - x_a}{d}, \qquad \frac{y_b - y_a}{d}, \qquad \frac{z_b - z_a}{d} \tag{5.6}$$

where

$$d = \sqrt{(x_b - x_a)^2 + (y_b - y_a)^2 + (z_b - z_a)^2}. \tag{5.7}$$

**Exercise 5.4** What are the direction cosines of $\mathbf{i}$? Work this out (a) by means of a diagram, and (b) purely algebraically, by means of Equation (5.2).

**Exercise 5.5** If $OP$ has direction cosines such that $l$ and $m$ are positive but $n$ is negative, into which octant does it point? What if $l$ and $m$ were negative and $n$ positive?

**Exercise 5.6** A line lies in the plane of $\mathbf{i}$ and $\mathbf{j}$, and makes equal angles with the lines containing $\mathbf{i}$ and $\mathbf{j}$. What are its three direction cosines? Express them in terms of $\alpha$ only [There are four possible cases—two pairs of opposite directions. Had we spoken of *vectors* instead of *lines*, there would have been only two cases. Why?]

**Exercise 5.7** A line makes equal angles with **i**, **j**, and **k**. What are its direction cosines if all are positive?

**Exercise 5.8** A vector of magnitude $V$ has direction cosines $l, m$, and $n$. What are its components? [*Ans.* $(lV, mV, nV)$.]

**Exercise 5.9** If two of the direction cosines of a vector are $\cos\theta \cos\varphi$ and $\cos\theta \sin\varphi$, what is the third direction cosine? [*Ans.* $\pm \sin\theta$.]

**Exercise 5.10** If the direction cosines of a line are positive and in the ratio $2:3:4$, what are their actual values? [By Equation (5.3), the sum of the squares of the three direction cosines must be unity. But $2^2 + 3^2 + 4^2 = 29$. So we divide 2, 3, and 4 by $\sqrt{29}$ to obtain the direction cosines $2/\sqrt{29}$, $3/\sqrt{29}$, and $4/\sqrt{29}$. Three numbers that are *proportional* to direction cosines $l$, $m$, and $n$ are called *direction numbers*.]

**Exercise 5.11** A vector has direction numbers 1, 2, and 4. What are its direction cosines? [Two possibilities.]

## 6. ORTHOGONAL PROJECTIONS

We can get an idea of the usefulness and convenience of direction cosines by deriving some fundamental formulas of analytical geometry. We shall later derive these formulas by more compact vectorial methods, but the latter tend to obscure what is going on behind the scenes.

In deriving these formulas by means of direction cosines, we must make use of the idea of an *orthogonal projection.* Given a point $P$ and a plane $\pi$, drop a perpendicular from $P$ to $\pi$ to meet $\pi$ at the point $P'$. Then $P'$ is called the orthogonal projection of $P$ on the plane $\pi$. Given point $P$ and a line $\lambda$, the foot, $P'$, of the perpendicular from $P$ to $\lambda$ is called the orthogonal projection of $P$ on the line $\lambda$.

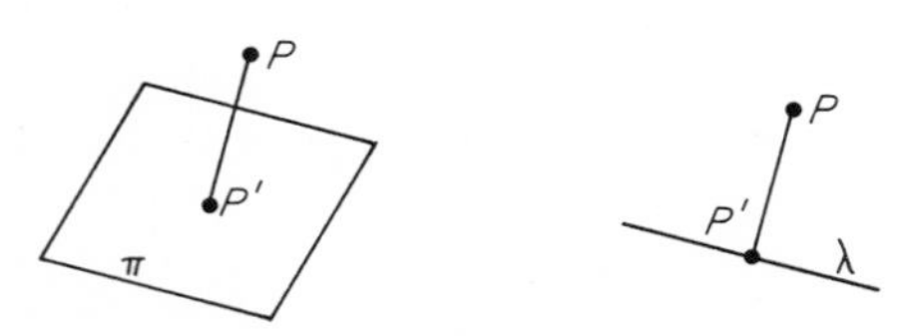


**Figure 6.1**

There are other types of projection, but it will be safe in this book for us to drop the word "orthogonal" when speaking of orthogonal projections. The idea of such projections seems so simple that one wonders how it could possibly be worth considering. Yet it is actually a powerful mathematical concept, as we shall see.

If $P$ traces out a curve, its projection, $P'$, on a plane $\pi$ traces out a curve called the projection on $\pi$ of the original curve. If the curve is a plane curve and closed, the area enclosed by its projection on $\pi$ is called the projection on $\pi$ of the area enclosed by the original curve.

**Exercise 6.1** What is the shape of the projection on $\pi$ of a straight line? What of a triangle? What of a circle? [*Ans.* A straight line or a point; a triangle or a line segment; a circle, or an ellipse, or a line segment.]

**Exercise 6.2** A rectangle $ABCD$ has $AB$ lying in the plane $\pi$ and $BC$ making an angle $\theta$ with the plane. If $K$ is the area of $ABCD$, show that the area of its projection on $\pi$ is $K \cos \theta$.

No matter how a point $P$ may move, its projection, $P'$, on a line $\lambda$ cannot move off the line. Two simple theorems form the basis of the applications of projections on a line.

**Theorem 1.** If a directed line segment $AB$ of length $d$ makes an angle $\theta$ with a directed line $\lambda$, the length and sign of its projection on $\lambda$ are given by $d \cos \theta$; the sign is positive if the projection points in the same direction as $\lambda$ and negative if it points in the opposite direction.

To prove this, pass planes through $A$ and $B$ that are perpendicular to $\lambda$ and let them cut $\lambda$ at $A'$ and $B'$. Then $A'$ and $B'$ are the projections of $A$ and $B$ whether the line $AB$ is coplanar with $\lambda$ or not. If the segment $AB$ is moved parallel to itself, with $A$ and $B$ remaining on the respective planes, the projections of $A$ and $B$ will be unaltered. So move $AB$ to the position $A'B_1$, $A'$ being the above-mentioned projection of A. Then from the right triangle $A'B'B_1$ we see that since $\cos \theta$ is negative when $\theta$ is obtuse, the length and sign of the projection, $A'B'$, are given by $d \cos \theta$.

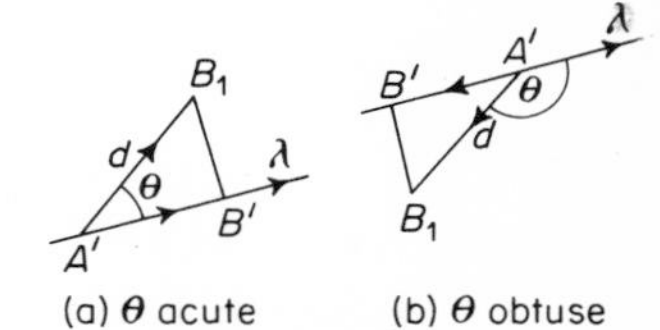


**Figure 6.2**

**Theorem 2.** If the projections of $A$ and $B$ on line $\lambda$ are $A'$ and $B'$, and a point $P$ starting at $A$ moves on a zigzag line ending at $B$, the algebraic sum of the projections on $\lambda$ of the line segments forming the zigzags is just $A'B'$.

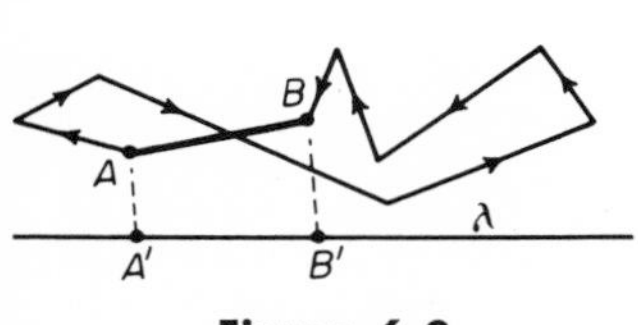


**Figure 6.3**

This powerful theorem may seem, at first glance, difficult to prove, especially when one realizes that the zigzag line need not lie in a plane but could even jut into four dimensions, or more. Yet the proof is almost a triviality, a fact that by no means lessens the theorem's importance. As $P$ traces out its zigzag path from $A$ to $B$, its projection, $P'$, moves to and fro on the line $\lambda$ starting at $A'$ and ending at $B'$; and when, for example, $P'$ retraces to the right ground previously traced out to the left, it cancels it. The *algebraic* sum of the projections is therefore $A'B'$. And that is that.

**Exercise 6.3** Let $O$ be the origin and $A$ the point $(x, 0, 0)$. Let $ON$ be a directed line segment having direction cosines $l$, $m$, and $n$. Show that the length and sign of the projection on $ON$ of the directed line segment $OA$ are given by $lx$ whatever the signs of $l$ and $x$. Also, if $K$ is the point $(x, y, 0)$, show that the length and sign of the projection on $ON$ of the directed line segment $AK$ are similarly given by $my$. [Use Theorem 1. Note how conveniently direction cosines enter.]

To apply these two theorems, we first consider a plane such that $ON$, the perpendicular to it from the origin, has length $p$ and direction cosines $l, m$, and $n$. Let any point $P$ on the plane have rectangular Cartesian coordinates $(x, y, z)$. We wish to find an equation that $x$, $y$, and $z$ must satisfy—an equation of the plane, as it is called. Drop a perpendicular from $P$ to the $xy$-plane meeting it at $K$. Draw $KA$ parallel to the $y$-axis to meet the $x$-axis at $A$. Then $OA = x$, $AK = y$, and $KP = z$. If and only if $P$ lies in the plane, $PN$ will be perpendicular to $ON$. Therefore:

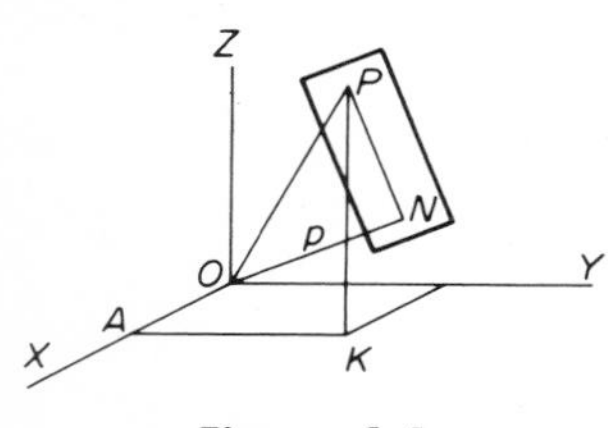

**Figure 6.4**

$$\text{projection of } OP \text{ on } ON = ON.$$

Using Theorem 2, we replace $OP$ by the zigzag $OAKP$. Then:

$$\text{sum of projections of } OA, AK, \text{ and } KP \text{ on } ON = ON. \tag{6.1}$$

Applying Exercise 6.3, and remembering that the length of $ON$ is $p$, we have immediately

$$lx + my + nz = p, \tag{6.2}$$

which is the equation we sought.

**Exercise 6.4** How far from the origin is the plane whose equation is $3x + 2y - 6z = 63$?

*Solution* The first step is to resist the temptation to say, by comparison with Equation (6.2), that $l = 3$, $m = 2$, $n = 6$ and $p = 63$. Obviously something is wrong with such a method, since we could have rewritten the given equation as, say, $300x + 200y - 600z = 6300$, and thus have come to the conclusion that $l$ is not 3 but 300, and similarly for $m$, $n$, and $p$. We have to remember that $l^2 + m^2 + n^2 = 1$. The numbers 3, 2, and $-6$ are not direction cosines but direction numbers (see Exercise 5.10). Since $\sqrt{3^2 + 2^2 + 6^2} = \sqrt{49} = 7$, we divide the given equation by 7 to obtain $\frac{3}{7}x + \frac{2}{7}y - \frac{6}{7}z = 9$. We may now make the identifications $l = \frac{3}{7}$, $m = \frac{2}{7}$, $n = \frac{6}{7}$, and therefore, also, $p = 9$.

**Exercise 6.5** Like Exercise 6.4 for the planes $x - 2y + 2z = 108$, $x + y + z = 9$, and $3y + 4z = 100$.

Consider next the problem of finding a formula for the angle between two directed lines having direction cosines $l, m, n$ and $l', m', n'$ respectively.

[If two lines do not intersect, the angle between them is defined as the angle between lines parallel to them that do intersect. Actually there are infinitely many angles, both positive and negative, between two given lines. When we talk of "*the* angle" between them, we presumably have some specific one in mind. Here, as on previous occasions, we mean the smallest positive angle between the positive directions of the lines regarded as directed lines.]

Denote the angle between the two lines by $\theta$, and for convenience (though it is not really necessary), imagine the lines emanating from the origin. Take points $P$ and $P'$ on the lines such that $OP$ and $OP'$ are of *unit* length. Then, by Theorem 1, the length and sign of the projection of $OP$ on $OP'$ will be given by just $\cos\theta$. But we can also compute the length and sign of the projection by using the zigzag path $OAKP$ instead of $OP$. Since $OP$ is of unit length, the coordinates of $p$ are just $(l, m, n)$, so that $OA = l$, $AK = m$, and $KP = n$. These line segments, being parallel to the coordinate axes, make with $OP'$ angles whose cosines are respectively $l'$, $m'$, and $n'$. Therefore the algebraic sum of the lengths of their projections on $OP'$ is $ll' + mm' + nn'$. So we must have:

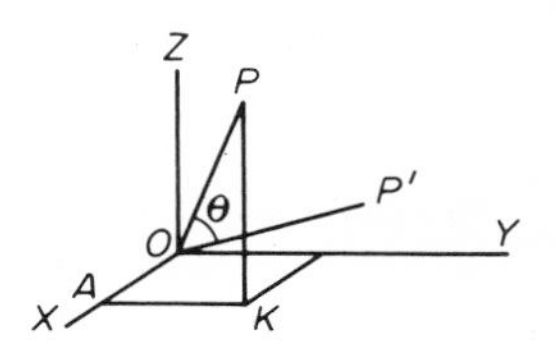

**Figure 6.5**

$$\cos\theta = ll' + mm' + nn'. \tag{6.3}$$

**Exercise 6.6** In the above, did we use the fact that $OP'$ was of unit length? Derive Equation (6.3) by projecting $OP'$ onto $OP$.

**Exercise 6.7** Derive Equation (6.3) by projecting $OP$ onto $OP'$, but taking $OP$ to be of arbitrary length $C$.

**Exercise 6.8** Find the cosine of the angle between two lines with direction numbers 1, 3, 5 and 2, $-$ 4, 3. $\left[\textit{Ans.}\ \dfrac{5}{\sqrt{35}\sqrt{29}}.\right]$

**Exercise 6.9** Prove that two lines are perpendicular if their direction cosines satisfy

$$ll' + mm' + nn' = 0. \tag{6.4}$$

**Exercise 6.10** Find the cosine of the angle between the planes $x + 3y + 5z = 10$ and $2x - 4y + 3z = 15$. [The angle is the same as that between the normals to the planes, and these normals have direction numbers 1, 3, 5 and 2, $-$ 4, 3 respectively.]

**Exercise 6.11** The $xy$-plane has an equation $z = 0$, and the $yz$-plane $x = 0$. Using Equation (6.4), verify that these planes are perpendicular.

**Exercise 6.12** What is the distance between the two parallel planes $x + 2y + z = 2$ and $x + 2y + z = 8$?

## 7. PROJECTIONS OF AREAS

In Exercise 6.2 we found that if rectangle $ABCD$ makes an angle $\theta$ with a plane $\pi$ through $AB$, the area of its projection is $\cos\theta$ times the area of $ABCD$. This is strongly reminiscent of the way projections of line segments behave, as given in Theorem 1. And the result is not confined to rectangles having one side in the plane of projection. It is not even confined to rectangles.

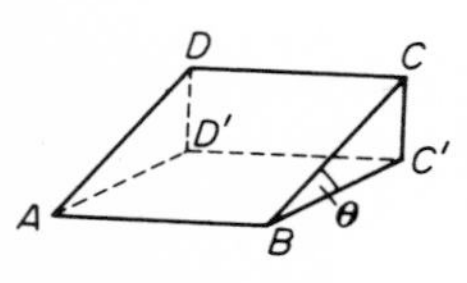


**Figure 7.1**

If we have a plane area of quite irregular shape, as in Figure 7.2, we can crisscross it with lines to break it up into a lot of rectangles—with some left-overs of irregular shape next to the boundary. Ignore the left-overs for the moment and consider the resulting serrated-edged region. Since the area of the projection of its constituent rectangles is $\cos\theta$ times the area of the original, the total area of the projection is $\cos\theta$ times that of the original serrated-edged region. So, for this region, the area of the projection is $\cos\theta$ times that of the original. The finer the crisscross mesh, the more nearly the serrated-edged region fills the over-all region, and it will hardly come as a surpise that, by methods belonging to the calculus, one can prove that the result actually holds for the over-all region too. Thus we have the important theorem that if a plane region of any shape has an area $K$, the area of its projection on a plane making an angle $\theta$ with it is $K\cos\theta$. We shall use this result in a later chapter.

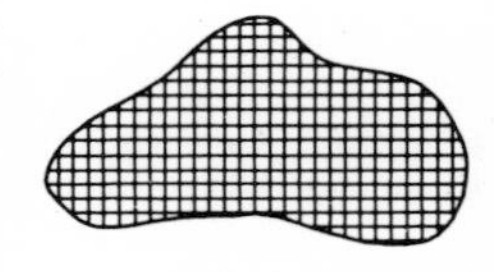

**Figure 7.2**

**Exercise 7.1** In Figure 7.1, take two points M and L on the line $DC$ such that $ML = DC$. Show that the area of the parallelogram $ABLM$ is equal to that of the rectangle $ABCD$. Hence, show that the area of the projection of this parallelogram is $\cos\theta$ times the area of the parallelogram.

**Exercise 7.2** A circular disc of radius 5 inches touches the ground, which is horizontal, and lies in a plane that is inclined at 45° to the ground. If the sun is vertically overhead, what is the area of the shadow of the disc, what is the time of day, and where on earth can the disc be?

# 4

# SCALARS. SCALAR PRODUCTS

## 1. UNITS AND SCALARS

The components of a vector **V** depend on the frame of reference. They change when we go from one frame to another. For example, if a displacement **V** lies in the plane of $\mathbf{e}_x$ and $\mathbf{e}_y$ as indicated on the left of Figure 1.1, its components $(\mathbf{V}_x, \mathbf{V}_y, \mathbf{V}_z)$ are such that $\mathbf{V}_x, \mathbf{V}_y \neq 0$, $\mathbf{V}_z = 0$. If we go over to a new reference frame in which $\mathbf{e}'_x$ and $\mathbf{e}'_z$ are the same as $\mathbf{e}_x$ and $\mathbf{e}_z$ but $\mathbf{e}'_y$ lies along **V**, as on the right of Figure 1.1, then the new components $(\mathbf{V}'_x, \mathbf{V}'_y, \mathbf{V}'_z)$ are such that $\mathbf{V}'_x = 0$, $\mathbf{V}'_y \neq 0$, and $\mathbf{V}'_z = 0$. Even when we merely change the magnitudes of the base vectors $\mathbf{e}_x$, $\mathbf{e}_y$, and $\mathbf{e}_z$ without altering their directions, the components of **V** are altered—unless **V** is the zero vector, of course. Nevertheless, no matter how we change the reference system, the vector itself remains the same [remember, here, the discussion we had earlier of the master reference frame], and in particular the magnitude of **V** remains the same.

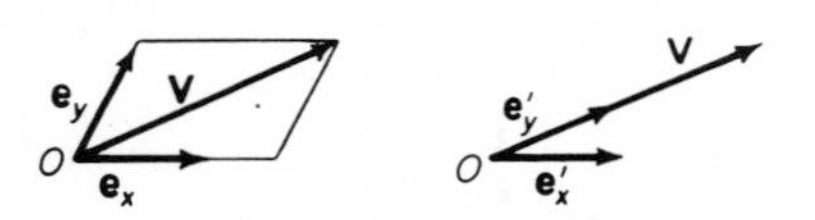

**Figure 1.1**

**Exercise 1.1** A displacement **V** points along the $\mathbf{e}_x$ direction and has magnitude $V$. Its components are therefore $(V/e_x, 0, 0)$, where $e_x$ is the magnitude of $\mathbf{e}_x$. What will its components be if we go over to a new reference frame in which $\mathbf{e}'_x$ has twice the magnitude of $\mathbf{e}_x$? [*Ans.* $(V/2e_x, 0, 0)$.]

We must now make a subtle distinction. Suppose that the displacement **V** has a magnitude of 108 inches, that the base vectors $\mathbf{e}_x$, $\mathbf{e}_y$, and $\mathbf{e}_z$ have each a magnitude of 1 inch, and that the components of **V** are (36, 24, 72). If we go over to a new reference frame in which the base vectors $\mathbf{e}'_x$, $\mathbf{e}'_y$, and $\mathbf{e}'_z$ differ from $\mathbf{e}_x$, $\mathbf{e}_y$, and $\mathbf{e}_z$ only in that their magnitudes are 12 inches instead of 1 inch, the components of **V** change to one-twelfth of their former values, i. e., to (3, 2, 6). Nevertheless the magnitude of **V** is still 108 inches. Its numerical value, 108, has not changed. Suppose, though, that we go back to the original reference frame and *without altering the base vectors or the vector* **V**, we use a different scale of measurement for all lengths, going from inches to feet. Then the components of **V** will still be (36, 24, 72) since they are the *ratios* of the magnitudes of the component vectors of **V** to the magnitudes of the **e**'s and thus are independent of the scale of measurement. But, though the magnitude of **V** is the same as before, its *numerical* value has changed from 108 to 9, because we are now measuring the magnitude in feet instead of inches.

**Exercise 1.2** What will the components of **V** be relative to the foregoing $\mathbf{e}'_x$, $\mathbf{e}'_y$, and $\mathbf{e}'_z$, when we go over from measuring in inches to measuring in feet? [*Ans.* (3, 2, 6), just as they were when we used inches in this reference frame.]

**Exercise 1.3** A force **F** has a magnitude of 6 lbs., and is represented by an arrow-headed line segment of length 6 inches. Its components relative to $\mathbf{e}_x$, $\mathbf{e}_y$, and $\mathbf{e}_z$ are (4, 1, 3). If the reference frame is changed to one with new base vectors $\mathbf{e}'_x$, $\mathbf{e}'_y$, and $\mathbf{e}'_z$ where $\mathbf{e}'_x = 10\mathbf{e}_y$, $\mathbf{e}'_y = 20\mathbf{e}_z$, and $\mathbf{e}'_z = 30\mathbf{e}_x$, what will be the new components of **F**? [*Ans.* $(\frac{1}{10}, \frac{3}{20}, \frac{2}{15})$.]

If the components of the displacement **V** are pure numbers, how can Equation (2.5) of chapter 3, namely,

$$V = \sqrt{V_x^2 + V_y^2 + V_z^2},$$

give us a length? How can it give us anything but a pure number? It is a standard equation, found in one guise or another in practically all textbooks on vectors. Let us consider it more carefully.

The fact is that we glossed over the question of units in deriving it. The legerdemain occurs in Equation (2.3) of Chapter 3. To keep the wording simple, let us discuss here the case in which all quantities involved are positive. If $OA$ means the length of the line segment $OA$, then it is not a pure number and so can not be equal to $\mathbf{V}_x$. It is really equal to $\mathbf{V}_x i$, where $i$ is the magnitude of **i** and thus a length. Since $i$ has the numerical value 1, it can be omitted—provided we realize that it is really there. Equation (2.5) of Chapter 3 can thus be regarded as yielding a length after all, though this would have been clearer had we written it as

$$V = \sqrt{V_x^2 i^2 + V_y^2 j^2 + V_z^2 k^2},$$

which corresponds (except for the taking of the square root) to Equation (2.1) of Chapter 3 for the case of a unit orthogonal triad. It was this sort of thing, among others, that we had in mind when we said there that the price of using such triads is loss of generality and loss of depth of understanding.

For a displacement **V**, which has a magnitude that is a length, the problem of units is relatively straightforward. What of a different sort of vector, such as a force **F**? The remark in Exercise 1.3 that the 6 lbs. is represented by 6 inches is less innocent than it seems. Forces combine by the parallelogram law with forces, not with displacements, whether the latter are thought of as localized or not. Therefore, a force cannot be equivalent to the vectorial sum of three such displacements. Strictly speaking, instead of writing an equation like

$$\mathbf{F} = F_x \mathbf{e}_x + F_y \mathbf{e}_y + F_z \mathbf{e}_z, \tag{1.1}$$

and thinking of the **e**'s as *geometrical* vectors having magnitudes that are lengths, we should write something like

$$\mathbf{F} = F_x \mathbf{f}_x + F_y \mathbf{f}_y + F_z \mathbf{f}_z, \tag{1.2}$$

where **F** and the **f**'s are all forces, and the components $F_x$, $F_y$, and $F_z$ are pure numbers. If we did this consistently with all types of vectors, the components would always be dimensionless numbers. But when working with vectors we like to draw arrow-headed line segments and parallelograms, and it is difficult to resist the temptation to think of the arrow-headed line segments as lengths. [We can hardly formulate the parallelogram law without thinking geometrically in this way. Compare page 9, where we explained the sense in which forces are vectors.] So we usually express **F** as in Equation (1.1) of this chapter. But, to clarify matters, let us here make a distinction between **F**, the force, and its geometrical counterpart, which we shall denote by $\mathscr{F}$. This $\mathscr{F}$ has a magnitude that is a length; for example, each pound of force could correspond to an inch of $\mathscr{F}$. Then if we write

$$\mathscr{F} = \mathscr{F}_x \mathbf{e}_x + \mathscr{F}_y \mathbf{e}_y + \mathscr{F}_z \mathbf{e}_z, \tag{1.3}$$

the components $\mathscr{F}_x$, $\mathscr{F}_y$, and $\mathscr{F}_z$ are pure numbers and are, in fact, the same as the $F_x$, $F_y$, and $F_z$ in Equation (1.2)—but not the same as those in Equation (1.1). For we can go from Equation (1.2) to Equation (1.1) by changing *all* lbs. into inches. But when we write Equation (1.1), we are apt to think of **F** as in lbs., not inches. In this case we have to say that the components $F_x$, $F_y$, and $F_z$ in Equation (1.1) are not pure numbers, but have the dimensions of a force divided by a length. When we then calculate the magnitude of **F** in terms of its components, we have to remember to bring in the magnitudes of the base vectors, the quantities $\mathscr{F}_x \mathbf{e}_x$, etc. being thus forces. Similar considerations apply to other vector quantities such as velocities, accelerations, and others we shall come upon later.

**Exercise 1.4** Force $\mathbf{F}$ has a magnitude of 6 lbs. and is represented by an arrow-headed line segment of length 6 inches. Its components $F_x$, $F_y$, and $F_z$ relative to $\mathbf{e}_x$, $\mathbf{e}_y$, and $\mathbf{e}_z$ are (4, 1, 3). (a) What are its components relative to corresponding vectors $\mathbf{f}_x$, $\mathbf{f}_y$, and $\mathbf{f}_z$ with magnitudes that are in lbs? (b) What are the components, $\mathscr{F}_x$, $\mathscr{F}_y$, and $\mathscr{F}_z$ of $\mathscr{F}$ relative to $\mathbf{e}_x$, $\mathbf{e}_y$, and $\mathbf{e}_z$? [*Ans.* (4, 1, 3) in each case.]

**Exercise 1.5** In Exercise 1.4, let the unit of force be changed from the lb. to the oz. (but with each lb. represented by 1 inch as before). What are the answers to (a) and (b) in this case? [*Ans.* The same as before. Which components *do* change?]

**Exercise 1.6** In Exercise 1.4, if we changed from one inch per lb. to one inch per oz., what would happen to the three sets of components mentioned therein? [*Ans.* Those in (a) and (b) would be unaltered. Those of $\mathbf{F}$ relative to $\mathbf{e}_x$, $\mathbf{e}_y$, and $\mathbf{e}_z$ would become (64, 16, 48).]

If a force has a magnitude of 6 lbs. and we change the unit of force from lbs. to oz., the *numerical* value of its magnitude changes to 96.

A magnitude that is objective in the sense that it does not change simply because we change the reference frame, though it *may* change its numerical measure when we change the scale of measurement, is called a *scalar*. Thus the magnitude of a vector is a scalar. But a scalar need not be associated with a vector. For example, the number of fingers on one's hand is a scalar. So are the mass of a body and the time interval between two events, according to Newtonian physics. A scalar need not be constant. For example, the height of a growing child is a scalar.

**Exercise 1.7** Name various other scalars. [Do not be content with "the number of fingers on two hands," "the number of fingers on three hands if one finger has been chopped off," and the like. Give various types of scalars including variable ones.]

**Exercise 1.8** Could the number of fingers on one's hand change because of a change of units? [This would make an excellent true-false question, because the answer is both yes and no. It depends on what we mean by *number*. If we mean that we can put the fingers in one-to-one correspondence with five pebbles, say, then the number will not change. But we can change the unit of counting: for example, we can count in dozens, in which case the *number* of fingers becomes 5/12. Thus, even a dimensionless number can be changed by a change of unit.]

**Exercise 1.9** Is the angle between two vectors having the same starting point a scalar? Is the area of the parallelogram defined by the two vectors a scalar?

Not everything that has magnitude is a scalar. For instance, the components of a vector are not scalars since they change when the reference system is changed. [But the "components" *relative to a particular* $\mathbf{e}_x$, $\mathbf{e}_y$, $\mathbf{e}_z$ are scalars

since these "components" are really related not to a reference frame regarded as such but to a particular set of fixed vectors $\mathbf{e}_x$, $\mathbf{e}_y$, $\mathbf{e}_z$.] The number of vertices of a given triangle, namely 3, is a scalar, but the number of vertices lying in coordinate planes is not.

**Exercise 1.10** Why not?

Note how long it has taken us to get to a place where we could try to define a scalar satisfactorily. Even so we have cheated. In Chapter 1 we pointed out the difficulty of giving satisfactory definitions that are brief. Have you noticed that though we have used the word *magnitude* freely, we have never defined it? We have no intention of doing so. But you might like to try for yourself.

**Exercise 1.11** Define *magnitude*. [You may come up with a definition along these lines: *magnitude is something that can be represented by a single number*. This is not satisfactory, though, for two opposite reasons. One is that we may need different numbers for the same magnitude in different units, so that the word "single" might need qualification despite the word "can." The other is that, by means of a code, we can represent pretty well anything by a single number. Anyway, what about such a thing as your telephone number? Perhaps we have to insist that numbers representing magnitudes of the same sort shall combine according to the usual arithmetical laws of addition. This sounds promising. It would exclude catalog numbers, telephone numbers, and other code numbers. But suppose we chose to measure weights by the *logarithms* of the numbers of pounds in them. Then, since for example $\log 2 + \log 5 = \log 2 \times 5 = \log 10$, which is not $\log 7$, would this fact mean that we could not regard the weight of a piece of matter as a magnitude? One would prefer that it did not. Evidently some rewording is necessary. Keep at it —if you are so inclined. And when you have overcome this particular difficulty, ask yourself whether temperature is a magnitude.]

**Exercise 1.12** Is the magnitude of a position vector a scalar? [This is not a question to be answered with a simple yes or no. It has facets. Think it through.]

Given a vector $\mathbf{V}$ of magnitude $V$, we can form a new vector by multiplying the magnitude by a scalar, say $h$, and keeping the same direction and location. This new vector is denoted by $h\mathbf{V}$, and we say that it is the result of multiplying the vector $\mathbf{V}$ by the scalar $h$. We have already discussed this for the case in which $h$ is a dimensionless number, and we have seen instances in which $h$ was not. In the latter case, $h\mathbf{V}$ cannot combine with $\mathbf{V}$ according to the parallelogram law since it is not of the same sort as $\mathbf{V}$. For example, if $\mathbf{a}$ is the acceleration vector of a particle of mass $m$, then $m\mathbf{a}$ is not an acceleration vector; by Newton's second law of motion, we find that it actually combines with forces (see Equation (6.1) of Chapter 2).

## 2. SCALAR PRODUCTS

We have seen how the arithmetical symbols + and − can be used to represent the combination of vectors according to the parallelogram law. Having seen in this way how we can talk of *adding* and *subtracting* vectors, we naturally ask ourselves how we can *multiply* and *divide* vectors. But actually there is no reason—except perhaps misguided optimism—why we should expect to be able to do so usefully. For example, though we can "add" colors (as on a color television screen) or "subtract" them (as in amateur color photography), it does not occur to us to ask how we could multiply or divide colors by colors. Again, we can add and subtract money; and if we are astute and lucky, we can multiply money—but by numbers, not by money. If we start with \$ 100 and triple our money, we end up with \$ 300 which is \$ 100 multiplied by 3, not multiplied by \$ 3. We *could* contemplate the idea of multiplying \$ 100 by \$ 3 and getting 300 "square dollars," but even apart from the name we would not feel that the idea was likely to be particularly useful. What right have we, then, to expect that there is even any meaning to "**U** times **V**" or "**U** divided by **V**"?

Let us see what we can think of that might be somehow worthy of the phrase "**U** times **V**." Given two vectors **U** and **V** with magnitudes $U$ and $V$, we could form the scalar $UV$ which would be a sort of product of the vectors **U** and **V**. But it is really just the product of the two scalars $U$ and $V$, and it does not involve the directions of the vectors at all.

We can bring in the directions to some extent by considering the angle between **U** and **V**. Let us call it $\theta$. The angle $\theta$ does not tell us everything about the directions of **U** and **V**. It tells us only one quarter of the total amount of information about the directions, since two angles are needed to specify each direction; thus, $\theta$ is one angle out of a needed four. It has the advantage, though, of being a scalar.

Using $\theta$ as well as the magnitudes $U$ and $V$, we can form all kinds of scalars that involve the product of the magnitudes. For example, $UV\theta$, $UV\theta^5$, $UV \sin\theta$, $UV \cos\theta$, $UV \tan\theta$, $UV \sec\theta$, $UV \sin^3\theta \cos\theta$, $UV \sin\theta/\theta^2$, and so on. Out of all these possibilities, there is one that has, shall we say, a pleasing geometrical significance: the quantity $UV \sin\theta$ gives the area of the parallelogram defined by **U** and **V**, as is easily seen. We might think, therefore, that this particular scalar quantity formed by multiplying various scalars connected with the vectors **U** and **V** would so merit our attention as to be worthy of some such title as the "scalar product" of the vectors **U** and **V**, a title that graces the name of this chapter. Why do we shrink from so regarding it? One reason, and it is a good one, is that we happen to know (as perhaps you do too) that the term "scalar product" is used for a different combination of $U$, $V$, and $\theta$. But that is hardly an enlightening reason. Let us see what makes us reluctant to give $UV \sin\theta$ the particular honor of the title "scalar product."

The trouble with $UV \sin\theta$ comes to light when we consider the parallelogram law. Suppose vectors **A**, **B**, and **C** are such that **C** is the resultant of **A** and **B**. We write this fact in the form

$$\mathbf{C} = \mathbf{A} + \mathbf{B}, \tag{2.1}$$

and we say that **C** is equivalent to **A** and **B** combined. Now if we really mean the word "equivalent," we will want the "scalar product" of a vector **V** with **C** to be the same as the scalar product of **V** with **A** + **B**. We can always ensure this by saying that **A** + **B** is just another way of saying **C**. But we would prefer to be able to say the more stringent, and more significant thing that the "scalar product" of **V** with the resultant (**A** + **B**) is the same as that of **V** with **A** plus that of **V** with **B**; that is, we would like the "scalar product" to obey the distributive law.

It is easy to see that $UV \sin \theta$ does not obey this law. Take, for simplicity, a vector **V** that is perpendicular to **A**, **B**, and **C**, as in Figure 2.1. Then, unless **A** and **B** have the same direction, the area of the rectangle formed by **V** and **C** is clearly smaller than the sum of the areas of the rectangle formed by **V** and **A** and the rectangle formed by **V** and **B**. This is because the magnitude of **C** is smaller than the sum of the magnitudes of **A** and **B**. Naturally, we are disappointed. The quantity $UV \sin \theta$ has such a pleasant geometrical significance that we are reluctant to see it fail our test. But it is not lost to us. We shall find a way to make it pass the test—at a price. That, however, comes later.

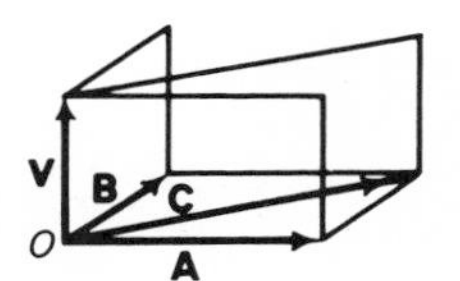


**Figure 2.1**

Meanwhile the foregoing gives us an important hint. Though the magnitudes of **A** and **B** do not add up to the magnitude of **C**, we recall Theorem 2 about projections, and we realize that it tells us, in effect, that the sum of the *projections* of **A** and **B**—on *any* line—is equal to the projection of **C** on that line. Let us see how we can profit from this fact. Take a vector **V** and project the vectors **A**, **B**, and **C** onto it. Then:

$$\text{projection of } \mathbf{C} = \text{projection of } \mathbf{A} + \text{projection of } \mathbf{B}.$$

If $\theta_{AV}$, $\theta_{BV}$, and $\theta_{CV}$ are the angles that **A**, **B**, and **C** make with **V**, this can be written, by Theorem 1, as

$$C \cos \theta_{CV} = A \cos \theta_{AV} + B \cos \theta_{BV}. \tag{2.2}$$

Therefore, if we define the scalar product of **V** with **U** as $U \cos \theta_{UV}$, and denote it by $\mathbf{V} \cdot \mathbf{U}$, we see that Equation (2.2) yields:

$$\mathbf{V} \cdot \mathbf{C} = \mathbf{V} \cdot (\mathbf{A} + \mathbf{B}) = \mathbf{V} \cdot \mathbf{A} + \mathbf{V} \cdot \mathbf{B}.$$

This is good as far as it goes, but such a quantity $\mathbf{V} \cdot \mathbf{U}$ would not be the same as $\mathbf{U} \cdot \mathbf{V}$: the former would be $U \cos \theta$, the latter $V \cos \theta$, where $\theta$ is the angle between **U** and **V**. There is no reason why we should insist that $\mathbf{V} \cdot \mathbf{U} = \mathbf{U} \cdot \mathbf{V}$, but it happens that there is a simple way of making this relation true, so we might as well make the most of our opportunity. We modify the above by regarding **U**·**V** not as $U \cos \theta$ but as $UV \cos \theta$. Let us make this official:

"Let **U** and **V** be two vectors with magnitudes $U$ and $V$, the angle between the vectors being $\theta$. The scalar $UV \cos \theta$ is denoted by the symbol **U**·**V** and is called the scalar product, or the dot product, of **U** and **V**."

Thus,

$$\mathbf{U}\cdot\mathbf{V} = UV \cos \theta. \tag{2.3}$$

In particular, we see that

$$\mathbf{V}\cdot\mathbf{V} = V^2. \tag{2.4}$$

**Exercise 2.1** Is $(\mathbf{A}\cdot\mathbf{B})\cdot\mathbf{C} = \mathbf{A}\cdot(\mathbf{B}\cdot\mathbf{C})$? [*Ans.* In a nonsensical sense, *yes*, since neither has any meaning. Why not?]

**Exercise 2.2** Prove that

$$\mathbf{U}\cdot\mathbf{V} = \mathbf{V}\cdot\mathbf{U}. \tag{2.5}$$

**Exercise 2.3** Prove that if $\mathbf{C} = \mathbf{A} + \mathbf{B}$, then $\mathbf{V}\cdot\mathbf{C} = \mathbf{V}\cdot\mathbf{A} + \mathbf{V}\cdot\mathbf{B}$, i. e., that

$$\mathbf{V}\cdot(\mathbf{A} + \mathbf{B}) = \mathbf{V}\cdot\mathbf{A} + \mathbf{V}\cdot\mathbf{B}. \tag{2.6}$$

**Exercise 2.4** What is the value of **U**·**V** if **U** and **V** are perpendicular? What if they have the same direction?

**Exercise 2.5** Prove that if two opposite edges of a tetrahedron are perpendicular, and two other opposite edges are perpendicular, then the remaining two are also perpendicular. [Denote the vertices by $P$, $Q$, $R$, and $S$. Then we are given that $\overrightarrow{PQ}\cdot\overrightarrow{RS} = 0$, and $\overrightarrow{PR}\cdot\overrightarrow{QS} = 0$, say, and we wish to prove that $\overrightarrow{PS}\cdot\overrightarrow{QR} = 0$. We express everything in terms of vectors emanating from $P$. Thus $\overrightarrow{RS} = \overrightarrow{PS} - \overrightarrow{PR}$, etc. So, given $\overrightarrow{PQ}\cdot(\overrightarrow{PS} - \overrightarrow{PR}) = 0$ and $\overrightarrow{PR}\cdot(\overrightarrow{PS} - \overrightarrow{PQ}) = 0$, we wish to prove that $\overrightarrow{PS}\cdot(\overrightarrow{PR} - \overrightarrow{PQ}) = 0$, but this is an immediate consequence, as you can see by adding the two preceding relations.]

As happens often, though not always, that which is mathematically elegant turns out to be that which is important in applications. In the field of physics, consider the work done by a constant force **F** with magnitude $F$ lbs. If it acts on a particle that moves through a distance $D$ ft. in its own direction, it does an amount of work $FD$ ft. lbs. Suppose that the displacement of the particle, **D**, is not along the line of action of the force. Then we can resolve the force **F** into two forces $\mathbf{F}_{//}$ and $\mathbf{F}_{\perp}$, the former lying along **D** and the latter perpendicular to it. The particle does not move in the direction of the

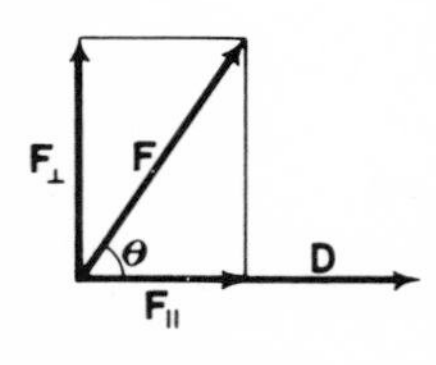

**Figure 2.2**

force $\mathbf{F}_\perp$ at all, and thus $\mathbf{F}_\perp$ does no work during the displacement. The force $\mathbf{F}_{//}$ lies along the displacement of the particle and since the magnitude of $\mathbf{F}_{//}$ is $F \cos \theta$, it does an amount of work $FD \cos \theta$ ft. lbs., which is thus the amount of work done by **F**. We see, then, that the work done is none other than the scalar product of **F** and **D**:

$$W = \mathbf{F} \cdot \mathbf{D}. \tag{2.7}$$

The above probably sounds like a convincing and enlightening argument. It may be enlightening, but it is not really convincing. It assumes, for example, that the work done by **F** can be found by adding the work done by $\mathbf{F}_\perp$ and the work done by $\mathbf{F}_{//}$, merely because $\mathbf{F} = \mathbf{F}_\perp + \mathbf{F}_{//}$; it also assumes that the work done by $\mathbf{F}_\perp$ is zero because the particle does not move in the direction of $\mathbf{F}_\perp$ at all.

Actually, the above argument is a valid *consequence* of the *definition* in Equation (2.7) of the work done by a constant force. It is not a derivation of that equation. From Equation (2.7) we can deduce the various assumptions made in the above argument. Without using calculus, it is difficult to show why physicists should be sufficiently interested in the quantity $\mathbf{F} \cdot \mathbf{D}$ to give it a name. The fact is that, by making certain manipulations on the Newtonian equation of motion of a particle, namely $\mathbf{F} = m\mathbf{a}$, we can derive from it, for the case of a constant force, the scalar equation,

$$\mathbf{F} \cdot \mathbf{D} = \tfrac{1}{2}mv^2 - \tfrac{1}{2}mv_o^2$$

where $v$ and $v_o$ are the magnitudes of the velocities of the particle at the end and the beginning of the displacement. [A more general equation, involving integration, results when **F** is not constant.] Because this equation is a mathematically simple consequence of the equation of motion and has various pleasant properties, it takes on important physical significance. Accordingly, physicists give names to the quantities therein. They call $\frac{1}{2}mv^2$ the *kinetic energy* of the particle, and, as we have already said, $\mathbf{F} \cdot \mathbf{D}$ the work done by **F** in the displacement **D**. This enables them to state the equation in words: the work done by the force is equal to the change it produces in the kinetic energy of the particle.

**Exercise 2.6** A constant force **F** of magnitude $F$ makes 45° with the $x$-axis. If it acts on a particle moving along the $x$-axis through a distance $D$, what is the work done?

**Exercise 2.7** Prove that if, in Exercise 2.6, **F** had been perpendicular to the $x$-axis, the work done would have been zero.

**Exercise 2.8** Show that if two constant forces $\mathbf{F}_1$ and $\mathbf{F}_2$ act on a particle, the work done by their resultant is the sum of the work done by $\mathbf{F}_1$ and the work done by $\mathbf{F}_2$.

**Exercise 2.9** A constant force having components $(\mathbf{F}_x, 0, 0)$ acts on a particle that undergoes a displacement with components $(D_x, 0, 0)$, all components being relative to **i**, **j**, and **k**. What is the work done by the force?

**Exercise 2.10** The same as Exercise 2.9, but with the force having components $(F_x, F_y, F_z)$. [*Ans.* $F_x D_x$.]

**Exercise 2.11** Show that the work done by a constant force **F** acting on a particle that moves through a displacement **D** can be regarded as the product of the magnitude of **F** and the length of the projection of **D** on the line of action of **F**.

## 3. SCALAR PRODUCTS AND UNIT ORTHOGONAL TRIADS

Using Equations (2.6) and (2.5), we have

$$\begin{aligned}(\mathbf{A} + \mathbf{B})\cdot(\mathbf{U} + \mathbf{V}) &= \mathbf{A}\cdot(\mathbf{U} + \mathbf{V}) + \mathbf{B}\cdot(\mathbf{U} + \mathbf{V}) \\ &= \mathbf{A}\cdot\mathbf{U} + \mathbf{A}\cdot\mathbf{V} + \mathbf{B}\cdot\mathbf{U} + \mathbf{B}\cdot\mathbf{V},\end{aligned}$$

and by repeated application of the same maneuvers, we see that a dot product of the type $(\mathbf{A} + \mathbf{B} + \ldots)\cdot(\mathbf{U} + \mathbf{V} + \ldots)$ is equal to the sum of all the individual dot products of each member of the first parenthesis with each member of the second.

Now consider two vectors **U** and **V** expressed in terms of their components relative to a unit orthogonal triad **i**, **j**, and **k**:

$$\mathbf{U} = U_x\mathbf{i} + U_y\mathbf{j} + U_z\mathbf{k} \tag{3.1}$$

$$\mathbf{V} = V_x\mathbf{i} + V_y\mathbf{j} + V_z\mathbf{k}. \tag{3.2}$$

How do we find the value of $\mathbf{U}\cdot\mathbf{V}$ in terms of the components? We could draw a diagram and work out the projection of **V** on **U** using a zigzag path. But there is an algebraic method that gives the result without any appeal to zigzag paths or even to a diagram, though, in fact, it is basically the same method.

We begin by noting that, from the definition of the scalar product, since the vectors **i**, **j**, and **k** have unit magnitude,

$$\left.\begin{aligned}\mathbf{i}\cdot\mathbf{i} = \mathbf{j}\cdot\mathbf{j} = \mathbf{k}\cdot\mathbf{k} = 1, \\ \mathbf{j}\cdot\mathbf{k} = \mathbf{k}\cdot\mathbf{i} = \mathbf{i}\cdot\mathbf{j} = 0, \\ \mathbf{k}\cdot\mathbf{j} = \mathbf{i}\cdot\mathbf{k} = \mathbf{j}\cdot\mathbf{i} = 0.\end{aligned}\right\} \tag{3.3}$$

We now write,

$$\mathbf{U}\cdot\mathbf{V} = (U_x\mathbf{i} + U_y\mathbf{j} + U_z\mathbf{k})\cdot(V_x\mathbf{i} + V_y\mathbf{j} + V_z\mathbf{k}),$$

and use the distributive law just discussed. There will be nine individual terms in the distributed product on the right, since each of the first three must multiply each of the second three. Of these nine terms, six will vanish because of the relations (3.3): for example, the term $(U_x\mathbf{i})\cdot(V_y\mathbf{j})$ equals $U_xV_y\mathbf{i}\cdot\mathbf{j}$ and this vanishes because $\mathbf{i}\cdot\mathbf{j} = 0$. Typical of the remaining three terms is $(U_x\mathbf{i})\cdot(V_x\mathbf{i})$ which equals $U_xV_x\mathbf{i}\cdot\mathbf{i}$, or just $U_xV_x$ since $\mathbf{i}\cdot\mathbf{i} = 1$. So we find in the end the neat-looking and powerful formula:

$$\mathbf{U}\cdot\mathbf{V} = U_xV_x + U_yV_y + U_zV_z. \tag{3.4}$$

**Exercise 3.1** In addition to the distributive law, we used the fact that $(U_x\mathbf{i})\cdot(V_x\mathbf{i}) = U_xV_x\mathbf{i}\cdot\mathbf{i}$, etc. Note that this is obviously valid.

In particular, using Equation (2.4), we see from this that

$$V^2 = \mathbf{V}\cdot\mathbf{V} = V_x^2 + V_y^2 + V_z^2, \tag{3.5}$$

a result we already knew. Also, if $\mathbf{U}$ and $\mathbf{V}$ are perpendicular, $\cos\theta$ will be zero and therefore $\mathbf{U}\cdot\mathbf{V}$ will be zero so that we have this condition for perpendicularity:

$$U_xV_x + U_yV_y + U_zV_z = 0. \tag{3.6}$$

**Exercise 3.2** Find the value of $\mathbf{U}\cdot\mathbf{V}$ if $\mathbf{U} = 2\mathbf{i} + 3\mathbf{j} - \mathbf{k}$ and $\mathbf{V} = 3\mathbf{i} + 4\mathbf{j} + 2\mathbf{k}$.

**Exercise 3.3** Find the value of $\mathbf{U}\cdot\mathbf{V}$ if $\mathbf{U} = 4\mathbf{i} + 2\mathbf{j} - \mathbf{k}$ and $\mathbf{V} = 3\mathbf{i} - 5\mathbf{j} + 2\mathbf{k}$, and so show that $\mathbf{U}$ and $\mathbf{V}$ are perpendicular. [Imagine trying to prove this without the aid of Equation (3.4).]

**Exercise 3.4** Do Exercises (2.9) and (2.10) using Equation (3.4).

**Exercise 3.5** For what value of $x$ will $3\mathbf{i} - 2\mathbf{j} + 3\mathbf{k}$ and $x\mathbf{i} - 2\mathbf{j} + 2\mathbf{k}$ be perpendicular?

**Exercise 3.6** In Exercise 3.2, find the values of $U$ and $V$, and thus, using Equation (2.3), show that for these vectors, $\cos\theta = 16/\sqrt{14}\sqrt{29}$.

**Exercise 3.7** Show that when $\mathbf{U}$ and $\mathbf{V}$ are vectors of unit magnitude, Equation (3.4) is just $\cos\theta = ll' + mm' + nn'$ in disguise. Then compare this method of derivation with the previous zigzag method and see if you can trace out the underlying identity of the two methods.

From Equations (2.3) and (3.4) we see that

$$UV\cos\theta = U_xV_x + U_yV_y + U_zV_z,$$

so that

$$\cos\theta = \frac{U_xV_x + U_yV_y + U_zV_z}{UV}. \tag{3.7}$$

Compare this with Exercise (3.7).

**Exercise 3.8** Using Equations (3.5) and (3.7), find $\cos\theta$ for the vectors $2\mathbf{i} + \mathbf{j} + \mathbf{k}$ and $3\mathbf{i} + 2\mathbf{j} + 2\mathbf{k}$. [*Ans.* $10/\sqrt{6}\sqrt{17}$.]

**Exercise 3.9** Find the angle (not just its cosine) between $\mathbf{j} + \mathbf{k}$ and $\mathbf{i} + \mathbf{k}$, and draw a diagram showing the two vectors in relation to $\mathbf{i}$, $\mathbf{j}$, and $\mathbf{k}$. [Note the bearing of this exercise on the following one.]

**Exercise 3.10** From a vertex, $O$, of a cube, diagonals are drawn of the three square faces that meet at $O$. Prove that these diagonals make angles of 60° with each other. [While this may seem like an excellent demonstration of the power of the formula (3.5), the angles are easily found by ordinary geometry. Find them in that way too.]

**Exercise 3.11** Prove that the vectors $\mathbf{i}$, $\mathbf{j} + \mathbf{k}$, and $\mathbf{i} + \mathbf{j} + \mathbf{k}$ are coplanar, and find the cosines of the angles between them. Also draw an appropriate diagram involving a cube, and from it check the angle between $\mathbf{i}$ and $\mathbf{j} + \mathbf{k}$ and the fact that the three vectors are coplanar. [To see without the diagram that the vectors are coplanar, note that one is the resultant of the other two.]

**Exercise 3.12** Would Equation (3.4) be valid if $(U_x, U_y, U_z)$ and $(V_x, V_y, V_z)$ had been the components of $\mathbf{U}$ and $\mathbf{V}$ relative to a reference frame defined by general base vectors $\mathbf{e}_x$, $\mathbf{e}_y$, and $\mathbf{e}_z$? [This is probably the nastiest question in the book. According to all that has been said here, the answer is *no*, so give yourself full credit if that is your answer. However, there is more to the concept of a scalar product than we are in a position to explain here, and in a significant sense it could actually be correct to say that Equation (3.4) is valid in all reference frame.]

Let us derive Equation (6.2) of Chapter 3 by using scalar products. Here is the relevant part of the diagram:

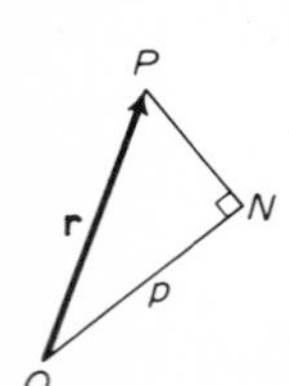


**Figure 3.1**

Denote by $\mathbf{r}$ the position vector of a general point $P$ on the plane through $N$ that is perpendicular to $ON$. Denote the unit vector along $ON$ by $\mathbf{n}$. Then, since the length of the projection of $OP$ on $ON$ is $p$, we have

$$\mathbf{r} \cdot \mathbf{n} = p, \tag{3.8}$$

which is the required equation of the plane. To put it in the less compact, nonvectorial form, we note that $\mathbf{r}$ has components $(x, y, z)$ and the components of the unit vector $\mathbf{n}$ are just the direction cosines $(l, m, n)$ of the normal to the plane. It follows at once that

$$lx + my + nz = p.$$

**Exercise 3.13** Derive this equation by expressing the fact that $PN$ is perpendicular to $ON$. [*Hint*. The displacement $\overrightarrow{PN}$ is given by $\overrightarrow{ON} - \overrightarrow{OP}$; therefore, the perpendicularity can be expressed as $\overrightarrow{ON} \cdot (\overrightarrow{ON} - \overrightarrow{OP}) = 0$. This gives $\overrightarrow{ON} \cdot \overrightarrow{OP} = \overrightarrow{ON} \cdot \overrightarrow{ON}$. Now use the fact that $\overrightarrow{ON}$ and $\overrightarrow{OP}$ have components $(lp, mp, np)$ and $(x, y, z)$.]

Equation (3.7) for $\cos \theta$ can be used to obtain proofs of familiar formulas in trigonometry and two-dimensional analytical geometry. By way of orientation do this exercise first.

**Exercise 3.14** Prove that the vectors $\mathbf{i} + 3\mathbf{j}$ and $4\mathbf{i} + 2\mathbf{j}$ make 45°

with each other. [Note that these vectors are in the $xy$-plane and can thus be regarded as belonging to two-dimensional analytical geometry in that plane. Note, too, how powerful the $\cos\theta$ formula is by contemplating the problem of trying to prove that the angle is 45° using only the methods of ordinary geometry. Draw the diagram and see for yourself.]

We shall now derive the well-known condition in two-dimensional analytical geometry that two lines be perpedicular. We recall that the slope of a line is defined as the tangent of the angle it makes with the $x$-axis, and that it is usually denoted by $m$ (which, of course, is not to be confused with the $m$ denoting a direction cosine). If a line in the $xy$-plane makes an angle $\alpha$ with the $x$-axis, its slope is given by $m = \tan\alpha$, and its direction cosines are easily seen to be $\cos\alpha$, $\sin\alpha$, and 0. Consider two lines making angles $\alpha$ and $\alpha'$ with the $x$-axis, and having slopes $m$ and $m'$. We have, for perpendicularity, by Equation (3.6)

$$\cos\alpha\cos\alpha' + \sin\alpha\sin\alpha' = 0.$$

Divide this by $\cos\alpha\cos\alpha'$ and we find that $1 + mm' = 0$, or

$$mm' = -1,$$

showing that the slopes are negative reciprocals.

**Exercise 3.15** Using the scalar product, derive the well-known trigonometric formula $\cos(A - B) = \cos A\cos B + \sin A\sin B$. [Consider the direction cosines of two lines in the $xy$-plane making angles $A$ and $B$ with the $x$-axis, and note that the angle between them is $A - B$.]

It is possible to derive the so-called law of cosines by means of the scalar product. In triangle $ABC$ let the sides be of length $a$, $b$, and $c$ as shown. Using displacements, we have $\overrightarrow{BC} = \overrightarrow{AC} - \overrightarrow{AB}$. So,

$$\begin{aligned} BC^2 &= \overrightarrow{BC}\cdot\overrightarrow{BC} = (\overrightarrow{AC} - \overrightarrow{AB})\cdot(\overrightarrow{AC} - \overrightarrow{AB}) \\ &= \overrightarrow{AC}\cdot\overrightarrow{AC} - 2\overrightarrow{AB}\cdot\overrightarrow{AC} + \overrightarrow{AB}\cdot\overrightarrow{AB} = AC^2 + AB^2 - 2(AB)(AC)\cos A. \end{aligned}$$

Therefore,

$$a^2 = b^2 + c^2 - 2bc\cos A.$$

# 5

# VECTOR PRODUCTS. QUOTIENTS OF VECTORS

## 1. AREAS OF PARALLELOGRAMS

In the preceding chapter we noted that the quantity $UV \sin \theta$ gives the area of the parallelogram defined by the vectors **U** and **V**. We were therefore tempted to regard it as a "scalar product" of these vectors. We were discouraged from doing so when we found that this product would not obey the distributive law, but that $UV \cos \theta$ would. One of the purposes of this chapter is to rescue the quantity $UV \sin \theta$ from the danger of vectorial oblivion.

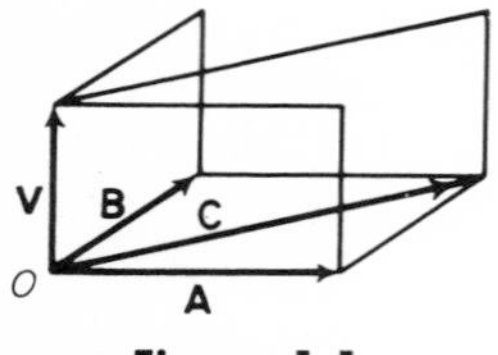

**Figure 1.1**

At this stage we do not know how to do this. But let us cross our fingers and agree to denote whatever it is we are seeking by the symbol $\mathbf{U} \times \mathbf{V}$ (to distinguish it from the $\mathbf{U} \cdot \mathbf{V}$ discussed in the preceding chapter) and to call it the *cross product* of **U** and **V**.

In Figure 1.1, the vectors **A**, **B**, and **C** are such that

$$\mathbf{C} = \mathbf{A} + \mathbf{B}. \tag{1.1}$$

To make our exploratory task easier, we consider first the special case in which vector **V** is perpendicular to the vectors **A**, **B**, and **C**.

By the symbol $\mathbf{V} \times \mathbf{A}$ we mean somehow to refer to the area of the parallelogram defined by **V** and **A**. [In the present special case, this and other parallelograms are actually rectangles, but let us use the word parallelogram nevertheless.] Here $\mathbf{V} \times \mathbf{A}$, the area of the parallelogram defined by **V** and **A**, is just $VA$. Similarly, the areas corresponding to the cross products $\mathbf{V} \times \mathbf{B}$ and $\mathbf{V} \times \mathbf{C}$ are $VB$ and $VC$. We recall from the previous chapter that if we take $\mathbf{V} \times \mathbf{A}$ to mean just the area of the parallelogram defined by **V** and **A**, then $\mathbf{V} \times \mathbf{C} = \mathbf{V} \times (\mathbf{A} + \mathbf{B}) \neq \mathbf{V} \times \mathbf{A} + \mathbf{V} \times \mathbf{B}$. This is because the magnitude of $\mathbf{A} + \mathbf{B}$ is rarely equal to the sum of the magnitudes of **A** and **B**; that is, despite Equation (1.1) we have, in general,

$$C \neq A + B. \tag{1.2}$$

**Exercise 1.1** What is wrong with the following argument? Since $V$, $A$, and $B$ are scalars, we have, by ordinary algebra,

$$V(A + B) = VA + VB.$$

So if $\mathbf{V} \times \mathbf{A}$ stands for just $VA$, etc. we can write this as

$$\mathbf{V} \times (\mathbf{A} + \mathbf{B}) = \mathbf{V} \times \mathbf{A} + \mathbf{V} \times \mathbf{B}.$$

[*Hint.* Does the + on the left in the second equation mean the same thing as the other three +'s?]

One of the most important things in life is the ability to extract success from failure. This usually involves a seemingly contradictory combination of stubbornness and flexibility: one must be stubborn in one's determination to reach a solution, but highly flexible in seeking it.

Our failure above gives us an idea. We have, in effect, been multiplying Equation (1.2) by $V$. Why not work on Equation (1.1) instead? We shall then get

$$V\mathbf{C} = V(\mathbf{A} + \mathbf{B}) = V\mathbf{A} + V\mathbf{B}. \tag{1.3}$$

So if we agree to look on $\mathbf{V} \times \mathbf{A}$, for example, not as a *scalar* representing the area of the parallelogram defined by **V** and **A** but as a *vector* lying along **A** and having this area as magnitude, we shall be able to rewrite Equation (1.3) in the form

$$\mathbf{V} \times \mathbf{C} = \mathbf{V} \times (\mathbf{A} + \mathbf{B}) = \mathbf{V} \times \mathbf{A} + \mathbf{V} \times \mathbf{B}, \tag{1.4}$$

and apparently our troubles will be over—at least for the special case we are considering here.

But there is something unsatisfying and inelegant about representing the area of a parallelogram by a vector pointing along one of its sides. Why one side rather than the other? Why should $\mathbf{V} \times \mathbf{A}$, for instance, point along **A** rather than along **V**? Is there some way by which we can avoid playing favorites? We had better try choosing a neutral direction—a direction *perpendicular* to both **A** and **V**. How does Equation (1.4) fare if we think of $\mathbf{V} \times \mathbf{A}$ as having the same magnitude as before, but as pointing in a direction perpendicular to the plane of the parallelogram? Surprising as it may seem, Equation (1.4) survives. You may think this will be hard to prove. But actually it is obvious. For a glance at Figure 1.1 shows that the new $\mathbf{V} \times \mathbf{A}$, $\mathbf{V} \times \mathbf{B}$, and $\mathbf{V} \times \mathbf{C}$ are merely the old ones turned as a rigid body through 90° about $\mathbf{V}$ as an axis. Since Equation (1.3) holds for the $\mathbf{A}$, $\mathbf{B}$, and $\mathbf{C}$ in Figure 1.1, it obviously has to hold for the rotated vectors. Therefore Equation (1.4) still holds.

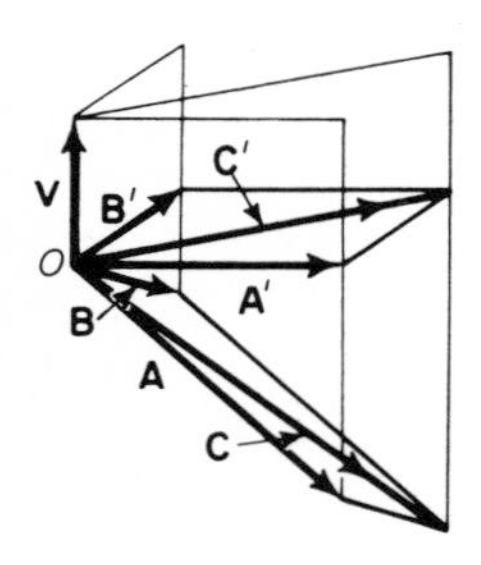


**Figure 1.2**

But we have been making things easy for ourselves by considering the special case in which $\mathbf{V}$ is perpendicular to $\mathbf{A}$, $\mathbf{B}$, and $\mathbf{C}$. Will our luck hold when $\mathbf{V}$ is not perpendicular to them? It is easy to see that it will. Consider Figure 1.2, in which, though the plane of the vectors $\mathbf{A}$, $\mathbf{B}$, and $\mathbf{C}$ is not perpendicular to $\mathbf{V}$, the plane of the vectors $\mathbf{A}'$, $\mathbf{B}'$, and $\mathbf{C}'$ is; and the lines that look parallel to $\mathbf{V}$ are, in fact, parallel to $\mathbf{V}$. The area of the (undrawn) parallelogram defined by $\mathbf{V}$ and $\mathbf{A}$ is equal to the area of the coplanar rectangle defined by $\mathbf{V}$ and $\mathbf{A}'$. Therefore $\mathbf{V} \times \mathbf{A}' = \mathbf{V} \times \mathbf{A}$. Similarly $\mathbf{V} \times \mathbf{B}' = \mathbf{V} \times \mathbf{B}$, and $\mathbf{V} \times \mathbf{C}' = \mathbf{V} \times \mathbf{C}$, and that is about all we need. For the upper part of Figure 1.2 apes Figure 1.1, with $\mathbf{A}'$, $\mathbf{B}'$, and $\mathbf{C}'$ replacing the $\mathbf{A}$, $\mathbf{B}$, and $\mathbf{C}$ in Figure 1.1. So relation (1.4) holds with $\mathbf{A}'$, $\mathbf{B}'$, and $\mathbf{C}'$ replacing $\mathbf{A}$, $\mathbf{B}$, and $\mathbf{C}$, and therefore, by the above equalities, it holds without the primes too, even though the present $\mathbf{A}$, $\mathbf{B}$, and $\mathbf{C}$ are not perpendicular to $\mathbf{V}$.

> **Exercise 1.2** Go through the above proof for the special case in which $\mathbf{V}$, $\mathbf{A}$, $\mathbf{B}$, and $\mathbf{C}$ are all coplanar, making the necessary adjustments in the diagram. [Since $\mathbf{V} \times \mathbf{A}$, $\mathbf{V} \times \mathbf{B}$, $\mathbf{V} \times \mathbf{C}$ have the same direction, Equation (4.1) reduces to a relation between the areas of three parallelograms in a plane. These areas are given by the products of a common base $\mathbf{V}$ with three altitudes the sum of the lengths of two of which is easily shown to be equal to the length of the third.]

What specially pleases us here is that when we look at the way Equation (1.4) was obtained, we see that it plays a dual role. Not only does it show that the distributive law is valid, but it also shows in what way we can regard the new $\mathbf{V} \times \mathbf{A}$ and $\mathbf{V} \times \mathbf{B}$ as combining according to the parallelogram law: if $\mathbf{C}$ is the vectorial resultant of $\mathbf{A}$ and $\mathbf{B}$, then $\mathbf{V} \times \mathbf{C}$ will be the vectorial resultant of $\mathbf{V} \times \mathbf{A}$ and $\mathbf{V} \times \mathbf{B}$. We begin to sense that despite the change to the perpendicular direction, cross products can still be regarded as vectors. But what about the combination not of $\mathbf{V} \times \mathbf{A}$ and $\mathbf{V} \times \mathbf{B}$ but of, say, $\mathbf{L} \times \mathbf{M}$ and $\mathbf{R} \times \mathbf{S}$, in which there is no common letter, the vectors $\mathbf{L}$, $\mathbf{M}$, $\mathbf{R}$, and $\mathbf{S}$, all starting at $O$? Our luck still holds. we can reduce this to the simpler case quite easily as follows. Let the plane defined by $\mathbf{L}$ and $\mathbf{M}$ cut the plane defined by $\mathbf{R}$ and $\mathbf{S}$ in the line $\lambda$. Take any nonzero vector $\mathbf{V}$ along $\lambda$. Then we can always find a vector $\mathbf{A}$ in the plane of $\mathbf{L}$ and $\mathbf{M}$ such that the area of the parallelogram defined by $\mathbf{V}$ and $\mathbf{A}$ is the same as that of the parallelogram defined by $\mathbf{L}$ and $\mathbf{M}$. Since $\mathbf{V}$ and $\mathbf{A}$ lie in the same plane as $\mathbf{L}$ and $\mathbf{M}$, the quantity $\mathbf{V} \times \mathbf{A}$ will be the same as $\mathbf{L} \times \mathbf{M}$. Similarly we can replace $\mathbf{R} \times \mathbf{S}$ by $\mathbf{V} \times \mathbf{B}$ *where* $\mathbf{V}$ *is the same* $\mathbf{V}$ *as before*. Consequently $\mathbf{L} \times \mathbf{M} + \mathbf{R} \times \mathbf{S}$ reduces to the case $\mathbf{V} \times \mathbf{A} + \mathbf{V} \times \mathbf{B}$ that we have already considered.

**Exercise 1.3** In the above, if $\mathbf{L}$ and $\mathbf{M}$ were $\mathbf{i}$ and $\mathbf{j}$ respectively, and $\mathbf{V}$ was $\mathbf{i} + \mathbf{j}$, what would $\mathbf{A}$ be? [*Ans.* It could be $\mathbf{i}$. It also could be $\mathbf{j}$. Or $\frac{1}{2}\mathbf{j} - \frac{1}{2}\mathbf{i}$. And there are infinitely many other possibilities. The fact is that the $\mathbf{A}$ and $\mathbf{B}$ above were by no means unique.]

**Exercise 1.4** Deduce from the above discussion that $\mathbf{L} \times \mathbf{M} + \mathbf{R} \times \mathbf{S} = \mathbf{R} \times \mathbf{S} + \mathbf{L} \times \mathbf{M}$.

By now we may be fairly well convinced that we can regard these cross products as vectors. But a minor trouble comes to mind: which way should $\mathbf{V} \times \mathbf{A}$ point? There are two opposite directions perpendicular to a given plane. If we are not consistent in our choice of direction, we shall lose Equation (1.4).

**Exercise 1.5** Why?

We shall therefore have to adopt a special convention. Let us agree to say that $\mathbf{U}$, $\mathbf{V}$, and $\mathbf{U} \times \mathbf{V}$, in that order, form a right-handed system. This then takes care of the problem. But it does so at a price:

$$\mathbf{U} \times \mathbf{V} = -\mathbf{V} \times \mathbf{U}. \tag{1.5}$$

**Exercise 1.6** Prove Equation (1.5). [Note that, by the convention, $\mathbf{V}$, $\mathbf{U}$, and $\mathbf{V} \times \mathbf{U}$, in that order, must also form a right-handed system; but the third member of a right-handed triad starting with $\mathbf{V}$, $\mathbf{U}$ points in the opposite direction to that of a triad starting with $\mathbf{U}$, $\mathbf{V}$.]

We are now ready for the official definition: *the cross product* $\mathbf{U} \times \mathbf{V}$ *denotes a vector having magnitude* $UV \sin \theta$ *and pointing in a direction perpendicular to* $\mathbf{U}$ *and* $\mathbf{V}$ *in such a way as to make* $\mathbf{U}$, $\mathbf{V}$, *and* $\mathbf{U} \times \mathbf{V}$, *in that order, a right-handed system.* The cross product is often also called the *vector product.* [We shall have more to say about the right-handed system later.]

Perhaps you think that this standard definition is long overdue. But let us go into this matter of the vector product with our eyes open. We have indeed shown how cross products can be regarded as vectors. But, as you have doubtless uneasily realized, in so doing we have had to agree to some curious things about the way areas are to be regarded as combining. We shall have to look into this in more detail later. Meanwhile, let us learn how to work with vector products.

**Exercise 1.7** Show that $\mathbf{A} \times \mathbf{A} = 0$, and that $\mathbf{A} \times (\alpha\mathbf{A}) = 0$, where $\alpha$ is a scalar.

**Exercise 1.8** If $\mathbf{U}$ is of magnitude 3 and points horizontally due east, and $\mathbf{V}$ is of magnitude 5 and points vertically upward, describe the vector $\mathbf{U} \times \mathbf{V}$. [*Ans.* Magnitude 15, and pointing—well, let's see—due south—I hope!]

**Exercise 1.9** If $\mathbf{U}$ is of magnitude 8 and points horizontally due west, and $\mathbf{V}$ is of magnitude 7 and points horizontally 45° north of west, describe the vector $\mathbf{U} \times \mathbf{V}$.

**Exercise 1.10** By applying Equation (1.5) to Equation (1.4) show that

$$(\mathbf{A} + \mathbf{B}) \times \mathbf{V} = \mathbf{A} \times \mathbf{V} + \mathbf{B} \times \mathbf{V}. \tag{1.6}$$

**Exercise 1.11** Using Equations (1.4) and (1.6) show that

$$(\mathbf{A} + \mathbf{B}) \times (\mathbf{U} + \mathbf{V}) = \mathbf{A} \times \mathbf{U} + \mathbf{A} \times \mathbf{V} + \mathbf{B} \times \mathbf{U} + \mathbf{B} \times \mathbf{V}.$$

and generalize this to the case in which each parenthesis contains the sum of any number of vectors.

**Exercise 1.12** Triangle $ABC$ has sides of length $a$, $b$, and $c$. Using vector products, prove that $a/\sin A = b/\sin B = c/\sin C$. [*Hint.* Using displacements, we have $\overrightarrow{BC} + \overrightarrow{CA} + \overrightarrow{AB} = 0$. Take the cross product of this with $\overrightarrow{BC}$, and use Equation (1.5). Then do the same with $\overrightarrow{CA}$.]

**Exercise 1.13** Show that $(\mathbf{A} - \mathbf{B})\cdot(\mathbf{A} + \mathbf{B}) = A^2 - B^2$ and that $(\mathbf{A} - \mathbf{B}) \times (\mathbf{A} + \mathbf{B}) = 2\mathbf{A} \times \mathbf{B}$, and interpret these results geometrically by means of appropriate diagrams.

## 2. CROSS PRODUCTS OF i, j, AND k

From the definition of a vector product, since $\sin 0° = 0$ and $\sin 90° = 1$, and $\mathbf{i}$, $\mathbf{j}$, and $\mathbf{k}$ are unit vectors, we have for a right-handed unit orthogonal triad the important relations

$$\left.\begin{aligned} \mathbf{i} \times \mathbf{i} = \mathbf{j} \times \mathbf{j} = \mathbf{k} \times \mathbf{k} &= 0, \\ \mathbf{j} \times \mathbf{k} = -\,\mathbf{k} \times \mathbf{j} &= \mathbf{i}, \\ \mathbf{k} \times \mathbf{i} = -\,\mathbf{i} \times \mathbf{k} &= \mathbf{j}, \\ \mathbf{i} \times \mathbf{j} = -\,\mathbf{j} \times \mathbf{i} &= \mathbf{k}. \end{aligned}\right\} \tag{2.1}$$

These relations are worth comparing with the corresponding relations for dot products given in Equation (3.3) of the preceding chapter.

**Exercise 2.1** If $\alpha$ and $\beta$ are scalars, show from the definition of the cross product that $(\alpha\mathbf{i}) \times (\beta\mathbf{i}) = 0$, $(\alpha\mathbf{j}) \times (\beta\mathbf{k}) = \alpha\beta\mathbf{i}$, $(\alpha\mathbf{k}) \times (\beta\mathbf{j}) = -\,\alpha\beta\mathbf{i}$, etc., and in general that $(\alpha\mathbf{U}) \times (\beta\mathbf{V}) = \alpha\beta\, \mathbf{U} \times \mathbf{V}$.

**Exercise 2.2** Using the relations (2.1), show that

$$2\mathbf{i} \times (3\mathbf{i} + 4\mathbf{j} + 5\mathbf{k}) = -\,10\mathbf{j} + 8\mathbf{k}.$$

**Exercise 2.3** As in Exercise 2.2, find the components relative to $\mathbf{i}$, $\mathbf{j}$, and $\mathbf{k}$ of the vector $(U_x\mathbf{i}) \times (V_x\mathbf{i} + V_y\mathbf{j} + V_z\mathbf{k})$.

**Exercise 2.4** Like Exercise 2.3 for $(U_x\mathbf{i} + U_y\mathbf{j}) \times (V_x\mathbf{i} + V_y\mathbf{j})$. Note that the result is perpendicular to $\mathbf{i}$ and $\mathbf{j}$, and explain why this was to have been expected. [Note that, for example, $U_x\mathbf{i} + U_y\mathbf{j}$ lies in the same plane as **i** and **j**.]

**Exercise 2.5** Using the scalar product, find the cosine of the angle between **i** and $\mathbf{i} + \mathbf{j} + \mathbf{k}$. Then, using the vector product, find the sine of this angle. Check that the sum of their squares comes to unity. [To find the sine, calculate the magnitudes of **i** and $\mathbf{i} + \mathbf{j} + \mathbf{k}$ and multiply by the unknown $\sin\theta$. The result must be equal to the magnitude of the cross product $\mathbf{i} \times (\mathbf{i} + \mathbf{j} + \mathbf{k})$. The sine comes to $\sqrt{2}/\sqrt{3}$ and the cosine to $1/\sqrt{3}$.]

**Exercise 2.6** Like Exercise 2.5 for $\mathbf{i} + \mathbf{j}$ and $\mathbf{i} + \mathbf{j} + \mathbf{k}$.

**Exercise 2.7** By drawing a cube and noting two congruent triangles, explain why the cosine in Exercise 2.5 equals the sine in Exercise 2.6.

## 3. COMPONENTS OF CROSS PRODUCTS RELATIVE TO i, j, AND k

If

$$\mathbf{U} = U_x\mathbf{i} + U_y\mathbf{j} + U_z\mathbf{k}, \tag{3.1}$$

$$\mathbf{V} = V_x\mathbf{i} + V_y\mathbf{j} + V_z\mathbf{k}, \tag{3.2}$$

we can find the components of $\mathbf{U} \times \mathbf{V}$ by much the same method we used for finding $\mathbf{U}\cdot\mathbf{V}$ under similar circumstances. We have

$$\mathbf{U} \times \mathbf{V} = (U_x\mathbf{i} + U_y\mathbf{j} + U_z\mathbf{k}) \times (V_x\mathbf{i} + V_y\mathbf{j} + V_z\mathbf{k}). \tag{3.3}$$

Of the nine terms that arise from the product on the right, three are zero, namely $U_xV_x\mathbf{i} \times \mathbf{i}$, $U_yV_y\mathbf{j} \times \mathbf{j}$, and $U_zV_z\mathbf{k} \times \mathbf{k}$, because of Equations (2.1). The other six terms fall naturally into three pairs. For example, $U_yV_z\mathbf{j} \times \mathbf{k}$ and $U_zV_y\mathbf{k} \times \mathbf{j}$ belong together because $\mathbf{j} \times \mathbf{k} = -\mathbf{k} \times \mathbf{j} = \mathbf{i}$. In fact, these two terms may be combined as $(U_yV_z - U_zV_y)\mathbf{i}$. Two other pairs may be similarly treated, and we ultimately find that

$$\mathbf{U} \times \mathbf{V} = (U_yV_z - U_zV_y)\mathbf{i} + (U_zV_x - U_xV_z)\mathbf{j} + (U_xV_y - U_yV_x)\mathbf{k}. \tag{3.4}$$

**Exercise 3.1** Check that the other two pairs do yield the two terms written in Equation (3.4).

Comparing this formula with that for $\mathbf{U}\cdot\mathbf{V}$, we see that the products $U_xV_x$, $U_yV_y$, $U_zV_z$ that survive in $\mathbf{U}\cdot\mathbf{V}$ are absent from $\mathbf{U} \times \mathbf{V}$, while the products $U_xV_y$, etc. that are absent from $\mathbf{U}\cdot\mathbf{V}$ are the ones that survive in $\mathbf{U} \times \mathbf{V}$. The formula for $\mathbf{U} \times \mathbf{V}$ is not as pleasant as the one for $\mathbf{U}\cdot\mathbf{V}$; but its structure is not at all as complicated as it seems, and writing it down from memory is a fairly simple matter. Note first that the coefficient of **i** does not contain $U_x$ and $V_x$, the components along **i** of the original vectors; and similarly for **j** and **k**. Next note that if we take $x$, $y$, and $z$, in cyclic order (that is, $x \to y \to z \to x \cdots$, as would be the case if $x$, $y$, and $z$ were written in a circle), then the positive terms have the suffixes in this order (when we write the components of the first vector first, as we would naturally do). The positive terms may thus be written down easily, and if gaps are left for the negative terms, we can fill them in later by simply reversing the order of the suffixes that appeared in the corresponding positive terms.

**Exercise 3.2** Study the above. Then, *without looking at Equation* (3.4), fill in the missing terms in the following:

$$\mathbf{A} \times \mathbf{B} = (A_y B_z - \qquad)\mathbf{i} + (A_z B_x - \qquad)\mathbf{j} + (A_x B_y - \qquad)\mathbf{k}.$$

**Exercise 3.3** Write the positive terms in the formula for $\mathbf{L} \times \mathbf{Q}$, leaving gaps for the negative terms. Then fill in the latter.

**Exercise 3.4** Write the formula for $\mathbf{V} \times \mathbf{U}$ and verify that it is the negative of that for $\mathbf{U} \times \mathbf{V}$.

**Exercise 3.5** Without looking at the full formula, write down the coefficient of $\mathbf{j}$ in $\mathbf{U} \times \mathbf{V}$ and that of $\mathbf{k}$ in $\mathbf{B} \times \mathbf{A}$.

There is another way of remembering how to write the formula for a vector product in terms of components. First write

$$\begin{array}{cccc} \mathbf{i} & \mathbf{j} & \mathbf{k} & \mathbf{i} \\ U_x & U_y & U_z & U_x \\ V_x & V_y & V_z & V_x . \end{array} \tag{3.5}$$

[The arrows may be omitted once you are used to the method.] For the component of $\mathbf{U} \times \mathbf{V}$ along $\mathbf{i}$, consult the $\mathbf{j}$ and $\mathbf{k}$ columns, and form the two products indicated by the arrows, counting the downward products positive and the upward ones negative. For the coefficient of $\mathbf{j}$, do the same with the $\mathbf{k}$ and $\mathbf{i}$ columns (the third and fourth), and for the coefficient of $\mathbf{k}$ use the $\mathbf{i}$ and $\mathbf{j}$ columns (the first and second).

If you have studied determinants, you will already have realized that, formally,

$$\mathbf{U} \times \mathbf{V} = \begin{vmatrix} \mathbf{i} & \mathbf{j} & \mathbf{k} \\ U_x & U_y & U_z \\ V_x & V_y & V_z \end{vmatrix} \tag{3.6}$$

and there in no reason why you should not find cross products by means of this formula.

Choose whichever method you find most congenial. When dealing with symbols having appropriate suffixes, I prefer the first since it lets one write down the formula or any part of it immediately without bothering with Equation (3.4), or schema (3.5), or Equation (3.6). In numerical problems involving $\mathbf{i}$, $\mathbf{j}$, and $\mathbf{k}$, we can use the distributive law and the relations (2.1). For example, to work out $(2\mathbf{i} + 3\mathbf{j} + \mathbf{k}) \times (3\mathbf{i} + 2\mathbf{j} + 5\mathbf{k})$, we first note that the $\mathbf{i} \times \mathbf{i}$, $\mathbf{j} \times \mathbf{j}$, and $\mathbf{k} \times \mathbf{k}$ terms will vanish. Then we group the $15\mathbf{j} \times \mathbf{k}$ and the $2\mathbf{k} \times \mathbf{j}$ to give $15\mathbf{i} - 2\mathbf{i}$ or $13\mathbf{i}$, and similarly for the other terms. The result is $13\mathbf{i} - 7\mathbf{j} - 5\mathbf{k}$.

**Exercise 3.6** Show that $(\mathbf{i} + 2\mathbf{j} + 3\mathbf{k}) \times (4\mathbf{i} + 5\mathbf{j} + 6\mathbf{k}) = 3(-\mathbf{i} + 2\mathbf{j} - \mathbf{k})$.

**Exercise 3.7** Show that $(\mathbf{i} - 2\mathbf{j} + 3\mathbf{k}) \times (-4\mathbf{i} + 5\mathbf{j} + 6\mathbf{k}) = -3(9\mathbf{i} + 6\mathbf{j} + \mathbf{k})$.

**Exercise 3.8** Find the sine of the angle between $\mathbf{i} + \mathbf{j} + \mathbf{k}$ and $\mathbf{i} + 2\mathbf{j} + 3\mathbf{k}$. Check the answer by finding the cosine from the scalar product and verifying that the sum of the squares of the sine and cosine is unity. [*Ans.* $1/\sqrt{7}$, $\sqrt{6}/\sqrt{7}$, the signs being correct for $0° \leq \theta \leq 180°$.]

**Exercise 3.9** Like Exercise 3.8 for $3\mathbf{i} + 4\mathbf{k}$ and $2\mathbf{i} - \mathbf{j} - 2\mathbf{k}$. [*Ans.* $\sqrt{221}/15$, $-2/15$]

**Exercise 3.10** Show that $\mathbf{i} \cos \alpha + \mathbf{j} \sin \alpha$ is a vector of unit magnitude in the **ij**-plane, making an angle $\alpha$ with **i**. Then, by considering the cross product of this vector with a similar one making an angle $\beta$ with **i**, derive the well-known trigonometrical formula $\sin(\beta - \alpha) = \sin \beta \cos \alpha - \cos \beta \sin \alpha$.

**Exercise 3.11** If $\mathbf{W} = \mathbf{U} \times \mathbf{V}$, then **W** is perpendicular to both **U** and **V**. Consequently $\mathbf{W} \cdot \mathbf{U} = 0$ and $\mathbf{W} \cdot \mathbf{V} = 0$. Express these two equations in terms of the components of **U**, **V**, and **W**, and by solving them, show that the components of **W** must be in the ratio

$$(U_y V_z - U_z V_y) : (U_z V_x - U_x V_z) : (U_x V_y - U_y V_x).$$

## 4. TRIPLE PRODUCTS

In ordinary algebraic multiplication of scalars, the expression *uvw* has a definite meaning. The situation is different when we are dealing with scalar products or vector products of vectors. For example, $\mathbf{U} \cdot \mathbf{V} \cdot \mathbf{W}$ has no meaning at all: if we form $\mathbf{U} \cdot \mathbf{V}$ first we obtain a scalar, and we can not form a dot product of this scalar with **W**; nor are we any better off if we form, say, $\mathbf{V} \cdot \mathbf{W}$ first. With cross products the situation is different. Consider, as a simple case, the expression $\mathbf{i} \times \mathbf{j} \times \mathbf{j}$. If we first form the product $\mathbf{i} \times \mathbf{j}$ we obtain **k**, and since this is a vector, we can certainly form its cross product with the remaining **j**: the result is $-\mathbf{i}$. But if we had first formed the product $\mathbf{j} \times \mathbf{j}$ we would have obtained the result zero. Let us write this in symbols:

$$(\mathbf{i} \times \mathbf{j}) \times \mathbf{j} = \mathbf{k} \times \mathbf{j} = -\mathbf{i} \quad \text{but} \quad \mathbf{i} \times (\mathbf{j} \times \mathbf{j}) = \mathbf{i} \times 0 = 0.$$

We see from this that an expression like $\mathbf{U} \times \mathbf{V} \times \mathbf{W}$ is ambiguous, its value depending on whether we first form $\mathbf{U} \times \mathbf{V}$ or $\mathbf{V} \times \mathbf{W}$. In general,

$$(\mathbf{U} \times \mathbf{V}) \times \mathbf{W} \neq \mathbf{U} \times (\mathbf{V} \times \mathbf{W}), \tag{4.1}$$

a fact that we express by saying that vector products do not obey the *associative* law. Therefore, whenever we have dealings with a quantity like

$\mathbf{U} \times \mathbf{V} \times \mathbf{W}$, we have to include parentheses, as $(\mathbf{U} \times \mathbf{V}) \times \mathbf{W}$ or $\mathbf{U} \times (\mathbf{V} \times \mathbf{W})$, to show what we mean. Notice that Equation (4.1) preserved the order of the vectors **U**, **V**, and **W**. We are not concerned here with the changes of sign that could arise if we altered the order.

**Exercise 4.1** Find the values of $(\mathbf{j} \times \mathbf{k}) \times \mathbf{k}$ and $\mathbf{j} \times (\mathbf{k} \times \mathbf{k})$.

**Exercise 4.2** Find the values of $(\mathbf{i} \times \mathbf{j}) \times \mathbf{i}$ and $\mathbf{i} \times (\mathbf{j} \times \mathbf{i})$ and show that they happen to be equal.

**Exercise 4.3** Find the values of all of the following that have meaning: $(\mathbf{i} + \mathbf{j} + \mathbf{k}) \times \mathbf{k}$, $(\mathbf{i} \cdot \mathbf{i}) \times \mathbf{j}$, $\mathbf{i} \cdot (\mathbf{i} \times \mathbf{j})$, $(\mathbf{i} \cdot \mathbf{i})\, \mathbf{j}$, $(\mathbf{i} \cdot \mathbf{i}) \cdot \mathbf{j}$, $(\mathbf{i} - \mathbf{j}) \cdot (\mathbf{i} + \mathbf{j})$, $[(\mathbf{i} - \mathbf{j}) \times (\mathbf{i} + \mathbf{j})] \times \mathbf{i}$, $[(\mathbf{i} - \mathbf{j}) \times (\mathbf{i} + \mathbf{j})] \times \mathbf{k}$, $(\mathbf{i} \cdot \mathbf{i})(\mathbf{i} \cdot \mathbf{i})$, $(\mathbf{i} \cdot \mathbf{i}) \times (\mathbf{i} \cdot \mathbf{i})$, $(2\mathbf{i}) \cdot (3\mathbf{i})$. [Three are meaningless.]

**Exercise 4.4** Find the values of $[(\mathbf{i} + \mathbf{k}) \times \mathbf{j}] \times (\mathbf{i} + \mathbf{k})$ and $(\mathbf{i} + \mathbf{k}) \times [\mathbf{j} \times (\mathbf{i} + \mathbf{k})]$ and show that they happen to be equal.

**Exercise 4.5** From Exercises, 4.2 and 4.4 we begin to suspect that $(\mathbf{U} \times \mathbf{V}) \times \mathbf{U} = \mathbf{U} \times (\mathbf{V} \times \mathbf{U})$. Using Equation (1.5) twice, prove that this relation is indeed true. [Use Equation (1.5) once on $(\mathbf{U} \times \mathbf{V})$ and then on the pair $(\mathbf{U} \times \mathbf{V}) \times \mathbf{U}$.]

**Exercise 4.6** Show that $\mathbf{i} \cdot (\mathbf{j} \times \mathbf{k}) = (\mathbf{i} \cdot \mathbf{j}) \times \mathbf{k} = \mathbf{i} \times (\mathbf{j} \cdot \mathbf{k}) = (\mathbf{i} \times \mathbf{j}) \cdot \mathbf{k} = 1$. [This is not a coincidence. It is a special case of what we are about to prove.]

Consider the parallelepiped defined by three vectors **U**, **V**, and **W** having a common starting point $O$.

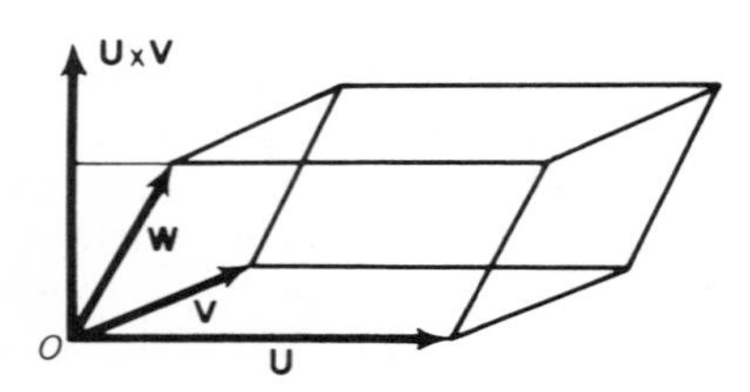


The cross product $\mathbf{U} \times \mathbf{V}$ is perpendicular to the parallelogram defined by **U** and **V**, and its magnitude is equal to the area of the parallelogram. To find the volume of the parallelepiped, we multiply the area of the base by the altitude. But the altitude is $W \cos \theta$, where $\theta$ is the angle that **W** makes with the normal to the base. The vector $\mathbf{U} \times \mathbf{V}$ happens conveniently to be normal to the base. Therefore $(\mathbf{U} \times \mathbf{V}) \cdot \mathbf{W}$ will give us the volume of the parallelepiped, provided **U**, **V**, and **W** form a right-handed set. If not, $(\mathbf{U} \times \mathbf{V}) \cdot \mathbf{W}$ will come out negative and will give minus the volume. Since the dot product is commutative, $(\mathbf{U} \times \mathbf{V}) \cdot \mathbf{W} = \mathbf{W} \cdot (\mathbf{U} \times \mathbf{V})$. Also, since the expression $\mathbf{U} \times (\mathbf{V} \cdot \mathbf{W})$ would be nonsense, we can omit the parentheses in $(\mathbf{U} \times \mathbf{V}) \cdot \mathbf{W}$ and write it without ambiguity as $\mathbf{U} \times \mathbf{V} \cdot \mathbf{W}$.

It is clear that we could also find the volume using different base parallelograms, namely, those defined by **V** and **W**, and by **W** and **U** (note the order of this last). We thus have, for a right-handed set of vectors **U**, **V**, and **W**, the following seemingly formidable array of equalities which in fact tells us something simple:

$$\text{Volume of parallelepiped defined by } \mathbf{U}, \mathbf{V}, \text{ and } \mathbf{W} = \mathbf{U} \times \mathbf{V} \cdot \mathbf{W} = \mathbf{V} \times \mathbf{W} \cdot \mathbf{U} = \mathbf{W} \times \mathbf{U} \cdot \mathbf{V} = \mathbf{U} \cdot \mathbf{V} \times \mathbf{W} = \mathbf{V} \cdot \mathbf{W} \times \mathbf{U} = \mathbf{W} \cdot \mathbf{U} \times \mathbf{V} = -\mathbf{V} \times \mathbf{U} \cdot \mathbf{W} = -\mathbf{W} \times \mathbf{V} \cdot \mathbf{U} = -\mathbf{U} \times \mathbf{W} \cdot \mathbf{V} = -\mathbf{U} \cdot \mathbf{W} \times \mathbf{V} = -\mathbf{V} \cdot \mathbf{U} \times \mathbf{W} = -\mathbf{W} \cdot \mathbf{V} \times \mathbf{U}. \tag{4.2}$$

What these relations tell us is that when we form a triple product of vectors using one dot and one cross, it does not matter where we put the dot and cross (within reason, of course—we would not want them on another page, nor even like this: · **UVW** ×). Moreover, it does not matter whether we start with **U**, or **V**, or **W**, just so long as we keep the letters in cyclic order. And if we should break the cyclic order, the only penalty would be a minus sign. Because of this flexibility, the six positive combinations in Equation (4.2) are often denoted by the single symbol [**UVW**].

**Exercise 4.7** Rewrite a sampling of the relations (4.2) for the case of a left-handed set of vectors **U**, **V**, and **W**. [$\mathbf{U} \times \mathbf{V} \cdot \mathbf{W}$ will now have a minus sign in front of it, and $\mathbf{V} \times \mathbf{U} \cdot \mathbf{W}$ a plus sign, for example. In practice, it is simplest to avoid left-handed sets **U**, **V**, and **W** by rearranging their order so as to form a right-handed set. If $\mathbf{U} \times \mathbf{V} \cdot \mathbf{W}$ comes out negative, we know that the set was left-handed. Let us assume in this section that our sets of vectors are right-handed. We shall have to look into the matter of handedness in more detail later.]

**Exercise 4.8** Show by means of Equation (4.2) that the volume of the cube defined by **i**, **j**, and **k** is unity. Work it out several ways.

**Exercise 4.9** Find the volume of the parallelepiped defined by $\mathbf{j} + \mathbf{k}$, $\mathbf{k} + \mathbf{i}$, and $\mathbf{i} + \mathbf{j}$. [Note how complicated this looks when one tries to prove the simple result geometrically by means of a diagram.]

**Exercise 4.10** Show that $\mathbf{i} \times (\mathbf{i} + \mathbf{j}) \cdot (\mathbf{i} + \mathbf{j} + \mathbf{k}) = 1$. Interpret the result geometrically in connection with a cube of unit edge, and then prove the theorem geometrically. [Once you see it, the geometrical proof is easy. But it may take you some time to see it.]

**Exercise 4.11** Find the volume of the parallelepiped defined by $\mathbf{i} + 2\mathbf{j} + 3\mathbf{k}$, $2\mathbf{i} + 3\mathbf{j} + 4\mathbf{k}$, and $\mathbf{i} + 3\mathbf{j} + \mathbf{k}$. [*Ans.* 4.]

**Exercise 4.12** If you have studied determinants show that

$$\mathbf{U} \cdot \mathbf{V} \times \mathbf{W} = \begin{vmatrix} U_x & U_y & U_z \\ V_x & V_y & V_z \\ W_x & W_y & W_z \end{vmatrix}.$$

**Exercise 4.13** What must be true geometrically of the three nonzero vectors **A**, **B**, and **C**, all starting at $O$, if $[\mathbf{ABC}] = 0$?

**Exercise 4.14** What must be true geometrically of the four nonzero vectors **A**, **B**, **C**, and **D**, all starting at $O$, if $(\mathbf{A} \times \mathbf{B}) \cdot (\mathbf{C} \times \mathbf{D}) = 0$? What if $(\mathbf{A} \times \mathbf{B}) \times (\mathbf{C} \times \mathbf{D}) = 0$?

**Exercise 4.15** Show that $[\mathbf{UVU}] = 0$. Using this and similar results, show that $[(\mathbf{V} + \mathbf{W})(\mathbf{W} + \mathbf{U})(\mathbf{U} + \mathbf{V})] = 2[\mathbf{UVW}]$, and interpret this relation geometrically. [Compare Exercise 4.9.]

As an example of the use of Equation (3.4), consider the not altogether pleasant problem of working out a formula for $\mathbf{U} \times (\mathbf{V} \times \mathbf{W})$. The result is clearly a vector. Let us therefore concentrate on the $\mathbf{i}$ component. This will come from $U_y\mathbf{j}$ multiplying the $\mathbf{k}$ term in $\mathbf{V} \times \mathbf{W}$, and $U_z\mathbf{k}$ multiplying the $\mathbf{j}$ term in $\mathbf{V} \times \mathbf{W}$. So it is:

$$U_y(V_x W_y - V_y W_x) - U_z(V_z W_x - V_x W_z).$$

Two of the terms have $V_x$ as a factor and two others have $W_x$. So we regroup this as:

$$V_x(U_y W_y + U_z W_z) - W_x(U_y V_y + U_z V_z).$$

Now we add $U_x V_x W_x - U_x V_x W_x$, which is zero, and obtain

$$V_x(U_x W_x + U_y W_y + U_z W_z) - W_x(U_x V_x + U_y V_y + U_z V_z),$$

which is the $\mathbf{i}$ component of $\mathbf{V}(\mathbf{U}\cdot\mathbf{W}) - \mathbf{W}(\mathbf{U}\cdot\mathbf{V})$ or $(\mathbf{U}\cdot\mathbf{W})\mathbf{V} - (\mathbf{U}\cdot\mathbf{V})\mathbf{W}$. By symmetry, the other components will fit the same scheme. So

$$\mathbf{U} \times (\mathbf{V} \times \mathbf{W}) = (\mathbf{U}\cdot\mathbf{W})\mathbf{V} - (\mathbf{U}\cdot\mathbf{V})\mathbf{W}. \tag{4.3}$$

**Exercise 4.16** Work out the $\mathbf{j}$ component by the same method.

## 5. MOMENTS

In Chapter 2, in connection with a seesaw, we became acquainted with the idea of the moment of a force. We now extend the discussion to the three-dimensional case.

Consider a rigid body acted on by a force $\mathbf{F}$ at a point $A$ in it as shown in Figure 5.1. Let $O$ be a pivotal point in the body. [The point $O$ need not be a pivotal point, nor need there be a rigid body. One can take moments about any point, with or without a rigid body connecting the point and the force. For example, one can take the moment about the center of the sun of a force acting on a planet. However, we shall more easily get the feel of what is going on if we think here of $O$ as an actual pivot in a rigid body acted on by the force $\mathbf{F}$.] The turning effect of the force $\mathbf{F}$ about $O$ is measured by the product of its magnitude $F$ and the perpendicular distance, $p$, between $O$ and its line of action. Thus the magnitude of the moment of $\mathbf{F}$ about $O$ is $pF$.

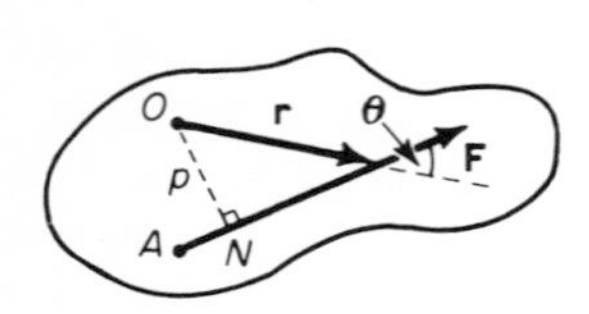

**Figure 5.1**

Now suppose we take *any* point $P$ on the line of action of $\mathbf{F}$. Let the position vector, $r$, of $P$ relative to $O$ make an angle $\theta$ with $\mathbf{F}$, as shown. Then $p = r \sin\theta$. Therefore, the magnitude of the vector product $\mathbf{r} \times \mathbf{F}$, which is $rF\sin\theta$, is just $pF$, the magnitude of the moment of $\mathbf{F}$ about $O$.

The relation between $\mathbf{r} \times \mathbf{F}$ and the moment of $\mathbf{F}$ about $O$ is even closer than this. It extends to direction as well as magnitude. For the effect of $\mathbf{F}$ is to turn the rigid body in such a way as not to tilt the plane containing $O$ and $\mathbf{F}$; thus the effect is not just to turn the body "about $O$" but rather to turn it about an axis through $O$ that is perpendicular to the plane of $\mathbf{r}$ and $\mathbf{F}$. So we see that the moment, or turning effect, of $\mathbf{F}$ about $O$ is, in this sense, given both as to its magnitude and the direction of its axis by $\mathbf{r} \times \mathbf{F}$. Moreover, if we reversed the direction of $\mathbf{F}$, thus producing a moment of opposite sign, the vector product $\mathbf{r} \times \mathbf{F}$ would automatically mirror this change of sign since, by the right-hand convention, $\mathbf{r} \times \mathbf{F}$ would now point in the direction diametrically opposite to its former direction.

Apparently, then, we may take $\mathbf{r} \times \mathbf{F}$ as representative of the moment of $\mathbf{F}$ about $O$. But before we do this, one thing remains to be verified. The quantity $\mathbf{r} \times \mathbf{F}$ is a vector. Do moments combine according to the parallelogram law? The answer is yes. But it is not an automatic yes. Simply agreeing to represent moments by vector products of the type $\mathbf{r} \times \mathbf{F}$ gives us no guarantee that the physical turning effect of an $\mathbf{r}_1 \times \mathbf{F}_1$ and an $\mathbf{r}_2 \times \mathbf{F}_2$ together is the same as the turning effect of their vectorial resultant. That it *is* the same can be deduced from equations $\mathbf{F} = m\mathbf{a}$ which are based on experiment. It can also be inferred directly from experiment. We omit the details.

This parallelogram-law behavior of moments of forces is a far-reaching thing. Consider, for example, the rigid cube shown in Figure 5.2, which we shall assume has edges of unit length. Let the cube be pivoted at $O$ and be acted on by two forces $\mathbf{F}_1$ and $\mathbf{F}_2$ as shown in Figure 5.2. What is the combined turning effect of the two forces? The forces act somewhat, but not entirely, against each other; and it is not easy to visualize offhand what sort of a twist they would tend to give to the cube when acting together. We can find their combined turning effect by forming the resultant of two cross products; a convenient way to do this is to use components and then stop visualizing while we perform some routine calculations.

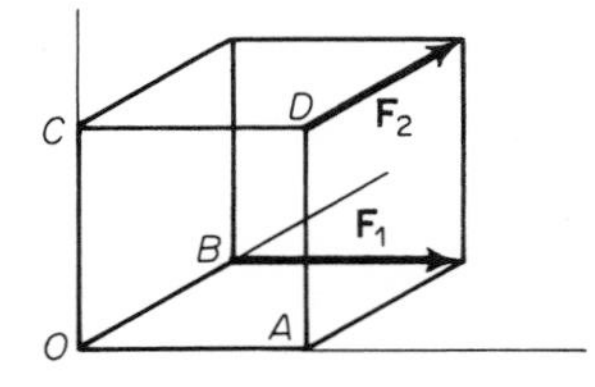


**Figure 5.2**

Let us introduce $\mathbf{i}$, $\mathbf{j}$, and $\mathbf{k}$ along $OA$, $OB$, and $OC$ respectively. Then the moment vector of $\mathbf{F}_1$ is easily seen to be $-F_1\mathbf{k}$. [Why the minus?] To find the moment vector of $\mathbf{F}_2$ about $O$, we note that the position vector, $\mathbf{r}$, of the point $D$ has components $(1, 0, 1)$, while the components of $\mathbf{F}_2$ are $(0, F_2, 0)$. So, by a standard procedure, we find that $\mathbf{r}_2 \times \mathbf{F}_2 = -\mathbf{F}_2\mathbf{i} + \mathbf{F}_2\mathbf{k}$. [The easiest procedure is to use the distributive law on $(\mathbf{i} + \mathbf{k}) \times (F_2\mathbf{j})$.] Thus the resultant moment is given by the vector $-F_2\mathbf{i} + (F_2 - F_1)\mathbf{k}$.

In particular, if the two forces have the same magnitudes, so that $F_1 = F_2$, we see that the resultant is just $-F_2\mathbf{i}$, a fact that may at first sight seem surprising. But if we look at the problem in a different way, this result becomes rather obvious. We simply ask ourselves what the combined turning effects of the forces are about the three axes $OA$, $OB$, and $OC$ individually. (The turning effect of a force **F** about $OA$, for example, is measured by the moment about $OA$ of the component of **F** perpendicular to $OA$.) About $OA$, the force $\mathbf{F}_1$ has zero moment but $\mathbf{F}_2$ has moment $-F_2$ because the perpendicular distance $AD$ from $OA$ to the line of action of $\mathbf{F}_2$ is unity. So the **i** component of the resultant moment is $-F_2$. About $OB$, the moments are both zero, because the line of action of $\mathbf{F}_1$ intersects $OB$, and $\mathbf{F}_2$ is parallel to $OB$. About $OC$, the moments are $-F_1$ and $F_2$, and when $F_1 = F_2$, the sum is zero.

**Exercise 5.1** Explain how the distributive law of vector products justfies the second method above.

**Exercise 5.2** Find the resultant moment about $O$ in Figure 5.2 if $\mathbf{F}_1$ acts along $CD$ and $\mathbf{F}_2$ along the diagonal $BC$. [*Ans.* $(F_2/\sqrt{2})\mathbf{i} + F_1\mathbf{j}$.]

**Exercise 5.3** Like Exercise 5.2, but with $\mathbf{F}_2$ acting along the diagonal $BD$. [*Ans.* $-(F_2/\sqrt{3})\mathbf{i} + F_1\mathbf{j} - (F_2/\sqrt{3})\mathbf{k}$.]

**Exercise 5.4** Show that if two nonzero forces act on a rigid body they will have zero resultant moment about a point $O$ only if they lie in a plane through $O$.

**Exercise 5.5** If the sum of the moments of $n$ forces about a point $O$ is zero, show that it will be zero about every other point, provided the resultant of the forces moved (as though they were free vectors) to act at a common point is zero. [Compare Exercise 6.6 of Chapter 2.]

We already know from Chapter 2 that a couple consists of two parallel forces having equal magnitudes and opposite directions but, in general, different lines of action. It is clear from the present discussion of the moment of a force about a point, that if $P_1$ and $P_2$ are any points on the respective lines of action of the forces $\mathbf{F}_1$ and $\mathbf{F}_2$ of a couple (these forces having equal magnitudes), the moment of the couple is given by the vector product $\overrightarrow{P_1P_2} \times \mathbf{F}_2$ (which is the same as $\overrightarrow{P_2P_1} \times \mathbf{F}_1$); and that if $F$ is the common magnitude of the forces and $d$ the perpendicular distance between them, the magnitude of the moment of the couple is $Fd$. Its direction, of course, is perpendicular to the plane of the couple.

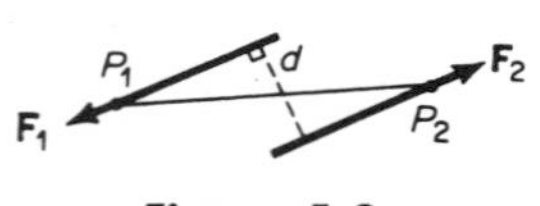

**Figure 5.3**

**Exercise 5.6** Given a couple and any point $O$ not necessarily in the plane of the forces comprising the couple, show that the resultant of the moments of the forces about $O$ is equal to the moment vector of the couple. [Use the fact tht $\overrightarrow{OP_2} = \overrightarrow{OP_1} + \overrightarrow{P_1P_2}$.]

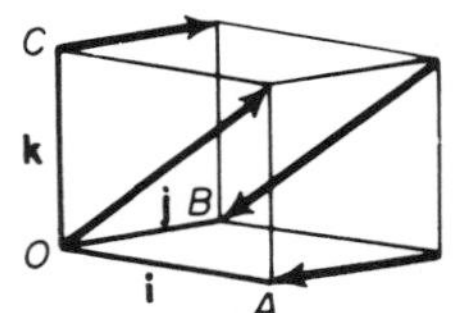


**Exercise 5.7** Find the moment of the resultant couple of the two couples shown in the diagram, if all forces have magnitude 5 and the square

has sides 2 units long. [*Ans.* The individual couples have moments $-10\mathbf{i} - 10\mathbf{k}$, $-10\mathbf{i} + 10\mathbf{k}$, so their resultant is $-20\mathbf{i}$.]

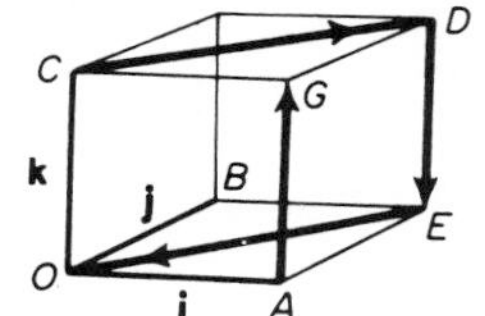


**Exercise 5.8** Like Exercise 5.7 but with the couples as shown, the forces $\overrightarrow{AG}$ and $\overrightarrow{DE}$ having magnitudes of 2 lbs, and $\overrightarrow{CD}$ and $\overrightarrow{EO}$ $2\sqrt{2}$ lbs. [Why do we get the wrong answer if we argue that the three vectors $\overrightarrow{CD}$, $\overrightarrow{DE}$, and $\overrightarrow{EO}$ are together equivalent to the vector $\overrightarrow{CO}$ so that we have, in effect, just a single couple with forces along $\overrightarrow{CO}$ and $\overrightarrow{AG}$?]

**Exercise 5.9** A force $\mathbf{F}$ acts at a point with position vector $\mathbf{r}$ relative to $O$. By introducing at $O$ a pair of cancelling forces parallel to $\mathbf{F}$ and having the same magnitude as $\mathbf{F}$, show that the original force is equivalent to a parallel force of equal magnitude through $O$ and a couple having a moment $\mathbf{r} \times \mathbf{F}$.

By applying the technique outlined in Exercise 5.9 to each force in succession, we can reduce any system of forces to a single resultant force at any point $O$ and a single resultant couple. Let us denote the single force at $O$ by $\mathbf{F}$ and the couple by $\mathbf{G}$. In general $\mathbf{F}$ and $\mathbf{G}$ will not be parallel. If we go to a new point $P$, the new force will be parallel to the old and, indeed, identical with it except for location. But the couple will in general be altered; for if the position vector of $P$ relative to $O$ is $\mathbf{r}$, there will be an additional couple $-\mathbf{r} \times \mathbf{F}$ arising from the change from $O$ to $P$, so the resultant couple will be $\mathbf{G} - \mathbf{r} \times \mathbf{F}$. We therefore have a chance of finding a point $P$ for which the resultant force and resultant couple of the system are parallel—in which case we say we have a *wrench*, this being the sort of influence one customarily exerts on a screwdriver. Let us see if we can find such a point $P$. We want $\mathbf{G} - \mathbf{r} \times \mathbf{F}$ to be parallel to $\mathbf{F}$. Therefore, we want $\mathbf{F} \times (\mathbf{G} - \mathbf{r} \times \mathbf{F}) = 0$, i.e.,

$$\mathbf{F} \times \mathbf{G} - \mathbf{F} \times (\mathbf{r} \times \mathbf{F}) = 0. \tag{5.1}$$

As it stands, this is not an easy equation to solve for $\mathbf{r}$. We therefore seek help from equation (4.3) which tells us that

$$\mathbf{F} \times (\mathbf{r} \times \mathbf{F}) = (\mathbf{F} \cdot \mathbf{F})\mathbf{r} - (\mathbf{F} \cdot \mathbf{r})\mathbf{F}.$$

This lets us write equation (5.1) in the form

$$\mathbf{F} \times \mathbf{G} - F^2\mathbf{r} + (\mathbf{F} \cdot \mathbf{r})\mathbf{F} = 0. \tag{5.2}$$

We are still not out of the woods. The next step is to decompose $\mathbf{r}$ into the sum of two vectors $\mathbf{n}$ and $\mathbf{p}$, $\mathbf{n}$ being parallel to $\mathbf{F} \times \mathbf{G}$ and thus normal to $\mathbf{F}$, and $\mathbf{p}$ being parallel to $\mathbf{F}$:

$$\mathbf{r} = \mathbf{n} + \mathbf{p}. \tag{5.3}$$

Then, since $\mathbf{F} \cdot \mathbf{n} = 0$, we have, from equation (5.2),

$$\mathbf{F} \times \mathbf{G} - F^2(\mathbf{n} + \mathbf{p}) + (\mathbf{F} \cdot \mathbf{p})\mathbf{F} = 0. \tag{5.4}$$

The component vectors of the vector 0 are zero in all directions. The component vectors of the resultant of the left hand side must therefore also be zero. Since $\mathbf{F} \times \mathbf{G}$ and $\mathbf{n}$ are perpendicular to $\mathbf{p}$ and $\mathbf{F}$, we see that we can extract two distinct equations from equation (5.4), one for the component vectors in the direction of $\mathbf{n}$, the other for the component vectors in the direction of $\mathbf{p}$. They are

$$\left.\begin{aligned} \mathbf{F} \times \mathbf{G} - F^2\mathbf{n} &= 0 \\ -F^2\mathbf{p} + (\mathbf{F}\cdot\mathbf{p})\mathbf{F} &= 0. \end{aligned}\right\} \qquad (5.5)$$

So

$$\mathbf{n} = \frac{\mathbf{F} \times \mathbf{G}}{F^2}, \quad \mathbf{p} = \frac{(\mathbf{F}\cdot\mathbf{p})\mathbf{F}}{F^2}. \qquad (5.6)$$

At first we are inclined to think that these equations (5.6), (5.3) fix the vector $\mathbf{r}$. But the second equation in (5.6) does not give $\mathbf{p}$; it gives only its direction —which we already knew was parallel to $\mathbf{F}$.

**Exercise 5.10** Why does it give only the direction? [*Hint:* Try to find the magnitude of $\mathbf{p}$ from the equation and see what happens.]

Since Equation (5.1) will hold for all vectors $\mathbf{p}$ satisfying Equation (5.6), we see that instead of finding a single point $P$ relative to which the forces of the system form a wrench, we have found a whole line of such points. This line is parallel to $\mathbf{F}$, and is called *Poinsot*'s *central axis*.

**Exercise 5.11** The above mathematics breaks down if $\mathbf{F} = 0$. Does it matter? What could one say about the whole idea of searching for a central axis when $\mathbf{F} = 0$?

**Exercise 5.12** Given one point relative to which a system of forces becomes a wrench, show that it is obvious that any point on the line through this point parallel to $\mathbf{F}$ will also be one relative to which the system is a wrench.

**Exercise 5.13** Show that the magnitude of the resultant moment of the forces of a system is a minimum when the moment is taken relative to points on the central axis. [*Hint*: $\mathbf{F}$ does not change its magnitude or direction as one goes from point to point. $\mathbf{G}$ changes by $-\mathbf{r} \times \mathbf{F}$ which is perpendicular to $\mathbf{F}$. So $\mathbf{G}\cdot\mathbf{F}$ does not change value. For given $\mathbf{F}$ and $\mathbf{G}\cdot\mathbf{F}$, what angle should $\mathbf{G}$ make with $\mathbf{F}$ if $G$ is to be a minimum?]

**Exercise 5.14** Find the central axis of the forces in Figure 5.2. [We have $\mathbf{F} = F_1\mathbf{i} - F_2\mathbf{j}$, and, taking moments about $O$, $\mathbf{G} = -F_2\mathbf{i} + (F_2 - F_1)\mathbf{k}$. So $F^2 = F_1^2 + F_2^2$, $\mathbf{F} \times \mathbf{G} = -F_2(F_2 - F_1)\mathbf{i} - F_1(F_2 - F_1)\mathbf{j} - F_2^2\mathbf{k}$. Then calculate $\mathbf{n}$ using equation (5.6). Note how complicated the result is for even so simple a case as this.]

**Exercise 5.15** Like Exercise 5.14 if the forces on the cube are those described in Exercise 5.2.

**Exercise 5.16** Relative to a point $O$, a wrench pointing along $\mathbf{i}$ has force and couple of magnitude 1 and 1. Another wrench, pointing along **j**, has force and couple of magnitude 1, 2 respectively. Find the resultant central axis. [*Ans.* A horizontal line through the point $(0, 0, \frac{1}{2})$ making 45° with the **i** and **j** directions.]

## 6. ANGULAR DISPLACEMENTS

If we rotate a rigid body about a fixed axis through an angle $\alpha$ radians, it undergoes an angular displacement. We can represent this angular displacement by means of an arrow-headed line segment of length $\alpha$ units (say inches) lying along the axis of rotation and pointing in the direction a right-handed corkscrew would move if similarly rotated.* Since the angular displacement has magnitude and direction and can be represented by an arrow-headed line segment, we begin to suspect that it behaves like a vector. Indeed, if we combine a rotation about a particular axis through an angle $\alpha$ radians with one about the same axis through an angle $\beta$ radians, the resulting angular displacement is one of $\alpha + \beta$ radians about the axis, and this is just what we would obtain by combining the corresponding arrow-headed line segments according to the parallelogram law—the parallelogram, in this special case being, of course, squashed into a straight line. Again, if we combine a rotation about an axis through an angle $\alpha$ radians with a rotation about the same axis through an angle $-\alpha$ radians (which is through an angle $\alpha$ radians in the opposite direction and would thus pertain to an arrow-headed line segment pointing in a direction opposite to that of the line segment representing the original angular displacement), we end up with a zero angular displacement, and this result, too, we would get from the parallelogram law.

$\alpha$ radians

$\alpha$ inches

**Figure 6.1**

Add to all this our vectorial experiences not only with moments but even with such unlikely things as areas of parallelograms, and we begin to feel confident that angular displacements can be treated as vectors. It seems hardly necessary to check further whether the arrow-headed line segments representing angular displacements obey the parallelogram law.

But suppose we rotated the rigid body through $\alpha$ radians about an axis $l_1$, and then through $-\alpha$ radians about a different, but parallel axis $l_2$. Then the two angular displacements together would not leave the body in its original position. They would displace it, without rotation, to a new position.

**Exercise 6.1** Check this assertion by considering the simple case of a rod $AB$ in a plane, first rotating it about $A$ and then rotating it through an equal angle in the opposite direction about the new position of $B$.

*Except, of course, that we were not contemplating any motion by the rigid body *along* the axis of rotation.

Evidently we have to tread carefully if we wish to regard angular displacements as vectors. Two ways of saving the situation seem open to us. One is to say that the resulting *rotation* above was zero, as predicted by the parallelogram law, and that the bodily displacement is not our present concern. The other is to say that then we combine angular displacements vectorially they must be angular displacements about the same axis. Clearly, the second is too drastic. If we combine only rotations having a common axis we shall never use the full parallelogram law but only the case in which the parallelogram is squashed into a line. While we would then be able to say correctly that rotations about a given axis behave like vectors, this would hardly be a worthwhile assertion.

We therefore try to combine the first possibility with a less stringent version of the second. We say that we want to look upon angular displacements not as free vectors but as bound (or at least sliding) vectors, and that we will limit ourselves to combining rotations about axes that have a common point. This immediately rules out the case of parallel axes that caused the trouble discussed earlier. Presumably, with this restriction, our luck will hold. Let us see.

Consider two rotations, each through $\pi/2$ radians,* about fixed perpendicular axes that, for convenience, we shall take to be the $x$-and $y$-axes of a three-dimensional rectangular Cartesian coordinate system. To simplify the discussion, take the rigid body to be a playing card, say the ace of spades. Place it face down in the position shown in Figure 6.2a. On rotating it about the $x$-axis through $\pi/2$ radians, we bring it into the position shown in Figure 6.2b. And on now rotating it about the $y$-axis through $\pi/2$ radians, we bring it finally to the position shown in Figure 6.2c.

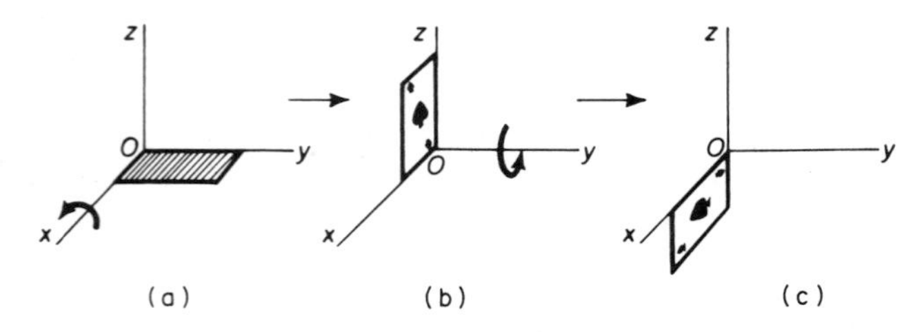


**Figure 6.2**

The two rotations, if treated as vectors, would be $(\pi/2)\mathbf{i}$ and $(\pi/2)\mathbf{j}$, and their resultant would be $(\pi/2)(\mathbf{i} + \mathbf{j})$, whose magnitude is $\pi/\sqrt{2}$ and whose direction bisects the angle between the positive $x$- and $y$-axes. But it is easy to see that if we had rotated the card in one move through the unpleasant angle $\pi/\sqrt{2}$ radians about the bisector of the foregoing angle we would not bring it into anything like the final position above. Therefore the arrow-headed line

*This is, of course, 90°. But we use radians here with an eye on the next section.

segments representing angular displacements have failed the parallelogram test, and they have done so even though we have limited ourselves to axes of rotation having a common point $O$. There is no avoiding the fact that we just cannot treat angular displacements about intersecting but nonidentical axes as vector quantities.

This fact becomes strikingly clear when we consider the effect of performing the above rotations in the reverse order. The sequence of positions of the card is shown in Figure 6.3, and the final position is quite different from

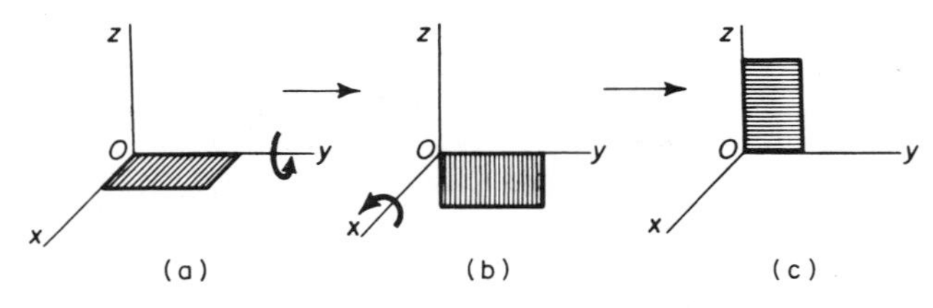


**Figure 6.3**

that in Figure 6.2. Thus, if we denote the rotations about the $x$- and $y$-axes respectively by $X$ and $Y$, we see that the addition of angular displacements does not even obey the commutative law:

$$X + Y \neq Y + X.$$

**Exercise 6.2** Starting with the card in the same position as above, rotate it first through $\pi/2$ radians about the $y$-axis and then through $\pi/2$ radians about the $z$-axis, and compare the resulting position with that obtained when these two rotations are perfomed in the reverse order. [The final positions are not the same.]

## 7. ANGULAR VELOCITY

Fresh from the chastening experience of the preceding section, let us consider a rigid body turning about a fixed axis at a constant rate of $\omega$ radians per second. Associated with the motion are a magnitude and a direction, the magnitude being $\omega$ and the direction that of the axis in, say, the right-hand corkscrew sense. Can it be properly regarded as a vector quantity?

**Exercise 7.1** Place your bet: yes or no? [Whichever you choose, you will be surprised—either in this section or later.]

There are two ways of thinking of the motion. One is as a single over-all rotation; the other is as the set of motions of all the individual particles making up the rigid body. Let us think of it in the latter sense.

Take a fixed point $O$ on the axis of rotation, and denote by $\mathbf{r}$ the position vector relative to $O$ of a general point $P$ of the rigid body. Draw $PN$ perpendicular to the axis of rotation. The rotation of the body causes $P$ to move in a circle perpendicular to the axis of rotation, the center of the circle being $N$. If $\mathbf{r}$ makes an angle $\theta$ with the axis, the radius of the circle, $PN$, is $r \sin \theta$.

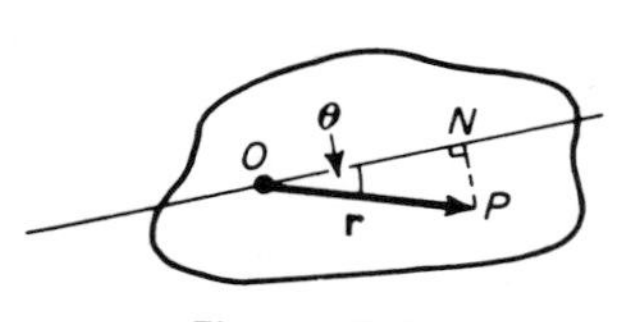


**Figure 7.1**

The speed of $P$ is easily calculated. In any time $t$ secs. the body turns through an angle of $\omega t$ radians. Therefore $P$ traverses an arc length of $(\omega t)(r \sin \theta)$. Dividing this arc length by the time taken, namely $t$, we find that the speed of $P$ is $\omega r \sin \theta$. [Note the advantage of using radians here. Had $\omega$ been given in deg./sec. the angle would have been $\omega t\pi/180$ radians, and the speed of $P$ would have been given by the ugly expression $(\pi\omega r \sin \theta)/180$.]

Though the speed of $P$ is constant, its velocity is not. At each instant it lies along the tangent at $P$ to the circle traversed by $P$, and its direction changes from instant to instant. The fact that the instantaneous velocity of $P$ has a continually changing direction is a considerable complication. To avoid trouble with the changing direction in what follows, we would like to talk about what happens at a particular instant. But a genuine instant affords no time for anything at all to happen. It was precisely this sort of dilemma that the calculus was designed to overcome; but the dilemma is of such subtlety that even when Newton and Leibnitz succeeded in constructing a calculus that gave formalized mathematical procedures for dealing with it, a century was to pass before a significant beginning was made to put these procedures on an acceptable mathematical basis. Nevertheless, during that century enormous progress was made by means of the calculus.

Taking a hint from the pioneers, then, let us cross our fingers and try to get the feel of what is going on without striving for mathematical rigor. We ask ourselves whether we can regard the above motion of a rigid body as a vector. Let us see what happens if we think of it as a vector starting at $O$, call it *angular velocity*, and denote it by $\boldsymbol{\omega}$.

The first thing we observe is that we may now represent the instantaneous velocity, $\mathbf{v}$, of the point $P$ by $\boldsymbol{\omega} \times \mathbf{r}$:

$$\mathbf{v} = \boldsymbol{\omega} \times \mathbf{r}. \tag{7.1}$$

**Exercise 7.2** Verify Equation (7.1) considering both the magnitude and direction of $\mathbf{v}$, and noting that $\mathbf{r} \times \boldsymbol{\omega}$ would have given the wrong sign for $\mathbf{v}$.

**Exercise 7.3** A rigid body is spinning about the $z$-axis with a constant angular speed of 10 radians/sec. At a particular instant, a point $P$ in the body has coordinates (2, 2, 5). What is its velocity at that instant? [*Ans.* $10\mathbf{k} \times (2\mathbf{i} + 2\mathbf{j} + 5\mathbf{k})$, i. e., $-20\mathbf{i} + 20\mathbf{j}$.]

**Exercise 7.4** Like Exercise 7.3 if the body was rotating about a line through the origin whose positive direction makes equal acute angles with the positive directions of the axes. [*Ans.* $10\sqrt{3}\,\mathbf{i} - 10\sqrt{3}\,\mathbf{j}$. Note that the instantaneous velocity has a horizontal direction. Try to visualize the situation without thinking of vector products to see why the velocity of $P$ at this instant has to be horizontal. You will then begin to realize the power of Equation (7.1).]

Now suppose that the rigid body was somehow subjected to two angular velocities $\boldsymbol{\omega}_1$ and $\boldsymbol{\omega}_2$ simultaneously, these having, in general, different axes but *axes with a common point O*. Since we are thinking of $\boldsymbol{\omega}_1$ and $\boldsymbol{\omega}_2$ as vectors starting at $O$, they will have a resultant, $\boldsymbol{\omega}$, also starting at $O$, that is given by:

$$\boldsymbol{\omega} = \boldsymbol{\omega}_1 + \boldsymbol{\omega}_2. \tag{7.2}$$

This equation is, of course, an expression of the fact that $\boldsymbol{\omega}_1$ and $\boldsymbol{\omega}_2$ yield $\boldsymbol{\omega}$ according to the parallelogram law.

Now $\boldsymbol{\omega}_1$ and $\boldsymbol{\omega}_2$ individually impart to $P$ the respective velocities $\mathbf{v}_1$ and $\mathbf{v}_2$, these being given by:

$$\mathbf{v}_1 = \boldsymbol{\omega}_1 \times \mathbf{r}, \qquad \mathbf{v}_2 = \boldsymbol{\omega}_2 \times \mathbf{r}. \tag{7.3}$$

The resultant angular velocity $\boldsymbol{\omega}$ would impart to $P$ the velocity $\mathbf{v}$ given by Equation (7.1). From Equation (7.2) we obtain

$$\boldsymbol{\omega} \times \mathbf{r} = \boldsymbol{\omega}_1 \times \mathbf{r} + \boldsymbol{\omega}_2 \times \mathbf{r}, \tag{7.4}$$

which is also an expression of the parallelogram law. But by Equations (7.1) and (7.2), this equation is just

$$\mathbf{v} = \mathbf{v}_1 + \mathbf{v}_2, \tag{7.5}$$

which tells us that if we represent angular velocities by vectors as above, the instantaneous velocities imparted to $P$ by $\boldsymbol{\omega}_1$ and $\boldsymbol{\omega}_2$ have a vectorial resultant that is just the velocity that would be imparted to $P$ by the vectorial resultant of $\boldsymbol{\omega}_1$ and $\boldsymbol{\omega}_2$. Since this holds for *every* point $P$ of the rigid body, we are justified in treating angular velocities as vectors.

Note in the above how important it is that the position vector $\mathbf{r}$ of $P$ should be the same throughout. We could not have obtained Equation (7.4) from Equation (7.2) otherwise. Therefore, we must here regard the angular velocities $\boldsymbol{\omega}$, $\boldsymbol{\omega}_1$, and $\boldsymbol{\omega}_2$ as having lines of action all passing through a common point $O$. As here regarded, they are, thus, not free vectors but sliding vectors.

That they are not free vectors is clear from the fact that if we spin a rigid body about an axis we do not cause the same motion as when we spin it with the same angular speed about a parallel axis that does not coincide with the first. Despite this, if we are willing to be content with the changes of orientation produced by angular velocities and to ignore the actual motions of individual points, we can think of angular velocities as free vectors.

**Exercise 7.5** Verify the above by considering the motion of a rigid rod $AB$ in a plane if the rod is spun about an axis through $A$ perpendicular to the plane, and if it is spun about a parallel axis through $B$.

**Exercise 7.6** You are probably familiar with *Foucault's pendulum.* It is merely a pendulum suspended in such a way that the support does not influence the direction in which the pendulum swings. What makes this simple piece of apparatus fascinating is that it reveals the rotation of the earth without our needing to look at the stars or other external objects to take our bearings. For example, if the pendulum were mounted at a pole, the vertical plane in which it swings would appear to rotate once every 24 hours. What really happens is that the pendulum swings in a plane that may be described as fixed in direction relative to the so-called fixed stars, and the earth rotates once every 24 hours beneath it. At the equator, the pendulum would not exhibit any such rotation relative to the earth. Why not?

**Exercise 7.7** At what rate relative to the earth would a Foucault pendulum seem to rotate if it were at a place of latitude $\lambda°$? [*Hint*: Let $O$ be the center of the earth and $P$ the point of support of the pendulum. Then we can resolve the angular velocity vector of the earth, say $\boldsymbol{\omega}$, into two components through $O$, one of magnitude $\omega \sin \lambda$ having its axis along $OP$, the other of magnitude $\omega \cos \lambda$ perpendicular to $OP$. Only the former will be revealed by the Foucault pendulum. Note how difficult it would be to visualize the situation without the concept of the angular velocity vector.]

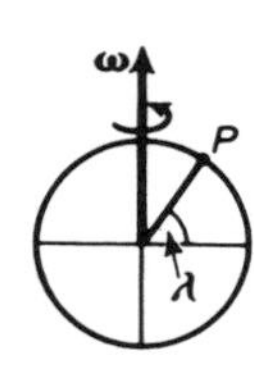


We must now return to the deliberately vague phrase "suppose that the rigid body was somehow subjected to two angular velocities $\boldsymbol{\omega}_1$ and $\boldsymbol{\omega}_2$ simultaneously" [see page 89]. What does it mean? There are two interpretations. Imagine the rigid body mounted on another body that is also rotating. Then if the first has angular velocity $\boldsymbol{\omega}_1$ relative to the second, and the second has angular velocity $\boldsymbol{\omega}_2$ relative to some master reference frame, the first is being subjected simultaneously to the two angular velocities $\boldsymbol{\omega}_1$ and $\boldsymbol{\omega}_2$, and its angular velocity relative to the master frame is the vectorial resultant of $\boldsymbol{\omega}_1$ and $\boldsymbol{\omega}_2$. We are using two reference frames here, just as we did for ordinary velocities when we considered a point moving on a moving platform.

But we can also take the mathematical instead of the physical point of view, just as we did when we looked upon ordinary velocity as the rate of change of displacement with respect to time and could thus combine velocities that were all measured relative to the same reference frame. With angular velocities the situation is as follows. Suppose a rigid body has angular velocity $\boldsymbol{\omega}$ about an axis through $O$. If we introduce a reference frame $\mathbf{e}_x$, $\mathbf{e}_y$, and $\mathbf{e}_z$ and write:

$$\boldsymbol{\omega} = \omega_x \mathbf{e}_x + \omega_y \mathbf{e}_y + \omega_z \mathbf{e}_z, \tag{7.6}$$

where the components $\omega_x$, $\omega_y$, and $\omega_z$ are found in the usual vectorial way, then the instantaneous velocity of each point of the rigid body, as given by

Equation (7.1), will also be correctly given by applying Equation (7.1) to the vectors $\omega_x \mathbf{e}_x$, $\omega_y \mathbf{e}_y$, and $\omega_z \mathbf{e}_z$ to obtain three velocity vectors, and then forming the resultant of these component velocity vectors.

**Exercise 7.8** Relative to **i**, **j**, and **k** a rigid body has angular velocity $\boldsymbol{\omega}$ having components $(\omega/\sqrt{2}, \omega/\sqrt{2}, 0)$. A point $P$ on the $x$-axis has a position vector with components $(r, 0, 0)$. Draw a picture to see what is going on. Then calculate the instantaneous velocity of $P$ (a) by working directly with $\boldsymbol{\omega}$ and forming $\boldsymbol{\omega} \times \mathbf{r}$ and (b) by finding the component velocities $(\omega_x \mathbf{i}) \times \mathbf{r}$ and $(\omega_y \mathbf{j}) \times \mathbf{r}$ and finding their resultant. In which direction does the instantaneous velocity of $P$ point?

**Exercise 7.9** Like Exercise 7.8, but with the components of the position vector of $P$ given by $(0, 0, r)$.

**Exercise 7.10** Like Exercises 7.8 and 7.9, but with the components of the position vector given by $(r/\sqrt{2}, 0, r/\sqrt{2})$. [Note the sudden increase in complexity here so far as visualization of the motions is concerned. The mathematical manipulations, on the contrary, are almost as easy as before.]

All this time we must have been wondering how it happens that angular velocities behave like vectors despite the fact that angular displacements do not. Here is the essential reason: If we make an angular displacement of a rigid body about a fixed axis, the points of the body move on arcs of circles. Suppose the point $P$ moves on the circle with center $N$ to the position $P'$, as shown in Figure 7.2. Then the displacement $PP'$ is not, in general, perpendicular to the position vector $\overrightarrow{OP}$, nor does it make a fixed angle with $\overrightarrow{OP}$ for all positions of $P'$—or for all positions of $P$.

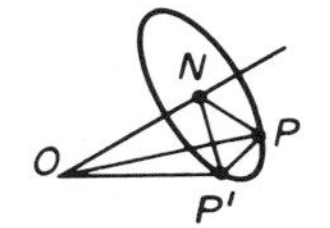


**Figure 7.2**

Look back at the discussion of the vectorial character of angular velocity and you will see that it depended crucially on the fact that the instantaneous velocity of $P$ was always perpendicular to the position vector of $P$. We might, if pressed, have managed even if the angle had not been a right angle, provided it had been a fixed angle. But with a varying angle we are beaten.

Even so, how does it happen that the instantaneous velocity of $P$, which is, after all, just a time rate of change of a displacement, is always perpendicular to $P$? The reason is that we are dealing with the *instantaneous* velocity. This is, roughly speaking, the time rate of change of an extremely small displacement. If we take $PP'$ extremely small in Figure 7.2 we see that it does indeed make approximately 90° with $OP$, being approximately tangent to the circle at $P$. And the fact is that extremely small angular displacements do indeed behave approximately like vectors, so that one often hears it said that infinitesimal angular displacements are vectors even though finite angular displacements are not.

## 8. MOMENTUM AND ANGULAR MOMENTUM

If a particle of mass $m$ has velocity $\mathbf{v}$, the vector $m\mathbf{v}$ is called its *momentum.* The combination $m\mathbf{v}$ is a rather natural one. Newton regarded it as measuring the quantity of motion. For example, if two particles of mass $m$ are moving with velocity $\mathbf{v}$, there is twice as much motion as with one; and the total momentum adds up to $2m\mathbf{v}$, accurately reflecting the fact.

In collisions momentum is *conserved*; that is, the total momentum vector of the system before the collision is equal to the total momentum vector after the collision.

**Exercise 8.1** A particle of mass $m$ moves in the positive direction along the $x$-axis with speed $v$. Another particle, of mass $2m$ moves in the positive direction along the $y$-axis with speed $v$. The two collide at the origin and coalesce into a single particle. What is the velocity of this particle?

*Solution* The momenta of the original particles are $mv\mathbf{i}$ and $2mv\mathbf{j}$ respectively. The resulting particle has mass $3m$. So if its velocity is $\mathbf{V}$, its momentum is $3m\mathbf{V}$. But the resultant of the original momenta is $mv\mathbf{i} + 2mv\mathbf{j}$. So, since the momentum is conserved in the collision, we must have $3m\mathbf{V} = mv\mathbf{i} + 2mv\mathbf{j}$; and from this we see that $\mathbf{V} = (v/3)\mathbf{i} + (2v/3)\mathbf{j}$. Thus, the resulting particle moves with speed $\sqrt{5}\,v/3$ in a direction making an angle $\tan^{-1} 2$ with the positive $x$-axis.

**Exercise 8.2** Like Exercise 8.1 but with a third particle, of mass $m$, moving with speed $2v$ in the positive direction along the $z$-axis, all three particles meeting simultaneously at $O$ and coalescing into a single particle. [*Ans.* Speed $3v/4$, and moving in a direction having direction cosines $(\frac{1}{3}, \frac{2}{3}, \frac{2}{3})$. Note that had the resulting particle in Exercise 8.1 later been struck by a particle of mass $m$ moving parallel to the $z$-axis in the positive direction with speed $2v$, and had this particle coalesced with the one it struck, the resulting speed and direction would be the same as before, though the velocity vector would have a different location.]

**Exercise 8.3** A particle of mass $m$ moving with speed $v$ in the $\mathbf{i}$ direction suddenly splits apart into two pieces. One piece, of mass $m_1$, moves with speed $v_1$ in the $\mathbf{j}$ direction. How does the other piece move?

[*Ans.* Its velocity is $\dfrac{mv}{m - m_1}\mathbf{i} - \dfrac{m_1 v_1}{m - m_1}\mathbf{j}$.]

**Exercise 8.4** Like Exercise 8.3 if the piece having mass $m$ moves with speed $v$ in a direction having direction cosines $\cos\alpha$, $\cos\beta$, and $\cos\gamma$.

The momentum vector is often denoted by the symbol $\mathbf{p}$. If a particle has momentum $\mathbf{p}$ and, relative to a point $O$, position vector $\mathbf{r}$, the vector product $\mathbf{r} \times \mathbf{p}$ is called the *moment of momentum* of the particle about $O$. It is often also called the *angular momentum* of the particle about $O$.

From the basic equation of motion $\mathbf{F} = m\mathbf{a}$, which is valid in an unaccelerated reference frame, we have:

$$\mathbf{r} \times \mathbf{F} = \mathbf{r} \times (m\mathbf{a}). \tag{8.1}$$

It happens that $\mathbf{r} \times (m\mathbf{a})$ is the rate of change of $\mathbf{r} \times (m\mathbf{v})$ with respect to time. This is not obvious, despite the fact that $\mathbf{a}$ is the rate of change of $\mathbf{v}$ with respect to time, since not only $\mathbf{v}$ but also $\mathbf{r}$ changes with time. However it does happen to be true (it can be easily proved using the calculus), so let us accept it as such here. Then Equation (8.1) tells us that the moment about an unaccelerated point $O$ of the force acting on a particle is equal to the rate of change of the angular momentum of the particle about $O$. In particular, if the force has zero moment about $O$ the angular momentum will be constant, since its rate of change is zero.

**Exercise 8.5** A particle acted on by zero force moves in a straight line with uniform speed. Verify that its moment of momentum about a fixed point $O$ is constant.

**Exercise 8.6** If a planet is acted on only by the force of gravity directed towards the center of the sun, show that its angular momentum about the center of the sun must be constant, and therefore, that it travels fastest in its orbit at the place where the tangent line to the orbit is closest to the sun.

A *gyroscope* is essentially just a top, though usually one of special shape and mounted in a special way. Consider the gyroscope shown, consisting of a massive squat cylinder spinning on its axis and mounted in such a way that its axis can turn freely in any direction about the center of symmetry $O$, which is kept in a fixed position. In practice, this requires a moderately elaborate mounting mechanism, but we can here think of the gyroscope as freely pivoted at $O$. Each particle of the gyroscope has a moment of momentum about $O$, and the sum of all these moments of momentum is the moment of momentum of the gyroscope about $O$. Denote it by vector $\mathbf{M}$ lying along the axis.

Equation (8.1) holds for each particle of the gyroscope, and by adding up the equations for all the particles (the mathematicians can do this rigorously) we find that the total moment of the forces acting on the gyroscope is equal to the rate of change of its total angular momentum.

Since the gyroscope is mounted at its center of symmetry, the gravitational forces on it have zero resultant moment about $O$. Suppose we point the axis of the gyroscope horizontally towards the right as shown in Figure 8.1. Since the resultant moment of the forces acting on it is zero, its angular momentum vector $\mathbf{M}$ will not change. Therefore, it will continue to point in the same direction. Now suppose we exert a constant pressure downward on the tip, $P$, of the axis, say by hanging a weight on it. How will the axis move? One's first guess—provided one has never played with a gyroscope—is that the tip of the axis will drop vertically. But it will not.

The weight on the tip of the axis exerts on the gyroscope a force whose moment vector about $O$ is in the direction of the horizontal line $OQ$ perpendicular to the axis $OP$, as shown. It will, therefore, cause the angular momentum vector of the gyroscope to change vectorially at a rate equal to this moment. We can think of the force as continually pumping into the gyroscope angular momentum that always points horizontally at right angles to the position of $OP$ as indicated in the figure by the arrow-headed line segment $OQ$. By forming the vectorial resultant of $\mathbf{M}$ and a small horizontal additional angular momentum perpendicular to it, we see that the effect of this pumping is to turn the axis of the gyroscope *horizontally* counterclockwise as viewed from above.

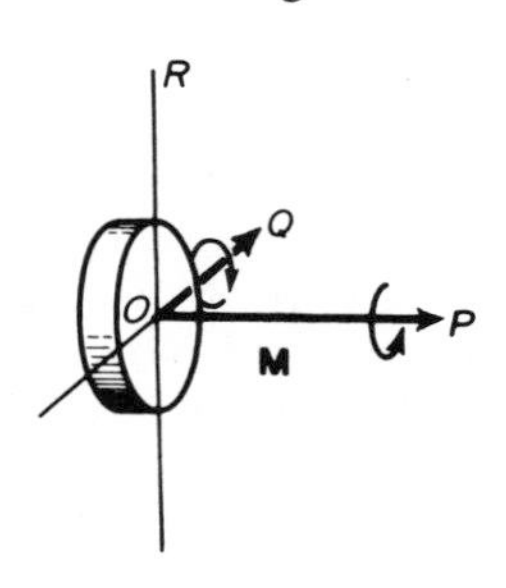

**Figure 8.1**

**Exercise 8.7** In what direction would the turning effect be if, instead of tying a weight to the tip of the axis, we exerted an upward pressure on it?

**Exercise 8.8** Like Exercise 8.7 if we pushed the tip of the axis horizontally into the page.

**Exercise 8.9** Assume that the propellers of an airplane are all turning in a clockwise sense as viewed from behind. If, in flight, the pilot turns the airplane towards the right, what is the gyroscopic effect of the propellers? [*Ans.* It tends to turn the nose of the airplane downward.]

We have seen that, when a weight is tied to the tip of the axis of the gyroscope in Figure 8.1, the axis will tend to rotate horizontally. When the weight is applied abruptly, the tip of the axis has a tendency to dip as well as to move horizontally and the resulting motion is somewhat complex, the tip of the axis bobbing up and down as the vertical plane containing it sweeps around like a revolving door. However, if the weight is applied gradually and the axis is allowed to find its appropriate motion one can manage, in principle, to end up with the axis turning steadily in a horizontal plane. Since the manifest cause of this horizontal motion of the axis is the *vertical* force exerted by the weight, this is likely to strike us as curious, to say the least. Let us therefore consider what is happening without making use of cross products. For simplicity, regard the gyroscope as a circular disc and consider the motion of a point $A$ on the rim of the gyroscope that is vertically above $O$, as in Figure 8.2. Because the gyroscope is spinning about its axis, the point $A$ has an instantaneous velocity tangent to the disc that we can represent by the arrow-headed line segment $\overrightarrow{AB}$. When we apply the weight at $P$ we tend to pull $OP$ downward; but this will tend to move the point $A$ *horizontally*, giving it an additional velocity that we can represent by $\overrightarrow{AC}$. The resultant velocity of $A$, given by $\overrightarrow{AD}$, thus lies outside the plane of the disc and shows that the disc tends to turn about a vertical axis in the manner already discussed. Had we taken a point diametrically opposite to $A$, both component velocities would be reversed in direction as, therefore, the resultant would be too, thus confirming the turning effect. Had we taken intermediate points the turning effect would be in the same general direction, except for points on the line $OQ$ which would seem to be unaffected. But since the gyroscope is reasonably rigid, even these points must join in the general turning. Indeed, the whole turning is a compromise among discordant local turning effects, and we see that when we tie a weight to the tip of the axis of a spinning gyroscope and thereby cause its axis to rotate in a horizontal plane we introduce considerable internal stresses of a quite complicated sort. But then, even if we did not tie a weight to the tip, the spinning of the gyroscope on its axis would still cause internal stresses, these being sometimes so great as to cause the gyroscope to explode into pieces.

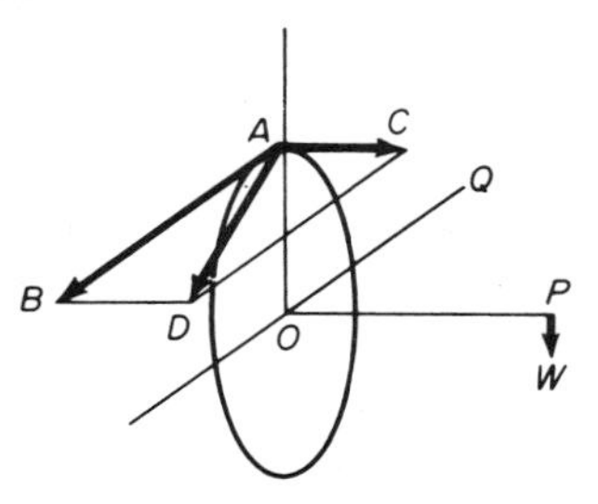

**Figure 8.2**

## 9. AREAS AND VECTORIAL ADDITION

When we consider vector products like $\mathbf{r} \times \mathbf{F}$ that yield moments, we have no qualms about accepting the physical fact that they combine according to the parallelogram law. But what about vector products that we think of as representing areas? Having decided that $\mathbf{U} \times \mathbf{V}$ represents the area of a parallelogram both in magnitude and, via the normal to the area, in direction, and having irrevocably agreed that cross products combine vectorially, we must face the fact that actual areas do not seem to fit into the vectorial scheme. For example, in Figure 9.1, if the cross product $\mathbf{U} \times \mathbf{V}$ represents the area of the parallelogram $OLMN$, then $\mathbf{V} \times \mathbf{U}$ represents the area of the parallelogram $ONML$, and we have to count this as the negative of the former area, even though the amount of paint we would need to paint the one is equal to and not the negative of the amount we would need to paint the other. But this is not too far removed from the idea of treating distances along a line as positive in one direction and negative in the opposite direction; and we have learned to live with that and even to approve of it as a valuable mathematical concept. So for the sake of mathematical harmony, and possible benefit, let us accept the official mathematical convention of counting an area as positive if, when "walking" around its rim (we may be somewhat upside down when doing so), we keep the area on our left, and negative if we keep it on our right.

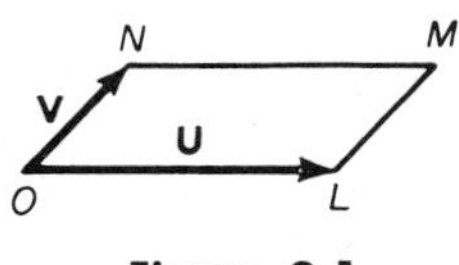

**Figure 9.1**

Immediately, we find ourselves rewarded by something pretty. For let us place two rectangles $LMNP$ and $MQRN$ side by side as in Figure 9.2 so that the sum of their areas is the area of the rectangle $LQRP$. Then consider the "sum" of the rims of the original rectangles. The part $\overrightarrow{MN}$ of the first can be regarded as cancelling the part $\overrightarrow{NM}$ of the second so that, in a reasonable sense, we can say that the rim of the sum of the areas is equal to the sum of their rims.

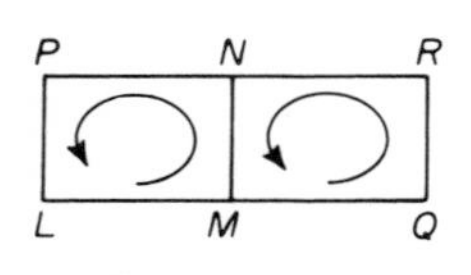

**Figure 9.2**

**Exercise 9.1** Show that it is not necessary for $LMNP$ and $MQRN$ to be rectangles, or even to have straight sides, so long as the areas are contiguous. [Actually they need not be contiguous, but then the prettiness is gone.]

**Exercise 9.2** Show that if an irregularly shaped region is crisscrossed by lines, as in Figure 9.3, and the areas of the small regions into which it is divided are all taken positively, then the sum of the rims of these small regions is just the rim of the whole region. Would this be true if all the areas were taken negatively?

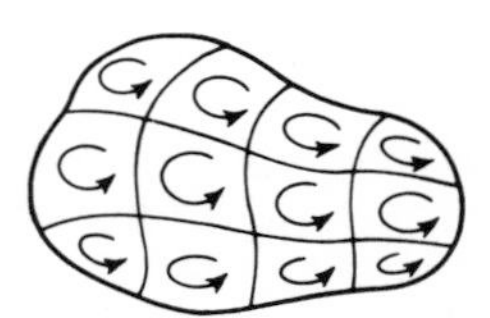
**Figure 9.3**

**Exercise 9.3** Observe that the sum of the areas of $OLMN$ and $ONML$ in Figure 9.1 is zero, and that the sum of their rims is too. Is this true of any two areas, one of which is counted as positive and the other, having the same number of square inches, is counted as negative? [Consider $\mathbf{U} \times \mathbf{V}$ and $(\frac{1}{2}\mathbf{V}) \times (2\mathbf{U})$. Note here, incidentally, that two areas of different shapes can nevertheless cancel each other vectorially.]

**Exercise 9.4** In Figure 9.1, if the parallelogram were rotated about $O$ while remaining in the plane, what would be the effect on $\mathbf{U} \times \mathbf{V}$?

Well, we have weathered the first crisis, and even benefited from it. But Exercises 9.3 and 9.4 have reminded us of something else that needs to be explored: the same vector can represent the areas of all sorts of parallelograms — a fact that we have already used, for example, in discussing $\mathbf{A} \times \mathbf{B} + \mathbf{C} \times \mathbf{D}$.

Suppose we have three parallelograms in a plane, as in Figure 9.4. Each of them can be represented by a vector through $O$ perpendicular to the plane. The resultant of the three vectors represents the area of the shaded jagged region—which is not a parallelogram.

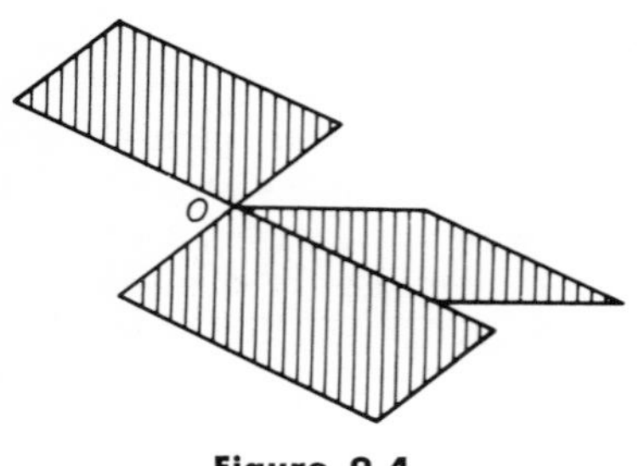

**Figure 9.4**

Indeed, if we are willing to treat the vector representing a plane area as a free vector, we can represent a plane area of quite general shape by means of a vector—simply by breaking the area up into a sufficient number of parallelograms and taking the sum of the corresponding vectors, all of which are parallel to one another. Infinitely many tiny parallelograms may be needed and a proper investigation would involve ideas belonging to the calculus. But the fact is that we can represent by means of a vector a plane area of any shape—even a curved one.

**Exercise 9.5** Must the plane area be all in one piece?

**Exercise 9.6** A circle of radius 10 ft. lies on the horizontal ground. What are the magnitude and direction of the vector by which a man would represent the circular area if he were walking around it with the area on his right? [*Ans.* Magnitude $100\pi$ ft$^2$., and pointing vertically downward.]

Let us recall that all this began in an innocent attempt to use $UV \sin \theta$ because it had geometrical interest: the area of the parallelogram defined by **U** and **V**. Soon we found ourselves looking on it as a vector, and now we are finding all sorts of unexpected consequences. We are no longer confined to parallelograms, nor even to cross products $\mathbf{U} \times \mathbf{V}$. Willy-nilly, we have had to give vectorial character in their own right to plane areas of all shapes, and this was forced on us by a special case only: the combination of cross products pointing in the same direction. We have still to consider what is forced on us by our combining areas vectorially in the general case.

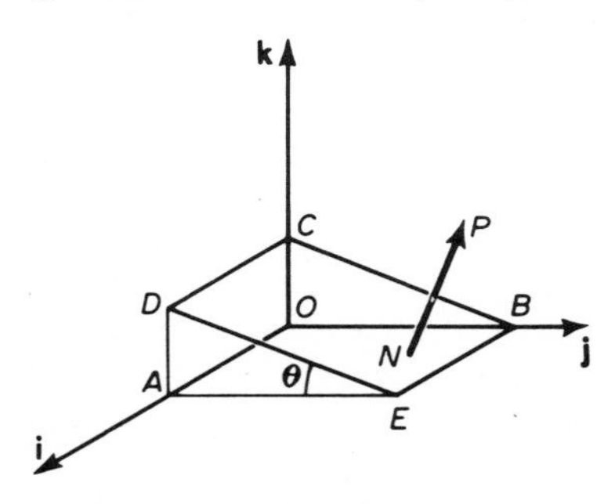


**Figure 9.5**

We start with the relatively simple case of the rectangle $CDEB$ shown in Figure 9.5. Let the area of the rectangle be $K$. If vector $\overrightarrow{NP}$ represents this area (we have drawn it as a free vector) its magnitude is $K$ and its direction cosines are 0, $\sin \theta$, and $\cos \theta$.

**Exercise 9.7** Prove that the direction cosines are as stated. [Note that $\overrightarrow{NP}$ is parallel to the **jk**-plane.]

Consequently the components of $\overrightarrow{NP}$ are $(0, K \sin \theta, K \cos \theta)$. But since the rectangle $OAEB$ is the projection of $CDEB$ on the **ij**-plane its area is $K \cos \theta$; so we see that the **k** component of $\overrightarrow{NP}$ is the area of the projection of $CDEB$ on the **ij**-plane. More, we see that the component *vector* of $\overrightarrow{NP}$ in the **k** direction, namely $(K \cos \theta)\mathbf{k}$, represents both in magnitude and direction the projection of $CDEB$ on the **ij**-plane. Similarly, the component vector of

$\overrightarrow{NP}$ in the **j** direction represents the projection of the original area $CDEB$ on the **ki**-plane.

**Exercise 9.8** Prove the last assertion.

**Exercise 9.9** Prove that a corresponding assertion is valid for the component vector in the **i** direction and the projection of the area on the **jk**-plane.

We can easily generalize this result to the case of any plane region oriented in any way relative to **i**, **j**, and **k**. The crucial fact that we have to use is that the angle between two planes is the same as the angle between their normals. Let the region have area $K$ and let a line normal to it have direction cosines $\cos \alpha$, $\cos \beta$, and $\cos \gamma$. Then the vector **V** representing the area of the plane region is given by:

$$\mathbf{V} = (K \cos \alpha)\mathbf{i} + (K \cos \beta)\mathbf{j} + (K \cos \gamma)\mathbf{k}. \tag{9.1}$$

Since the normal makes an angle $\gamma$ with **k** (which is normal to the **ij**-plane), the plane region makes the same angle $\gamma$ with the **ij**-plane. So its projection on the **ij**-plane has area $K \cos \gamma$, and we see that THE COMPONENT VECTOR IN THE **k** DIRECTION OF THE VECTOR **V**, NAMELY $(K \cos \gamma)\mathbf{k}$, REPRESENTS IN MAGNITUDE AND DIRECTION THE PROJECTION OF THE REGION ON THE **ij**-PLANE. SIMILAR RESULTS HOLD FOR THE OTHER COMPONENT VECTORS.

While this is an important theorem, and a useful one, it is stated in terms of **i**, **j**, and **k**, and it applies, at most, to reference frames in which the base vectors $\mathbf{e}_x$, $\mathbf{e}_y$, and $\mathbf{e}_z$ are mutually orthogonal. (For more general reference frames it can be restated in terms of oblique instead of orthogonal projections, but then it takes on a somewhat different character.) Meanwhile, Equation (9.1) confronts us with a stark fact that we have been trying to avoid: when we treat areas vectorially we commit ourselves to agreeing, for example, that three areas lying respectively in the **jk**-plane, the **ki**-plane, and the **ij**-plane are somehow to be regarded as together equivalent to a single area that lies in none of these planes, and that contains fewer square inches than the three do together.

Our first reaction is that this is preposterous. But we recall that when we said of ordinary vectors that $\overrightarrow{OA} + \overrightarrow{OB} = \overrightarrow{OC}$, the corresponding parallelogram diagram seemed equally preposterous. By a process of abstraction we came, via the idea of shifts, to the idea of displacements, and then at last we could accept the vectorial law of combination geometrically. We have to try to do something of the same sort with areas.

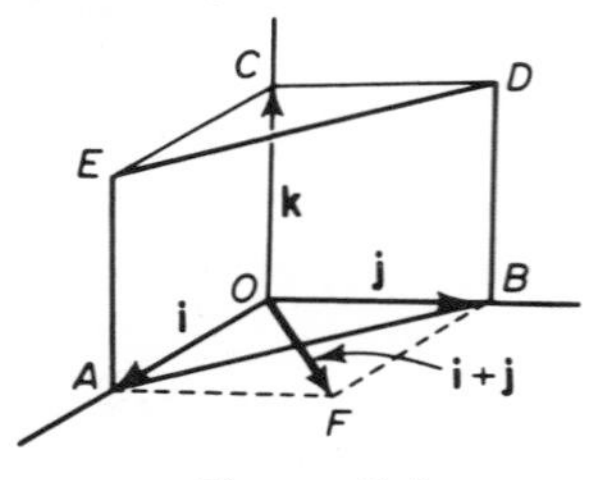


**Figure 9.6**

Instead of returning to Figure 9.5, let us avoid page-turning by considering a similar but even simpler situation that lends itself better to what we wish to discuss. The cross products $\mathbf{j} \times \mathbf{k}$ and $\mathbf{k} \times \mathbf{i}$ are respectively **i** and **j** and they can represent the areas of the squares $OBDC$ and $OCEA$

in Figure 9.6, though they could also represent differently shaped areas. The resultant of $\mathbf{i}$ and $\mathbf{j}$ is $\mathbf{i} + \mathbf{j}$, which is shown as $\overrightarrow{OF}$. This latter vector can represent the area of the rectangle $ABDE$ if we think of it as a free vector.

**Exercise 9.10** Show that the magnitude of $\mathbf{i} + \mathbf{j}$ does indeed give the area of $ABDE$, that $\mathbf{i} + \mathbf{j}$ is normal to $ABDE$, and that it points in the direction appropriate to $ABDE$ and not to $AEDB$.

When we say that $\mathbf{j} \times \mathbf{k} + \mathbf{k} \times \mathbf{i} = \mathbf{i} + \mathbf{j}$, we say that, vectorially speaking, the areas of the squares $OBDC$ and $OCEA$ are together somehow equivalent to the area of the rectangle $ABDE$ (though not to that of $AEDB$). We notice that triangles $DCE$ and $BOA$ have the same area. Since they are parallel, we can represent them by equal free vectors. Then if we reverse the sense of one of them, their sum will be zero. So let us add to the two squares $OBDC$ and $OCEA$ the pair of self-cancelling triangles $DEC$ and $BOA$. [Note the order: $DEC$, not $DCE$.] Then, from the vectorial point of view, we have to regard these four patches taken together as equivalent to the area $ABDE$. And what we now notice with the sudden delight that attends discovery, is that *the four patches taken together have the same rim as the area $ABDE$.*

**Exercise 9.11** Prove that the rims are as stated. [Note the various cancellations: for example, the $\overrightarrow{DC}$ of the square $OBDC$ cancels the $\overrightarrow{CD}$ of the triangle $DEC$.]

Now the question arises: is this just a coincidence or an instance of a general theorem? The matter is worth exploring. Consider a tetrahedron with vertices $P$, $A$, $B$, and $C$, as shown in Figure 9.7. Working with displacements, we see that $\overrightarrow{BC} = \overrightarrow{PC} - \overrightarrow{PB}$ and $\overrightarrow{CA} = \overrightarrow{PA} - \overrightarrow{PC}$. So, using the fact that $\overrightarrow{PC} \times \overrightarrow{PC} = 0$, we have

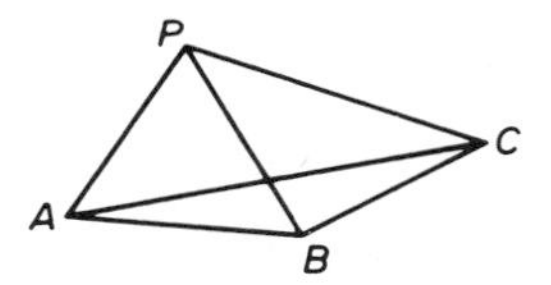

**Figure 9.7**

$$\tfrac{1}{2}\overrightarrow{BC} \times \overrightarrow{CA} = \tfrac{1}{2}(\overrightarrow{PC} - \overrightarrow{PB}) \times (\overrightarrow{PA} - \overrightarrow{PC}) = \tfrac{1}{2}\overrightarrow{PB} \times \overrightarrow{PC} + \tfrac{1}{2}\overrightarrow{PC} \times \overrightarrow{PA} + \tfrac{1}{2}\overrightarrow{PA} \times \overrightarrow{PB}.$$

[Note the change of sign when we reversed the order of the factors in the term $-\tfrac{1}{2}\overrightarrow{PB} \times \overrightarrow{PA}$.] This relation has an immediate geometrical interpretation. The cross product $\tfrac{1}{2}\overrightarrow{BC} \times \overrightarrow{CA}$ represents vectorially the area of the triangle $ABC$, the vector pointing upward if we think of $ABC$ as horizontal.

**Exercise 9.12** Why upward? [Note that $\overrightarrow{BC} = -\overrightarrow{CB}$.]

Also $\tfrac{1}{2}\overrightarrow{PB} \times \overrightarrow{PC}$ represents vectorially the area of the triangle $PBC$, the vector pointing in an upwardish direction—out of, not into the tetrahedron. Similarly for $\tfrac{1}{2}\overrightarrow{PC} \times \overrightarrow{PA}$ and $\tfrac{1}{2}\overrightarrow{PA} \times \overrightarrow{PB}$. Therefore the above relation, based on the distributive law, here tells us that we must regard the area $ABC$ as vectorially equivalent to the three areas $PBC$, $PCA$, and $PAB$ taken together. And we note that the rim of the former is the same as the rim of the latter three taken together.

**Exercise 9.13** Prove that the rims are as stated.

**Exercise 9.14** Prove the corresponding theorem for a pyramid $PABCD$ with a four-sided base. [Do it in two ways: (a) without cross products, by regarding the pyramid as two contiguous tetrahedra and applying the above theorem, and (b) by means of cross products as above. In method (b) you will, in effect, be considering two tetrahedra, since you will have to regard the area $ABCD$ as the sum of two triangles, such as $ABC$ and $BCD$.]

**Exercise 9.15** By method (a), extend the result in Exercise 9.14 to the case of a pyramid with an $n$-sided base.

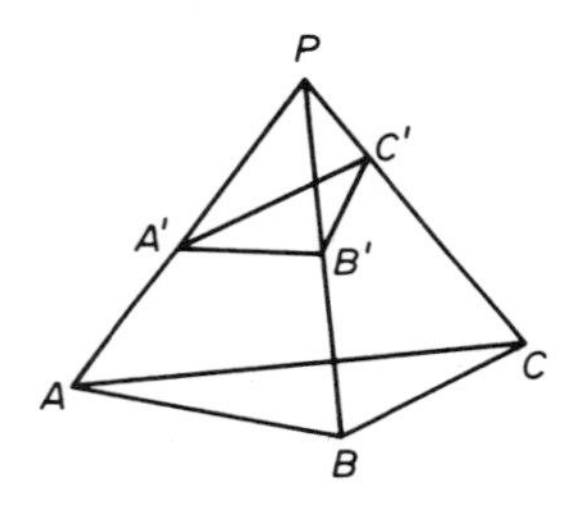


**Figure 9.8**

**Exercise 9.16** In Figure 9.8, the plane $A'B'C'$ is not necessarily parallel to the plane $ABC$. By "subtracting" the tetrahedron $PA'B'C'$ from the tetrahedron $PABC$, show, without resorting to cross products, that the area $ABC$ is vectorially equivalent to the sum of the areas of $ABB'A'$, $BCC'B'$, $CAA'C'$, and $A'B'C'$, and that the rim of $ABC$ is the rim of the four latter areas taken together. [*Hint*: We have, using a self-explanatory notation, $ABC = PBC + PCA + PAB$, and $A'B'C' = PB'C' + PC'A' + PA'B'$. Subtract, using the vectorial fact that, for example, $PBC - PB'C' = BCC'B'$ because these areas are coplanar.]

Now consider any plane area bounded by a simple closed curve $\Gamma$, as in Figure 9.9, and think of it as made of a ductile metal. By means of an appropriately shaped tool and a hammer, beat a tetrahedral dent $PABC$ into the surface. Then, vectorially, the area is equivalent to what it was before, because area $ABC$ is vectorially equivalent to the three areas $PBC$, $PCA$, and $PAB$ taken together.

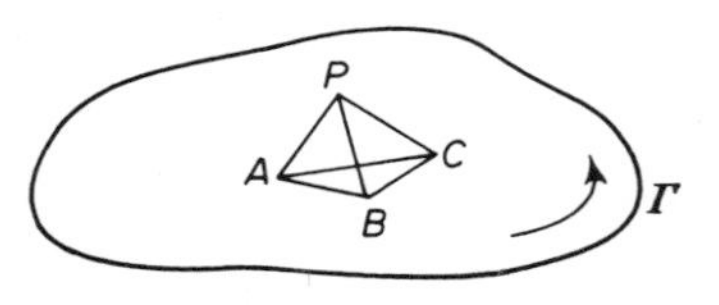


**Figure 9.9**

What about the rim of the distorted surface? In a sense it is obviously the same as before since the dent, not being a perforation, hardly qualifies as part of the rim. But let us convince ourselves of this. Had we punched out a triangular hole $ABC$, the rim would have been $\Gamma$ plus the rim of the triangle *taken in the order* $ACB$—remember, we have to walk the rim with the area on our left. When we add the cap $PABC$ we, in effect, add the rim $ABC$, and this just cancels the rim of the punched hole that we had to take in the opposite order.

So we have the important result that by punching the tetrahedral dent we changed neither the vectorial area nor the rim. We now become "punch drunk," making tetrahedral dents all over the place, and we end up with the realization that we can grossly distort the surface without altering its vectorial area, provided we leave its rim undistorted, which means, incidentally, that we may not punch holes in the surface. And we are ready to accept without proof the fact that the result holds even if the original surface is distorted into the shape of a bulging balloon, or a convoluted, semi-inflated one, or worse. The proof involves ideas belonging to the calculus.

At last we seem to understand the situation: areas are to be counted as vectorially equivalent if they have the same boundary. This is not the whole story, however.

**Exercise 9.17** Show that *the boundary curve need not be a plane curve*. [Consider, for simplicity, a boundary lying in a plane and in the form of a closed polygon. Now replace one side $AB$ by two lines $AP$ and $PB$ not in the plane. Consider two surfaces having this distorted rim. Add to each the triangle $APB$ and they will now have the same flat rim and, so, will be equivalent. So they must have been equivalent before.]

**Exercise 9.18** In Figure 9.6, what is the vector that would represent the area of a surface whose boundary was $OBDCEAO$? [*Ans.* $\mathbf{i} + \mathbf{j}$.]

This is the crux of the matter, but not the whole of it. In our discussion we had to treat areas as *free* vectors. So we must be ready to agree that two areas with congruent, similarly oriented rims are equivalent even if the rims do not coincide. Also we have to remember that, for example, a triangle and a parallelogram of the same area and in parallel planes are represented by the same vector and are thus regarded as equivalent; and that each could be equivalent to an aggregate of several isolated patches of area of suitable size and orientation. Since the exploration of all these possibilities frightens me, I shall pretend I have said enough about them.

There is a remarkable parallelism between the vectorial concept of area and the idea of a displacement. A displacement $\overrightarrow{AB}$ is equivalent to any sequence of zigzag displacements starting at $A$ and ending at $B$ (or, indeed, starting at $C$ and ending at $D$ if $\overrightarrow{CD} = \overrightarrow{AB}$; but for simplicity let us stick to the points $A$ and $B$ here). The points $A$ and $B$ can be regarded as the rim or boundary of the displacement $\overrightarrow{AB}$. So when we wish to treat line segments vectorially we have to agree that all zigzag line segments—or indeed any curved segments—having the same boundary are to be counted as equivalent. Analogously, when we wish to treat surface areas vectorially, we have to agree that, no matter how greatly convoluted they may be, all surfaces having the same boundary are to be counted as equivalent. [But see below.] When we stay with the same end points $A$ and $B$ rather than allowing also equivalent pairs of end points, like $C$ and $D$, we are dealing with shifts rather than vectors. When we correspondingly keep to areas having the same rim we are dealing with an areal analogue of shifts. In a restricted case, this "*areal shift*" has a simple pictorial significance. Consider a closed curve $\Gamma$ that is not necessarily plane. Imagine it capped by a suitably bent iron plate, and consider how we would go about protecting *one side* of the plate from rusting because of exposure to rain. We could attach a tarpaulin to the rim $\Gamma$ in a watertight way, and no matter how large we made the tarpaulin or how we arranged its billows and folds, it would protect the specified side of the iron from the rain. If we wished to protect the other side of the plate instead, we would attach the tarpaulin differently, though to the same rim. How can we specify the side we wish to protect? In conformity with what has gone before, we can associate with each side a sense of walking around the rim. If we walk with the immediate area of the iron cap on our left, we will agree to protect the side on which our feet tread; if on our right, the other side.

**Exercise 9.19** Show that the above is consistent in that, if we walk around $\Gamma$ in the opposite sense, the criterion will yield the opposite side in either case.

From all tarpaulin surfaces that protect a given side of the iron plate, we abstract everything except their common rim $\Gamma$ and the sense in which we go around $\Gamma$. The resulting abstraction is the areal shift, and all these surfaces yield the same areal shift.

The difficulty with this picture is that the surfaces we have been likening to tarpaulins can cross the cap. To save the analogy we could make the cap bulge so that none of the protecting surfaces crosses it. But since we have to contemplate all surfaces having the given rim, the bulge would have to be infinitely great. Now that the idea of protection from rusting has served its pictorial purpose for us, we might as well abandon the idea of an iron cap and say that the areal shift is basically the common rim of all the surfaces, but with a particular sense of going around it. Thus it is A RIM AND A SENSE; but with it goes a faint reminiscence of the idea of a covering area, just as an ordinary shift is a pair of points and a sense, but with a faint reminiscence of journeying.

When we go from areal shifts to areal vectors we encounter the frightening pictorial complications already mentioned. But these are analogous to those that arise in the transition from ordinary shifts to displacements, which require us to agree that

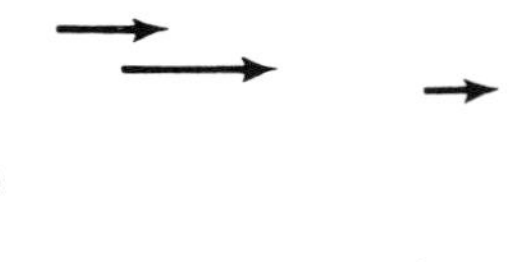

are together equivalent to

and if the situation here is less complex than it is for areas, it is nonetheless analogous to the areal situation.

It is interesting to speculate on the likelihood of our coming to the idea of a vector product had we started out with this concept of areal shifts.

**Exercise 9.20** In Figure 9.7, show that if one adopts the convention of representing the areas by normals that point outward from the tetrahedron, the vectorial sum of the four faces is zero. Show also that with this convention the total rim of the four faces together is zero.

**Exercise 9.21** Show that VECTORIALLY THE AREA OF A CLOSED POLYHEDRAL SURFACE IS ZERO IF THE VECTORS ALL POINT OUTWARD FROM THE REGION ENCLOSED BY THE FACES. [A simple method is to slice the region into two parts by means of a plane. Then the surface area of each half is equivalent vectorially to the plane area of the polygon of intersection, but the two halves are equivalent to this plane area with opposite signs.]

## 10. VECTOR PRODUCTS IN RIGHT- AND LEFT-HANDED REFERENCE FRAMES

We have been hinting for some time that there was something yet to be told about the direction of the vector product of two vectors. Suppose we say, as we have been saying all along, that the vector product $\mathbf{A} \times \mathbf{B}$ points in the direction such that $\mathbf{A}$, $\mathbf{B}$, and $\mathbf{A} \times \mathbf{B}$, in that order, form a right-handed system. Let us take a right-handed reference frame—for simplicity, a unit orthogonal triad, $\mathbf{i}$, $\mathbf{j}$, $\mathbf{k}$—and consider the components relative to it of $\mathbf{A}$, $\mathbf{B}$, and $\mathbf{C}$ where

$$\mathbf{C} = \mathbf{A} \times \mathbf{B}. \qquad (10.1)$$

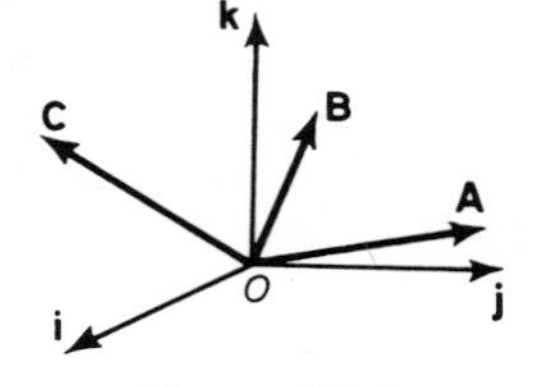


**Figure 10.1**

If $\mathbf{A}$ and $\mathbf{B}$ have components $(A_x, A_y, A_z)$ and $(B_x, B_y, B_z)$ respectively, we have seen that $\mathbf{C}$ has the components $(A_yB_z - A_zB_y, A_zB_x - A_xB_z, A_xB_y - A_yB_x)$.

Now suppose that we reflect Figure 10.1 in a mirror, so that it looks like Figure 10.2—except for the fact that Figure 10.2 has, for example, **B** instead of ᗺ. Clearly, the components of the **A** and **B** in Figure 10.2 with respect to **i**, **j**, and **k** in Figure 10.2 will still be $(A_x, A_y, A_z)$ and $(B_x, B_y, B_z)$ respectively. We would therefore like the components of their cross product **C** relative to this **i**, **j**, and **k** frame to be the same as before. But they obviously will not be, because the vector we have labelled **C** in Figure 10.2 is not really $\mathbf{A} \times \mathbf{B}$ if we adhere to the requirement that $\mathbf{A}$, $\mathbf{B}$, and $\mathbf{A} \times \mathbf{B}$, in that order, shall form a right-handed system. The reflection in the mirror has turned **A**, **B**, and **C** into a left-handed system, and if we adhered to our right-handed convention for the direction of the cross product, we should say that the true $\mathbf{A} \times \mathbf{B}$ in Figure 10.2 points in the direction opposite to that of the vector labelled **C** in that figure.

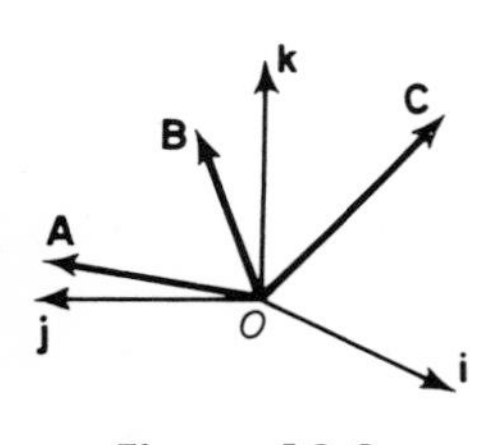


**Figure 10.2**

We can confirm the existence of the dilemma by forming the cross product $(A_x\mathbf{i} + A_y\mathbf{j} + A_z\mathbf{k}) \times (B_x\mathbf{i} + B_y\mathbf{j} + B_z\mathbf{k})$. At first we imagine we get the same components as before. But that is because we are using the relations $\mathbf{i} \times \mathbf{j} = \mathbf{k}$, etc. Those relations do not hold here, though, if, in calculating $\mathbf{i} \times \mathbf{j}$ etc. we use the convention that **A**, **B**, and $\mathbf{A} \times \mathbf{B}$ form a right-handed system. Instead of these relations between the **i**, **j**, and **k** in Figure 10.2, we should, by this convention, use $\mathbf{i} \times \mathbf{j} = -\mathbf{k}$, etc; and when we do this we find that the components come out to be the negatives of what they were before.

Now this is very strange behavior for a vector: that when *both it and its reference frame* are reflected in a mirror its components change sign. The vectors **A** and **B** did not behave in this unseemly way.

To preserve some semblance of order we appear to be driven to say that $\mathbf{A} \times \mathbf{B}$ is a vector perpendicular to **A** and **B** and pointing in a direction that makes **A**, **B**, and $\mathbf{A} \times \mathbf{B}$ a right-handed system when we are working in terms of a right-handed reference frame, but that it points in a direction that makes **A**, **B**, and $\mathbf{A} \times \mathbf{B}$ a left-handed system when we are working in terms of a left-handed reference frame. At least, under this convention, if **A** has the same components relative to both frames and **B** has the same components relative to both reference frames, then $\mathbf{A} \times \mathbf{B}$ will also. But is it wholly satisfactory? It means that if we draw **A** and **B** flat on the page as in Figure 10.3, we cannot say whether $\mathbf{A} \times \mathbf{B}$, which is perpendicular to **A** and **B**, points upward toward us or downward away from us until we decide whether, on future occasions, we shall use right-handed reference frames or left-handed ones. And if we simultaneously consult two bystanders, one of whom prefers right-handed frames, and the other left-handed frames we shall be in a grave quandary as to the direction of $\mathbf{A} \times \mathbf{B}$ in Figure 10.3. Equation (10.1), if true in a right-handed frame, would be false in a left-handed one, unless we said that though **C** was ↗ in a right-handed frame it had to switch to ↙ in a left-handed one, and this would hardly be the sort of vectorial behavior we have hitherto regarded as proper.

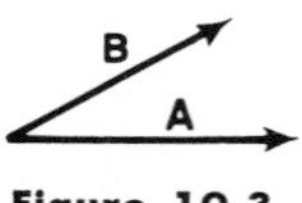

**Figure 10.3**

## 11. LOCATION AND CROSS PRODUCTS

Now that our suspicions have been aroused, we begin to think of other curious things about cross products. There is the problem of location, for instance, and while this problem is not confined to cross products, we might as well bring it in here while we are on the warpath.

When we began by thinking of $\mathbf{A} \times \mathbf{B}$ as representing the area of a parallelogram we had **A** and **B** starting at a common point $O$. But in the end, we found ourselves regarding the vectors representing areas as free vectors despite the frightening pictorial consequences. Contrast this with the situation in which we deal with moments. Physically, the moment $\mathbf{r} \times \mathbf{F}$ of a force $\mathbf{F}$ about a point $O$ would seem to have to be a vector passing through $O$. Yet if cross products are free vectors, the moment should presumably be a free vector. In contemplating $\mathbf{A} \times \mathbf{B}$ as an area, we could at least think of $\mathbf{A}$ and $\mathbf{B}$ as displacements, and thus as free vectors, thereby consoling ourselves for having to take $\mathbf{A} \times \mathbf{B}$ as a free vector. But with moments, the situation is different, because $\mathbf{r}$ and $\mathbf{F}$ are not free vectors. Indeed, they do not start from a common point—except in the case of zero moment. If we overlook this and still form $\mathbf{r} \times \mathbf{F}$, which is what we have been doing up to now, we can not assign a reasonable location to $\mathbf{r} \times \mathbf{F}$ because the point $P$, which is the only reasonable candidate as a localizing point, is *any* point on the line of action of $\mathbf{F}$. If we wish to use $O$, we may be tempted to add two cancelling forces at $O$, thus replacing $\mathbf{F}$ by a parallel force $\mathbf{F}_1$, of equal magnitude passing through $O$, and a couple of moment $\mathbf{r} \times \mathbf{F}$. Then we can form $\mathbf{r} \times \mathbf{F}_1$ and reasonably insist that *its* line of action pass through $O$. But what shall we then do with the extra, and seemingly superfluous couple? There does not seem to be a satisfactory way out of the dilemma. We would be much better off not talking about it.

**Exercise 11.1** Consider the problem of location for the relation $\mathbf{v} = \boldsymbol{\omega} \times \mathbf{r}$.

## 12. DOUBLE CROSS

Are cross products really vectors? By now we may be beginning to wonder. They have magnitude and direction, they combine according to the parallelogram law, and they obey all the other addenda to the definition that we have thought of so far. Yet they behave queerly in relation to right-handed and left-handed reference frames. Perhaps we need to add another amendment to the definition. Let us not try to do so, though. Making good definitions is far too difficult. Instead, let us use our general feeling for what is appropriate and what is not, without seeking to formulate sharp criteria.

OUR SUSPICIONS SHOULD HAVE BEEN AROUSED LONG AGO BY THE SEEMINGLY INNOCENT RELATIONS $\mathbf{i} \times \mathbf{j} = \mathbf{k}$, ETC. There is something obviously wrong with them. If the magnitudes of $\mathbf{i}$, $\mathbf{j}$, and $\mathbf{k}$ are lengths measured in, say, feet, then $\mathbf{i} \times \mathbf{j}$ has a magnitude of one square foot and so cannot be the same sort of thing as $\mathbf{k}$, whose magnitude is one foot. Perhaps we counter by arguing that since $\mathbf{i}$, $\mathbf{j}$, and $\mathbf{k}$ are of *unit* length, the magnitudes of $\mathbf{i} \times \mathbf{j}$ and $\mathbf{k}$ have the same numerical value, namely 1. But this will not do. If we change to inches, the former becomes 144 square inches and the latter 12 inches. Clearly, $\mathbf{i} \times \mathbf{j}$ cannot really be $\mathbf{k}$, even though we have been able to get extremely important equations by assuming it is. [Look back at Exercise 9.6 in this connection.]

Again, suppose we consider cross products in two dimensions instead of in three. Then, given two vectors $\mathbf{U}$ and $\mathbf{V}$, we say that their cross product has magnitude $UV \sin \theta$ and that it points—where? Why, nowhere. There is no third dimension into which it can point when we have only two dimensions in all. Clearly the cross product is not a vector in two dimensions.

What about four dimensions? Here again we start by forming the parallelogram defined by two vectors **U** and **V** starting at $O$. But when we seek a line through $O$ perpendicular to both **U** and **V**, instead of finding no possible directions as in the two-dimensional case, we now find far too many—infinitely many, in fact, and all of them on an equal footing. For in four dimensions, the lines through $O$ perpendicular to both **U** and **V** in general fan out into a plane. So here too the cross product is not a vector. And our conclusion is that in the three-dimensional case too, the cross product is not really a vector, even though by some sort of good fortune it has many of the characteristics of a vector and can be handled much as regular vectors are handled, provided it is handled with care.

Since the cross product in three dimensions has so many of the characteristics of a vector, a good name for an entity of this sort would be *pseudovector*, and this is a particularly apt choice since it happens to be the official technical term for it. [Such entities are sometimes referred to as *axial vectors*.]

## 13. DIVISION OF VECTORS

*The discussion in this section is confined to the three-dimensional case.*

Having explored the multiplication of vectors, we naturally wonder how to divide one vector by another. But here too we have to realize that there is no reason why such division should be possible. We have to see if we can find a useful meaning for the quotient of two vectors, and if we cannot, we shall probably wish to keep mum about the matter. There is, of course, no guarantee that the quotient, if it exists, will be a vector: let us recall that we found two reasonable types of "products" of vectors, one yielding scalars and the other pseudovectors. We must therefore be prepared for almost anything here.

Let **U** and **V** be two vectors starting at a common point $O$. Denote their quotient, if it exists and whatever it may be, by $Q$, so that we may write:

$$\frac{\mathbf{V}}{\mathbf{U}} = Q. \tag{13.1}$$

Then, since we want to keep as close as possible to our usual ideas of what constitutes the operation of division, we ask that Equation (13.1) mean that

$$\mathbf{V} = Q\mathbf{U}. \tag{13.2}$$

So the quantity $Q$ is to be something that acts on the vector **U** so as to convert it into the vector **V**.

If **U** and **V** have the same line of action, $Q$ is obviously just the ratio of their magnitudes. But we have to consider the case in which they have different directions. To convert **U** to **V** we have to change its magnitude by a factor $V/U$ and also to rotate it so that its direction coincides with that of

**V**. We find, therefore, that $Q$ involves a ratio of magnitudes and an angular displacement about an axis through $O$ perpendicular to **U** and **V**. Since the specification of the latter requires three numbers, we see that $Q$ requires four. It was therefore called a *quaternion* by W. R. Hamilton, who originated the concept.

This book is not the place to develop the quaternion calculus in any detail, but an extremely sketchy indication of the general idea as it was developed by Hamilton will not be without interest. In what follows, we make various reasonable assertions that, in a fuller treatment, would be listed as axioms.

If triangles $OAB$ and $OLM$ are coplanar, and similar in such a way that $OB/OA$ equals $OM/OL$ rather than $OL/OM$, then the operation that converts $\overrightarrow{OA}$ into $\overrightarrow{OB}$ is the same as that that converts $\overrightarrow{OL}$ into $\overrightarrow{OM}$. So the quaternions $\overrightarrow{OB}/\overrightarrow{OA}$ and $\overrightarrow{OM}/\overrightarrow{OL}$ are equal. Thus, given a quaternion, we can express it in a form that has as denominator any vector we wish in the plane of the two vectors involved. This being so, we can add two quaternions by first putting them on a common denominator, this latter being a vector lying in the line of intersection of their planes. We then have something of the form $(\overrightarrow{OB}/\overrightarrow{OA}) + (\overrightarrow{OC}/\overrightarrow{OA})$; and since $\overrightarrow{OB}$ and $\overrightarrow{OC}$ are vectors, these quaternions combine by the parallelogram law to yield $\overrightarrow{OR}/\overrightarrow{OA}$, where $\overrightarrow{OR}$ is the resultant of $\overrightarrow{OB}$ and $\overrightarrow{OC}$.

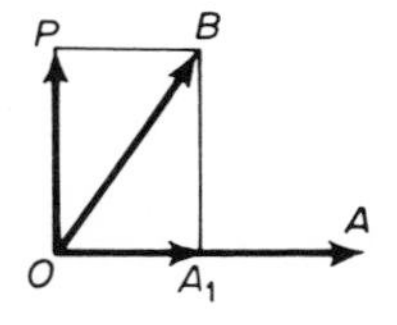


**Figure 13.1**

We now reverse the idea. Given a quaternion $\overrightarrow{OB}/\overrightarrow{OA}$, we split $\overrightarrow{OB}$ up into two components: $\overrightarrow{OA_1}$ lying along $\overrightarrow{OA}$, and $\overrightarrow{OP}$ perpendicular to it, as in Figure 13.1. Then,

$$\frac{\overrightarrow{OB}}{\overrightarrow{OA}} = \frac{\overrightarrow{OA_1}}{\overrightarrow{OA}} + \frac{\overrightarrow{OP}}{\overrightarrow{OA}}. \tag{13.3}$$

The quantity $\overrightarrow{OA_1}/\overrightarrow{OA}$ was called by Hamilton the *scalar* of the quaternion $\overrightarrow{OB}/\overrightarrow{OA}$. Note that, in general, it is not the same as the ratio of the magnitudes of the vectors $\overrightarrow{OB}$ and $\overrightarrow{OA}$. It is, in fact, their scalar product divided by the square of the magnitude of $\overrightarrow{OA}$. [Hamilton called the ratio of the magnitudes the *tensor*, a word that is now used in a quite different sense, as will be seen in the next chapter.]

**Exercise 13.1** Prove the above assertion about the scalar product.

**Exercise 13.2** Prove that the scalar of $\overrightarrow{OB}/\overrightarrow{OA}$ plus the scalar of $\overrightarrow{OC}/\overrightarrow{OA}$ is equal to the scalar of $\overrightarrow{OR}/\overrightarrow{OA}$, where $\overrightarrow{OR}$ is the resultant of $\overrightarrow{OB}$ and $\overrightarrow{OC}$.

The quaternion $\overrightarrow{OP}/\overrightarrow{OA}$ in Equation (13.3) is of a special type since it involves a rotation through a right angle. Such quaternions are called *right quaternions*. They turn out to be of special importance.

Hamilton introduced three basic right quaternions which he denoted by $i$,

$j$, and $k$. These are *not* the same as the vectors $\mathbf{i}$, $\mathbf{j}$, and $\mathbf{k}$, though they have distant kinship with them. They are defined in terms of three mutually perpendicular vectors $\mathbf{X}$, $\mathbf{Y}$, and $\mathbf{Z}$, starting at $O$, having unit magnitudes, and forming a right-handed system. The right quaternion $\mathbf{Z}/\mathbf{Y}$ converts $\mathbf{Y}$ into $\mathbf{Z}$. It is denoted by $i$, and it involves a rotation through 90° about $\mathbf{X}$. From a diagram, one easily sees that $i$ also converts $\mathbf{Z}$ into $-\mathbf{Y}$. So we may write:

$$i = \frac{\mathbf{Z}}{\mathbf{Y}} = \frac{-\mathbf{Y}}{\mathbf{Z}}.$$

Similarly, $j$ and $k$ are defined by:

$$j = \frac{\mathbf{X}}{\mathbf{Z}} = \frac{-\mathbf{Z}}{\mathbf{X}}, \qquad k = \frac{\mathbf{Y}}{\mathbf{X}} = \frac{-\mathbf{X}}{\mathbf{Y}}.$$

**Exercise 13.3** Show that the scalars of $i$, $j$, and $k$ are zero.

By $i^2$ we mean the result of applying $i$ twice. Since $i$ converts $\mathbf{Y}$ into $\mathbf{Z}$, and the second $i$ converts $\mathbf{Z}$ into $-\mathbf{Y}$, $i^2$ converts $\mathbf{Y}$ into $-\mathbf{Y}$ and we therefore write $i^2 = -1$.

**Exercise 13.4** Prove, similarly, that $j^2 = -1$ and $k^2 = -1$.

By $ij$ we mean the result of applying $j$ and then $i$. Since $j$ converts $\mathbf{X}$ into $-\mathbf{Z}$, and $i$ converts $\mathbf{Z}$ into $-\mathbf{Y}$ (and thus, as is easily seen, $-\mathbf{Z}$ into $\mathbf{Y}$), $ij$ converts $\mathbf{X}$ into $\mathbf{Y}$, which is just what $k$ does. So we write $ij = k$. However, since $i$ converts $\mathbf{Y}$ into $\mathbf{Z}$, and $j$ converts $\mathbf{Z}$ into $\mathbf{X}$, $ji$ converts $\mathbf{Y}$ into $\mathbf{X}$. But $k$ converts $\mathbf{Y}$ into $-\mathbf{X}$. Therefore we write $ji = -k$.

**Exercise 13.5** Prove, similarly, that $jk = i$, $kj = -i$, $ki = j$, and $ik = -j$.

**Exercise 13.6** Prove that $ijk = -1$; also, that $i(ij) = (ii)j = -j$, $i(ji) = (ij)i = j$, and $i(jj) = (ij)j = -i$. [Results such as these show that $i$, $j$, and $k$ obey the associative law of multiplication, even though they do not obey the commutative law.]

Consider now a right quaternion, such as $\overrightarrow{OP}/\overrightarrow{OA}$ in Equation (13.3). We can associate with it an entity called its *index*. This consists of an arrow-headed line segment $\overrightarrow{OP'}$ that lies, in the right-handed sense, along the axis of the rotation performed by the quaternion $\overrightarrow{OP}/\overrightarrow{OA}$, and it has a length representing the number $OP/OA$. Its direction is thus perpendicular to the plane of of $\overrightarrow{OP}$ and $\overrightarrow{OA}$. Since all right quaternions involve rotations through the same angle, 90°, their indices combine according to the parallelogram law.

**Exercise 13.7** Prove this. [If $\overrightarrow{OB} + \overrightarrow{OC} = \overrightarrow{OR}$, then $\overrightarrow{OB}/\overrightarrow{OA} + \overrightarrow{OC}/\overrightarrow{OA} = \overrightarrow{OR}/\overrightarrow{OA}$. We obtain the indices of these three quaternions by multiplying $\overrightarrow{OB}$, $\overrightarrow{OC}$, and $\overrightarrow{OR}$ by the same quantity, $1/OA$, and rotating them in their plane through 90° about an axis through $O$ perpendicular to their plane. Compare the analogous discussion in connection with cross products of perpendicular vectors.]

**Exercise 13.8** Show that the index of $i$ is the vector $\mathbf{X}$.

Thus, provided we stay with right-handed systems, the indices behave like vectors. (They are actually pseudovectors.) So we can express the index $\overrightarrow{OP}'$ of the quaternion $\overrightarrow{OP}/\overrightarrow{OA}$ in the form,

$$\overrightarrow{OP}' = x\mathbf{X} + y\mathbf{Y} + z\mathbf{Z}, \tag{13.4}$$

where $(x, y, z)$ are the coordinates of $P'$ relative to the base vectors $\mathbf{X}$, $\mathbf{Y}$, and $\mathbf{Z}$. Since the indices $\mathbf{X}$, $\mathbf{Y}$, and $\mathbf{Z}$ belong respectively to the quaternions $i$, $j$, and $k$, we have, corresponding to Equation (13.4), the quaternion equation:

$$\frac{\overrightarrow{OP}}{\overrightarrow{OA}} = xi + yj + zk, \tag{13.5}$$

and, therefore, such quaternion quantities obey the same algebraic rules as their vectorial indices. If we denote the scalar of the quaternion $\overrightarrow{OB}/\overrightarrow{OA}$ in Equation (13.3), namely $\overrightarrow{OA}_1/\overrightarrow{OA}$, by $w$, we see from Equation (13.5) that this quaternion, which is any quaternion $Q$, can be expressed in the form:

$$Q = w + xi + yj + zk. \tag{13.6}$$

The significant thing about this is that the quantities on the right, and thus quaternions in general, obey all the usual rules of algebra except the commutative law of multiplication. Quaternions are thus of considerable theoretical importance quite apart from their relationship to vectors. They constitute what algebraists call a noncommutative *field* (not to be confused with what physicists call a field—when they are talking physics or agriculture rather than algebra). Thus quaternions, like complex numbers, are a natural extension of the idea of number, and are often called *hypercomplex numbers.*

It is worthwhile contemplating Hamilton's achievement. You and I, if the idea had ever occurred to us of discussing the quotient of two vectors, would have noted, perhaps, that it involved the ratio of their magnitudes and also something akin to an angular displacement. The former we would have thought of as a scalar of no particular charm, and the latter would have scared us away from pursuing the idea further. Hamilton, by defining a new scalar, and focusing on *right* quaternions and relating them to their indices, not only created the essentials of scalar and vector products, but also constructed an algebra of extreme elegance.

Scalar and vector products were conceived by Hamilton and H. Grassmann independently in the 1840's. Since Hamilton's quaternion calculus proved unwieldy for applications to the physics of the nineteenth century, J. W. Gibbs and, independently, O. Heaviside, extracted from it the much more convenient vector calculus that we have been studying in this book. They did so in a creative way, and in the face of bitter opposition from the supporters of quaternionic methods.

When we tried to regard angular displacements as vectors, we were shocked and thwarted by their properties of noncommutativity and worse. We may well wonder, then, that these properties do not destroy Hamilton's structure. One reason is that, for example, the quaternions $i$, $j$, and $k$ are not angular displacements but merely relations between pairs of vectors; and another is that Hamilton made the noncommutativity apply to multiplication rather than addition. Even so he was, for a long time, frustrated by the lack of commutativity of angular displacements, and it came to him as a staggering discovery that the key to progress lay in the heretical renunciation of the commutative law of multiplication. When, finally, the idea did come to him he realized immediately that it was a momentous one. But he could not know that almost a century later noncommutative multiplication would prove to be of crucial importance in the quantum theory, and that the spin of atomic particles would actually be represented by symbols satisfying equations corresponding to:

$$ij + ji = 0,$$

as would other fundamental quantities of atomic physics.

**Exercise 13.9** Imagine unit vectors $\mathbf{X}$, $\mathbf{Y}$, and $\mathbf{Z}$ along the respective coordinate axes in Figures 6.2 and 6.3. Show that the vector $\mathbf{X}$ in Figure 6.3c occupies the position of $\mathbf{Y}$ in Figure 6.3a. Denote this by $\mathbf{X} \to \mathbf{Y}$. Show similarly that $\mathbf{Y} \to \mathbf{Z}$, and $\mathbf{Z} \to \mathbf{X}$. Show that the corresponding results for Figure 6.2 are $\mathbf{X} \to -\mathbf{Z}$, $\mathbf{Y} \to \mathbf{X}$, and $\mathbf{Z} \to -\mathbf{Y}$. Note that $\mathbf{X} \to \mathbf{Y}$ is the effect of the quaternion $k$, while $\mathbf{Y} \to \mathbf{X}$ is that of $-k$, these results being in conformity with $ij = -ji = k$; but note that $\mathbf{X} = -\mathbf{Z}$ is not the effect of $k$ but of $j$. Consider the other effects similarly. [This exercise indicates the subtlety and boldness of Hamilton's concept: though $i$, $j$, and $k$ are *related* to angular displacements, $ij$ does not yield an angular displacement about the $z$-axis. Despite this, Hamilton, with profound mathematical instinct, did not shy away from quaternion relations $ij = k$ and the like, as you and I would have. He later found a way to represent general angular displacements by means of quaternions.]

# 6

# TENSORS

## 1. HOW COMPONENTS OF VECTORS TRANSFORM

The time has come for a new approach. To make further progress with vectors, we concentrate on their components. This may seem an unpromising tactic, since the components change whenever we change the reference frame, whereas relations among vectors are independent of the choice of the frame. But by studying how such changeable things as components can embody this objectivity of vectors, we come to an important generalization: the concept of a *tensor*.

We do not propose to discuss location. For this and other reasons we use a fixed origin $O$ and assume that all vectors and the like start at $O$. To get the feel of how components of vectors behave, we start with a simple numerical example in two dimensions. Suppose we are given

$$\mathbf{V} = V_x \mathbf{e}_x + V_y \mathbf{e}_y, \tag{1.1}$$

and we change to a new reference frame with base vectors $\bar{\mathbf{e}}_x$ and $\bar{\mathbf{e}}_y$ where

$$\left.\begin{aligned} \bar{\mathbf{e}}_x &= 3\mathbf{e}_x + 4\mathbf{e}_y \\ \bar{\mathbf{e}}_y &= \mathbf{e}_x + 2\mathbf{e}_y. \end{aligned}\right\} \tag{1.2}$$

To find how the components change, we first solve Equations (1.2) for $\mathbf{e}_x$ and $\mathbf{e}_y$ in terms of $\bar{\mathbf{e}}_x$ and $\bar{\mathbf{e}}_y$; the result is:

$$\left.\begin{aligned} \mathbf{e}_x &= \bar{\mathbf{e}}_x - 2\bar{\mathbf{e}}_y \\ \mathbf{e}_y &= -\tfrac{1}{2}\bar{\mathbf{e}}_x + \tfrac{3}{2}\bar{\mathbf{e}}_y. \end{aligned}\right\} \tag{1.3}$$

Then from Equations (1.1) and (1.3) we have:

$$\begin{aligned} \mathbf{V} &= V_x(\bar{\mathbf{e}}_x - 2\bar{\mathbf{e}}_y) + V_y(-\tfrac{1}{2}\bar{\mathbf{e}}_x + \tfrac{3}{2}\bar{\mathbf{e}}_y) \\ &= (V_x - \tfrac{1}{2}V_y)\bar{\mathbf{e}}_x + (-2V_x + \tfrac{3}{2}V_y)\bar{\mathbf{e}}_y. \end{aligned} \tag{1.4}$$

But the components $\bar{V}_x$ and $\bar{V}_y$ in the new reference frame must satisfy an equation corresponding to Equation (1.1), namely,

$$\mathbf{V} = \bar{V}_x\bar{\mathbf{e}}_x + \bar{V}_y\bar{\mathbf{e}}_y. \tag{1.5}$$

So Equation (1.4) shows that the new components of $\mathbf{V}$ are given by

$$\left.\begin{aligned} \bar{V}_x &= V_x - \tfrac{1}{2}V_y \\ \bar{V}_y &= -2V_x + \tfrac{3}{2}V_y. \end{aligned}\right\} \tag{1.6}$$

Compare the coefficients in Equations (1.6) and (1.3).

**Exercise 1.1** If $\mathbf{V} = V_x\mathbf{e}_x + V_y\mathbf{e}_y$, find its components relative to $\bar{\mathbf{e}}_x$ and $\bar{\mathbf{e}}_y$, where $\bar{\mathbf{e}}_x = 4\mathbf{e}_x + \mathbf{e}_y$ and $\bar{\mathbf{e}}_y = 2\mathbf{e}_x + \mathbf{e}_y$. [*Ans.* From $\mathbf{e}_x = \tfrac{1}{2}\bar{\mathbf{e}}_x - \tfrac{1}{2}\bar{\mathbf{e}}_y$ and $\mathbf{e}_y = -\bar{\mathbf{e}}_x + 2\bar{\mathbf{e}}_y$, we obtain $\bar{V}_x = \tfrac{1}{2}V_x - V_y$ and $\bar{V}_y = -\tfrac{1}{2}V_x + 2V_y$. Explain why the coefficients of the $V$'s automatically come out to be the same as those of the $\mathbf{e}$'s, though in different places.]

**Exercise 1.2** Like Exercise 1.1 for $\bar{\mathbf{e}}_x = \mathbf{e}_x - 2\mathbf{e}_y$ and $\bar{\mathbf{e}}_y = -\tfrac{1}{2}\mathbf{e}_x + \tfrac{3}{2}\mathbf{e}_y$, and compare with the example worked out in the text.

## 2. THE INDEX NOTATION

In three dimensions, we have been denoting the base vectors by $\mathbf{e}_x$, $\mathbf{e}_y$, and $\mathbf{e}_z$, and the components of a vector by $V_x$, $V_y$, and $V_z$. In four dimensions we already find ourselves in a squeeze when we try to think of what letter to add to $x, y$, and $z$, and in $n$ dimensions, especially when $n > 26$, the inconvenience of using different letters becomes considerable. So we use numbers instead, denoting the base vectors, for example, by $\mathbf{e}_1, \mathbf{e}_2, \ldots, \mathbf{e}_n$. Then we denote all these $\mathbf{e}$'s collectively by the single symbol $\mathbf{e}_a$, where $a$ takes on the values $1, 2, \ldots, n$, in turn. Thus, in three dimensions, $a$ takes on the values 1, 2, and 3, and $\mathbf{e}_a$ stands for $\mathbf{e}_1$, $\mathbf{e}_2$, and $\mathbf{e}_3$, which are our new symbols for $\mathbf{e}_x$, $\mathbf{e}_y$, and $\mathbf{e}_z$. Note that $\mathbf{e}_b$, with $b$ running over the same range as $a$, stands for the same set of $\mathbf{e}$'s as $\mathbf{e}_a$ did.

For a reason that will become apparent later, we use an upper index when denoting the components of a vector $\mathbf{V}$ [but more about this later]. Thus we denote the components collectively by $V^a$, which stands for $V^1$, $V^2$, $\ldots$, $V^n$. Note that, for example, $V^2$ does not mean "$V$ squared." It means the second component of $\mathbf{V}$, and is what we have been calling $V_y$ in three dimensions.

Using the index notation, we can write:

$$\mathbf{V} = \sum_a V^a \mathbf{e}_a, \tag{2.1}$$

where the symbol $\sum_a$ means that we must let $a$ go from 1 to $n$ and must then add the terms that result. For $n = 2$, Equation (2.1) yields essentially Equation (1.1).

The index notation plays an important role in the theory of relativity. Einstein noticed that, when an index letter occurred in an upper position and the same index letter occurred in a lower position in the same term, then almost always a $\sum$ occurred too. So he introduced the *summation convention* according to which an index repeated in this way automatically implies a summation. This removes the unsightly $\sum$'s and makes the work look neater and simpler.

Using the summation convention, we write Equation (2.1) in the form:

$$\mathbf{V} = V^a \mathbf{e}_a, \tag{2.2}$$

which stands for

$$\mathbf{V} = V^1 \mathbf{e}_1 + V^2 \mathbf{e}_2 + \cdots + V^n \mathbf{e}_n.$$

**Exercise 2.1** If $V^a = (3, 2)$ and $W_a = (1, 5)$, show that $V^a W_a = 13$.

**Exercise 2.2** If $V^b = (1, 3, 6)$ and $W_a = (2, 7, -3)$, show that $V^a W_a = V^b W_b = 5$.

**Exercise 2.3** If $T^a_b$ is such that $T^1_1 = 5$, $T^1_2 = -1$, $T^2_1 = 3$, and $T^2_2 = 4$, show that $T^a_a = 9$. What is the value of $T^b_b$?

**Exercise 2.4** How many quantities are there in $T^a_b$ if $a$ and $b$ go from 1 to 3? [*Ans.* 9.] How many if they go from 1 to $n$?

A linear change from one set of base vectors $\mathbf{e}_a$ to another set $\bar{\mathbf{e}}_a$ with the same origin can be written in the form:

$$\bar{\mathbf{e}}_a = \mathscr{A}^b{}_a \mathbf{e}_b. \tag{2.3}$$

For example, if $a$, $b$ have the range 1, 2, and

$$\mathscr{A}^b_a = \begin{array}{c|c|c} a \backslash b & 1 & 2 \\ \hline 1 & 3 & 4 \\ \hline 2 & 1 & 2 \end{array}, \tag{2.4}$$

then equation (2.3) is essentially the two equations in (1.2); for when $a = 1$ it yields

$$\bar{\mathbf{e}}_1 = \mathscr{A}^b_1 \mathbf{e}_b = \mathscr{A}^1_1 \mathbf{e}_1 + \mathscr{A}^2_1 \mathbf{e}_2 = 3\mathbf{e}_1 + 4\mathbf{e}_2,$$

and similarly for $a = 2$.

**Exercise 2.5** Find the values of $\mathscr{A}^b_a$ for Exercises 1.1 and 1.2. [*Ans.*

In Exercise 1.1, $\mathscr{A}_1^1 = 4, \mathscr{A}_1^2 = 1, \mathscr{A}_2^1 = 2, \mathscr{A}_2^2 = 1.$]

**Exercise 2.6** Write out Equation (2.3) for the three-dimensional case. [*Ans.* One of the equations is $\bar{\mathbf{e}}_1 = \mathscr{A}_1^1\mathbf{e}_1 + \mathscr{A}_1^2\mathbf{e}_2 + \mathscr{A}_1^3\mathbf{e}_3$.]

When we solve Equation (2.3) for the **e**'s in terms of the $\bar{\mathbf{e}}$'s, we get an equation of the form,

$$\mathbf{e}_a = A_a^b \bar{\mathbf{e}}_b. \tag{2.5}$$

Thus if $a$, $b$ have the range 1, 2, and if

$$\mathbf{A}_a^b = \begin{array}{c|c|c} {}_a\diagdown^b & 1 & 2 \\ \hline 1 & 1 & -2 \\ \hline 2 & -\frac{1}{2} & \frac{3}{2} \end{array}, \tag{2.6}$$

Equation (2.5) becomes essentially the pair of Equations (1.3).

**Exercise 2.7** Write down the values of $A_a^b$ for Exercises 1.1 and 1.2 and save the results for later.

Substituting from Equation (2:5) into Equation (2.2), we have

$$\mathbf{V} = V^a \mathbf{e}_a = V^a(A_a^b \bar{\mathbf{e}}_b) = (A_a^b V^a)\bar{\mathbf{e}}_b. \tag{2.7}$$

But if the components of **V** relative to $\bar{\mathbf{e}}_a$ are $\bar{V}^a$ we must have (compare Equation (1.5))

$$\mathbf{V} = \bar{V}^a \bar{\mathbf{e}}_a. \tag{2.8}$$

When we compare this with Equation (2.7) we are puzzled by the fact that $\bar{\mathbf{e}}_b$ appears there but $\bar{\mathbf{e}}_a$ here. But then we remember that we can equally well write Equation (2.8) as

$$\mathbf{V} = \bar{V}^b \bar{\mathbf{e}}_b. \tag{2.9}$$

And comparing this with Equation (2.7) we see that

$$\bar{V}^b = A_a^b V^a. \tag{2.10}$$

This equation means exactly the same thing as do $\bar{V}^a = A_b^a V^b$, $\bar{V}^c = A_a^c V^a$, etc. For the indices that occur on both sides of the equality signs merely label $n$ equations one after the other, while the summed indices indicate the same summation in each case. Indices that are summed according to the summation convention are called *dummy indices.*

**Exercise 2.8** In deriving Equation (2.7), we moved parentheses and changed the order of terms in a product. Our experiences with cross products make us wary of such maneuvers. Show, for the two-dimensional case, that $V^a(A_a^b\bar{\mathbf{e}}_b) = (V^a A_a^b)\bar{\mathbf{e}}_b = V^1 A_1^1 \bar{\mathbf{e}}_1 + V^1 A_1^2 \bar{\mathbf{e}}_2 + V^2 A_2^1 \bar{\mathbf{e}}_1 + V^2 A_2^2 \bar{\mathbf{e}}_2$, and similarly that $V^a A_a^b = A_a^b V^a$. Then extend the results to the $n$-dimensional case.

From Equations (2.10) and (2.5), we see that $\bar{V}^a$ is given in terms of $V^a$ by means of the same $A_b^a$ that gives $\mathbf{e}_a$ in terms of $\bar{\mathbf{e}}_a$ (not $\bar{\mathbf{e}}_a$ in terms of $\mathbf{e}_a$).

Accordingly $\mathbf{V}$ is called a *contravariant vector*. Note, though, that in Equation (2.10) the summation is over the lower index of $A_a^b$ while in Equation (2.5) it is over the upper index. It is this that causes the displacement of the numbers noted when we compared Equations (1.3) and (1.6).

How are $\mathscr{A}_a^b$ and $A_a^b$ related? To find out, we introduce an index $c$ having the same range as $a$ and $b$, and we write Equation (2.5) in the form:

$$\mathbf{e}_b = A_b^c \bar{\mathbf{e}}_c.$$

Applying this to the right-hand side of Equation (2.3) we see that

$$\bar{\mathbf{e}}_a = \mathscr{A}_a^b \mathbf{e}_b = \mathscr{A}_a^b A_b^c \bar{\mathbf{e}}_c. \tag{2.11}$$

For convenience, let us write,

$$\mathscr{A}_a^b A_b^c = \delta_a^c. \tag{2.12}$$

Then Equation (2.11) takes the form:

$$\bar{\mathbf{e}}_a = \delta_a^c \bar{\mathbf{e}}_c. \tag{2.13}$$

Since this is summed over $c$, we must have

$$\delta_a^c = \begin{cases} 1 \text{ if } a = c \\ 0 \text{ if } a \neq c. \end{cases} \tag{2.14}$$

Because of its importance, this quantity $\delta_a^c$ is called the *Kronecker delta*, in honor of a famous mathematician who did not invent it. Equation (2.12), with $\delta_a^c$ defined by Equation (2.14), gives us one set of the relations we seek between $\mathscr{A}_a^b$ and $A_a^b$. It can be written, of course, in the equivalent form:

$$A_b^c \mathscr{A}_a^b = \delta_a^c. \tag{2.15}$$

**Exercise 2.9** Why are Equations (2.15) and (2.12) equivalent?

To obtain another relation, we repeat the process in reverse order, applying Equation (2.3) to Equation (2.5). Thus,

$$\mathbf{e}_a = A_a^b \bar{\mathbf{e}}_b = A_a^b \mathscr{A}_b^c \mathbf{e}_c,$$

from which it follows that

$$A_a^b \mathscr{A}_b^c = \delta_a^c. \tag{2.16}$$

While this can be written in the equivalent form

$$\mathscr{A}_b^c A_a^b = \delta_a^c, \tag{2.17}$$

it is not the same as Equations (2.12) and (2.15). Note that the summation is over differently situated indices.

**Exercise 2.10** Show that in two dimensions $\delta_b^a = \begin{array}{|c|c|}\hline 1 & 0 \\ \hline 0 & 1 \\ \hline\end{array}$. What is it in $n$ dimensions?

**Exercise 2.11** Show that $\delta_a^b T_b = T_b \delta_a^b = T_a$. [*Hint*: $\delta_1^b T_b = \delta_1^1 T_1 + \delta_1^2 T_2 + \cdots + \delta_1^n T_n$, and only the first term survives. Similarly, in $\delta_2^b T_b$ only the second term survives. And so on.]

**Exercise 2.12** Show that $\delta^c_a V^a = V^a \delta^c_a = V^c$, and that $\delta^c_a T_{bc} = T_{bc}\delta^c_a = T_{ba}$.

**Exercise 2.13** Show that $\delta^a_a = \delta^b_b = n$.

**Exercise 2.14** Show that $\delta^a_b T^{bcd}_{ef} = T^{acd}_{ef}$ and note that *the effect of a Kronecker $\delta$ is merely to change one index letter on a $T$ to a new letter.*

**Exercise 2.15** Using the remark in Exercise 2.14, show that $\delta^a_b \delta^b_c \delta^c_d = \delta^a_d$. What would be the value of $\delta^a_b \delta^b_c \delta^c_d \delta^d_a$?

Working with the base vectors $\mathbf{e}_a$, we found that $\mathscr{A}$ cancels the effect of $A$, and $A$ cancels the effect of $\mathscr{A}$. Because of this, each is said to be the *inverse* of the other. This cancelling is not confined to the **e**'s. For example, we can easily see that if we solve Equation (2.10) for $V^a$ in terms of $\bar{V}^a$ the result must be:

$$V^a = \mathscr{A}^a_b \bar{V}^b. \tag{2.18}$$

For, using Equation (2.17), we have:

$$\mathscr{A}^a_b \bar{V}^b = \mathscr{A}^a_b A^b_c V^c = \delta^a_c V^c = V^a.$$

**Exercise 2.16** Starting with Equation (2.10) and applying Equation (2.18) to the $V^a$ there, show that the $A$ cancels the $\mathscr{A}$, and note that this follows from Equation (2.15), with appropriate indices, rather than from Equation (2.17).

## 3. THE NEW CONCEPT OF A VECTOR

[In this book we do not consider coordinate systems that are curvilinear or marked off nonuniformly. What follows would need significant modification before it could apply to such systems.]

Take a reference frame with origin $O$ and base vectors $\mathbf{e}_a$. The components of the position vector, $\mathbf{r}$, of a point $P$ are the coordinates of $P$ relative to this frame. In the index notation, we denote the coordinates by $x^a$, and this lets us write the compact equation:

$$\mathbf{r} = x^a \mathbf{e}_a. \tag{3.1}$$

Equation (2.10) holds for the components of any contravariant vector. Applying it to $\mathbf{r}$, we see that if we go over to a new reference frame having the same origin but different base vectors $\bar{\mathbf{e}}_a$ given, as in Equation (2.3), by

$$\bar{\mathbf{e}}_a = \mathscr{A}^b_a \mathbf{e}_b,$$

the coordinates of $P$ change to $\bar{x}^a$ where

$$\bar{x}^a = A^a_b x^b. \tag{3.2}$$

This suggests a new way of thinking about vectors. We do not harp on arrow-headed line segments. Instead, we say first that a contravariant vector is an entity that has a different set of $n$ components in each coordinate system

—one set per system. And then we add that, if the components are $V^a$ when the coordinates are $x^a$, and $\bar{V}^a$ when the coordinates are $\bar{x}^a$, and if $\bar{x}^a$ is related to $x^a$ by Equation (3.2), then

$$\bar{V}^a = A^a_b V^b, \tag{3.3}$$

this last being essentially none other than Equation (2.10).

As thus expressed, this is disconcertingly unpictorial, but therein lies its strength, as we shall see. Inasmuch as it deals with components, it embodies an important aspect of the parallelogram law. But for full embodiment, we must add that if, in a particular coordinate system, the components of two vectors are $U^a$ and $V^a$, then the components of their resultant are $(U^a + V^a)$.

We have now to settle an important question: how do we *know* that we can get away with our statement that a vector, as here regarded, has a *unique* set of components in a given coordinate system? Not simply because we asserted as much: the law of transformation (3.3) might betray us. For example, if we went from coordinates $x^a$ to coordinates $\bar{x}^a$ and from $\bar{x}^a$ to $\hat{x}^a$, the components would go from $V^a$ to $\bar{V}^a$ and then to $\hat{V}^a$. But if we went directly from $x^a$ to $\hat{x}^a$, would the corresponding $\hat{V}^a$ be the same as the previous $\hat{V}^a$? The answer is obviously yes, because $V^a$ transforms exactly as $x^a$ transforms. So the uniqueness of the components in a given coordinate system is assured.

Now comes the crucial theorem: *If $U^a$ and $V^a$ are the components of two contravariant vectors in a given coordinate system, and*

$$U^a = V^a, \tag{3.4}$$

*then a corresponding relation holds in all coordinate systems.*

This is easily proved. For since

$$\bar{U}^a = A^a_b U^b \quad \text{and} \quad \bar{V}^a = A^a_b V^b, \tag{3.5}$$

we have, by subtraction,

$$\bar{U}_a - \bar{V}_a = A^a_b(U^b - V^b). \tag{3.6}$$

But the right-hand side of Equation (3.6) vanishes, by Equation (3.4), since $U^b - V^b$ is zero for each value of $b$. So the left-hand side also vanishes and, therefore,

$$\bar{U}^a = \bar{V}^b. \tag{3.7}$$

We now understand how the changeable components can embody objectivity. For if $U^a = V^a$, this equation tells us something that holds for all coordinate systems and, thus, something that is actually independent of the coordinate system used.

> **Exercise 3.1** By $cU^a$ we mean $(cU^1, cU^2, \ldots, cU^n)$, $c$ being a number. Prove that if $U^a$ are the components of a contravariant vector, $cU^a$ are components of one also. [Simply check that the components of $cU^a$ in any given coordinate system are unique, that they transform

according to the law of transformation (3.3), and that they combine correctly with other contravariant vector components. The proofs are extremely easy.]

**Exercise 3.2** Let $l$, $m$, and $n$ be numbers, and $U^a$, $V^a$, and $W^a$ components of contravariant vectors in a given coordinate system. Show that if $lU^a + mV^a - nW^a = 0$, a corresponding relation holds in every coordinate system.

Our new way of looking at vectors suggests new possibilities, for we can think of other ways of attaining uniqueness of components and objectivity. Thus, suppose we have an entity that has components $V_a$ when the coordinates are $x^a$, and suppose that when we go to new coordinates $\bar{x}^a$ by means of Equation (3.2), the components transform according to the law:

$$\bar{V}_a = \mathscr{A}^b_a V_b. \tag{3.8}$$

Then the components in a given coordinate system are obviously unique, because $V_a$ transforms as the base vectors $\mathbf{e}_a$ do. Also, if $U_a$ and $V_a$ are components of two such entities in the $x^a$ coordinate system, and $U_a = V_a$, then a corresponding relation holds in every other coordinate system. The proof so closely parallels that for $U^a = V^a$ that we shall not bother to write it out.

Clearly, this objective entity with components $V_a$ is entitled to the name vector. It is called a *covariant vector.*

We have only begun, and already we have learned something surprising: there are two types of vectors. Do not make the mistake of thinking that covariant vectors are vectors of a sort that we knew about before, say, pseudovectors. They are not. If we confine ourselves to **ijk** reference frames, including both right-handed and left-handed kinds, it turns out that contravariant and covariant vectors transform in the same way and are thus indistinguishable. But consider, for example, how $V^a$ and $V_a$ behave under the simple change of reference frame $\bar{\mathbf{e}}_a = 12\mathbf{e}_a$. From the laws of transformation, we easily see that $\bar{V}^a = \frac{1}{12}V^a$, but $\bar{V}_a = 12V_a$. [Note that $\mathbf{e}_a$ are contravariant vectors, not covariant vectors. They are printed in boldface type and the index $a$ does not label *their* components.]

**Exercise 3.3** In the $x^a$ coordinate system in two dimensions, a contravariant vector and a covariant vector have the same components: $V^a = (3, 5), V_a = (3, 5)$. The coordinates are changed to $\bar{x}^a$ where $\bar{x}^1 = 3x^1 + 2x^2$ and $\bar{x}^2 = x^1 + x^2$. Find $\bar{V}^a$ and $\bar{V}_a$. [*Ans.* $\bar{V}^a = (19, 8)$, $\bar{V}_a = (-2, 9)$. Note that because Equation (2.18) applies to the $x$'s one can obtain $\mathscr{A}^a_b$ by solving for the $x$'s in terms of the $\bar{x}$'s. One finds $\mathscr{A}^1_1 = 1, \mathscr{A}^1_2 = -2, \mathscr{A}^2_1 = -1, \mathscr{A}^2_2 = 3$. Note also that $V^a V_a = \bar{V}^a \bar{V}_a = 34$. This is no accident.]

**Exercise 3.4** Like Exercise 3.3 if $V^a = (2, 3)$, $V_a = (2, 3)$, and $\bar{x}^1 = 4x^1 + 5x^2, \bar{x}^2 = 3x^1 + 4x^2$. [*Ans.* $\bar{V}^a = (23, 18), \bar{V}_a = (-1, 2)$. If you mix up the values of $\mathscr{A}^1_2$ and $\mathscr{A}^2$ you will get the wrong answer $\bar{V}_a = (-7, 6)$. Note that $V^a V_a = \bar{V}^a \bar{V}_a = 13$.]

## 4. TENSORS

Later we shall tell how to visualize a covariant vector. Meanwhile we find other ideas crowding in on us. For example, if $U^a$ and $V^a$ are components of two contravariant vectors, we can form their various products two at a time, namely $U^1 V^1, U^1 V^2, U^1 V^3, \ldots, U^1 V^n, U^2 V^1, U^2 V^2, \ldots$, up to $U^n V^n$, there being $n^2$ of them. The index notation lets us represent all $n^2$ of them by the single symbol $U^a V^b$. Since $U^a$ and $V^a$ are unique in each coordinate system, the same must be true of $U^a V^b$, and since $U^a$ and $V^a$ have the objective property, there is a chance that $U^a V^b$ will too. The crucial thing is the way $U^a V^b$ transforms, and we easily see that

$$\bar{U}^a \bar{V}^b = A^a_c A^b_d U^c V^d. \tag{4.1}$$

From this, much as before, we find that if $R^a$ and $S^a$ are also components of contravariant vectors, and $U^a V^b = R^a S^b$, then a corresponding relation holds in every coordinate system.

**Exercise 4.1** Prove this.

But now the thought strikes us that we do not need products of vector components in order to have uniqueness and objectivity. We could imagine a a more general entity having $n^2$ components $T^{ab}$ in the $x^a$ coordinate system, these components transforming as $U^a V^b$ did, namely, according to the law:

$$T^{ab} = A^a_c A^b_d T^{cd}. \tag{4.2}$$

Such an entity will have only one set of components per coordinate system. Also, if two such entities have components $S^{ab}$ and $T^{ab}$ in the $x^a$ coordinate system, and $S^{ab} = T^{ab}$, then a corresponding relation will hold in every coordinate system. For from Equation (4.2) and a similar one for $S^{ab}$ we obtain:

$$\bar{S}^{ab} - \bar{T}^{ab} = A^a_c A^b_d (S^{cd} - T^{cd}).$$

But the $n^2$ quantities $S^{cd} - T^{cd}$ are all zero. Hence $\bar{S}^{ab} = \bar{T}^{ab}$.

Obviously, an analogous entity with components $T_{ab}$ that transform as the quantities $U_a V_b$ do will have similar objectivity. But why stop here? We can have analogous objective entities with components $T^{ab\ldots}_{cd\ldots}$. All such entities are called *tensors.*

*A tensor is an entity having a unique set of components* $T^{ab\ldots}_{cd\ldots}$ *in a given coordinate system* $x^a$ *and such that these components transform according to the law:*

$$\bar{T}^{ab\ldots}_{cd\ldots} = A^a_r A^b_s \ldots \mathscr{A}^t_c \mathscr{A}^u_d \ldots T^{rs\ldots}_{tu\ldots} \tag{4.3}$$

*when the coordinates are changed as in Equation (3.2).**

Unfortunately we shall have scant opportunity in this book to indicate the

*Strictly speaking. it is incorrect to say that the components in a given coordinate system are unique. If the components are not pure numbers, they will change when the units of length, time, and mass are changed.

crucial role of tensors. Though the above definition is abstract to the point of aridness, tensors are of major importance since they are objective entities. If we wish to represent objective phenomena mathematically, we can hardly escape them.

The tensor law of transformation is easy to remember. For each contravariant index write an $A$, and for each covariant index an $\mathscr{A}$. Then first put in indices as shown:

$$\bar{T}^{ab\cdots}_{cd\cdots} = A^a A^b \ldots \mathscr{A}_c \mathscr{A}_d \ldots T,$$

and after that, fill in the dummy indices in pairs; note that if, for example, a dummy is placed on $\mathscr{A}_c$, the corresponding dummy goes on $T$ in the position that $c$ occupied on $\bar{T}$.

**Exercise 4.2** Cover up Equation (4.3) and write down the laws of transformation for $T^b_a$, $T^a_{bc}$, and $T^{abc}$.

Since it becomes tedious to say "the tensor having components $T^{ab\cdots}_{cd\cdots}$ in a given coordinate system $x^a$," mathematicians say simply "the tensor $T^{ab\cdots}_{cd\cdots}$," and use similar contractions on other occasions, as in the following exercise.

**Exercise 4.3** If $S^a_{bc}$ and $T^a_{bc}$ are tensors [note the language], and $S^a_{bc} = T^a_{bc}$ [again, note the language], prove that a corresponding relation holds in every coordinate system. Also prove that if $R^a_{bc}$, $S^a_{bc}$, and $T^a_{bc}$ are tensors, $l$, $m$, and $n$ are numbers, and $lR^a_{bc} + mS^a_{bc} - nT^a_{bc} = 0$, a corresponding relation holds in every coordinate system. [The meaning of $lR^a_{bc}$ is reasonably clear.]

**Exercise 4.4** Show that the results in Exercise 4.3 would not hold for $R^b_{ac}$, $S^a_{bc}$, and $T^a_b$.

**Exercise 4.5** Generalize Exercises 4.3 and 4.4 to the case of tensors having any number of indices.

**Exercise 4.6** If $S^a_b$ and $T^{cd}$ are tensors, show that $S^a_b T^{cd}$ is a tensor. [*Hint:* From the laws of transformation of $S^a_b$ and $T^{cd}$ we see at once that $\bar{S}^a_b \bar{T}^{cd} = A^a_r \mathscr{A}^s_b A^c_t A^d_u S^r_s T^{tu}$, and that is practically all there is to it.]

**Exercise 4.7** Generalize Exercise 4.6 to the general case of $S^{a\cdots}_{b\cdots}$ and $T^{a\cdots}_{b\cdots}$.

## 5. SCALARS. CONTRACTION

The number of indices of a tensor is called the *order* of the tensor, and sometimes its *valence*. Thus $T^{ab}_c$ is of order 3. Often we say that such a tensor has contravariant order 2 and covariant order 1, or that it is of the second contravariant order and the first covariant order. Vectors are tensors of order 1. What about tensors of order zero? Since they have no indices we write them as $T$ (or $S$, or $Q$, or some other index-free symbol). The corresponding law of transformation, by Equation (4.3), is just

$$\bar{T} = T.$$

Thus, tensors of zero order are scalars, a fact that probably heightens our respect for the concept of a tensor.

If $U_a$ and $V^a$ are vectors, what can we say about $U_a V^a$? By Exercise (4.7) we know that $U_a V^b$ is a second-order tensor; but $U_a V^a$ has a pair of dummy indices and thus, in a sense, no indices: $U_a V^b$ has $n^2$ components, but $U_a V^a$ has only one. The notation suggests, as did Exercises (3.3) and (3.4), that $U_a V^a$ is a scalar, and we can easily prove it is one. For we may write the laws of transformation of $U_a$ and $V^a$ as $\bar{U}_a = \mathscr{A}^b_a U_b$ and $\bar{V}^a = A^a_c V^c$. Therefore, using Equation (2.17), we have:

$$\bar{U}_a \bar{V}^a = \mathscr{A}^b_a A^a_c U_b V^c = \delta^b_c U_b V^c = U_c V^c = U_a V^a.$$

**Exercise 5.1** If $T^a_b$ is a tensor, prove that $T^a_a$ is a scalar.

**Exercise 5.2** If $T^{ab}_c$ is a tensor, prove that $T^{ab}_a$ is a contravariant vector.

Exercises 5.1 and 5.2 are special cases of the following general theorem:

*If $T^{\cdots a \cdots}_{\cdots b \cdots}$ is a tensor of order r, then $T^{\cdots a \cdots}_{\cdots a \cdots}$ is a tensor of order r — 2.*

We omit the proof. It is the same as that in Exercise 5.1, except for the presence of dots all over the place. The process of summing over a repeated index, one contravariant and the other covariant, to obtain a tensor of lower order is called *contraction.*

The proof that contraction yields a tensor depends crucially on the presence of an $\mathscr{A}$ and an $A$ that combine to form a $\delta$. So, even if an expression like $T^{aa}$ or $T_{aa}$ were summed over $a$, the result would not be a scalar. We can easily verify this by referring to Exercise (3.3). There $\sum_a V^a V^a = 9 + 25 = 34$, but $\sum_a \bar{V}^a \bar{V}^a = 19^2 + 8^2$ which is too unpleasant to be worth evaluating but is obviously not 34. Also, $\sum_a V_a V_a$ was equal to 34 there too, but $\sum_a \bar{V}_a \bar{V}_a$ comes to 85. There is something disquieting about this, as we shall see later.

## 6. VISUALIZING TENSORS

The title of this section almost constitutes fraudulent labelling. The fact is that, with a few exceptions, tensors cannot be readily visualized.

Scalars we are familiar with, and therefore, perhaps we feel that we can visualize them. Contravariant vectors we picture as arrow-headed line segments, albeit with a curious mode of combination. But what of covariant vectors?

If $V_a$ is a covariant vector, the quantity $V_a x^a$ is a scalar, since $x^a$ is a contravariant vector. So if we write the equation,

$$V_a x^a = 1, \tag{6.1}$$

it will mean the same thing in all coordinate systems. In two dimensions it is, in the $x$, $y$ notation, just

$$V_x x + V_y y = 1,$$

which represents a straight line. So we can visualize $V_a$ in two dimensions as this straight line. But do not jump to conclusions. It is not a line *segment*—with or without an arrowhead. It is a whole line. Moreover, in three dimensions Equation (6.1) becomes

$$V_x x + V_y y + V_z z = 1,$$

which represents a plane, not a line. So we can there picture $V_a$ as this plane. For the record we mention, without elaboration, that in $n$ dimensions we can picture $V_a$ as an $(n - 1)$-dimensional hyperplane.

A second-order covariant tensor $T_{ab}$ can be pictured with the aid of the equation:

$$T_{ab} x^a x^b = 1. \tag{6.2}$$

In two dimensions this represents an ellipse, a hyperbola, or a pair of lines.* In three dimesions it represents an ellipsoid, a hyperboloid, a cylinder, or a pair of planes.* As we see, the game of visualization becomes more complicated the greater the number of indices, and we shall pursue it no further except to remark that the above process does not apply to contravariant indices.

**Exercise 6.1** In two dimensions if $T_{11} = T_{22} = 1$, $T_{12} = T_{21} = 0$, in rectangular Cartesian coordinates, show that Equation (6.2) represents a circle.

## 7. SYMMETRY AND ANTISYMMETRY. CROSS PRODUCTS

If a tensor $T^{ab}$ is such that

$$T^{ab} = T^{ba} \tag{7.1}$$

in one coordinate system, a similar relation holds in every coordinate system. For we have

$$\bar{T}^{ab} = A^a_c A^b_d T^{cd} \tag{7.2}$$

$$\bar{T}^{ba} = A^b_c A^a_d T^{cd} \tag{7.3}$$

But in Equation (7.3) we can replace $T^{cd}$ by $T^{dc}$, because of Equation (7.1). Therefore,

$$\bar{T}^{ba} = A^b_c A^a_d T^{dc}.$$

Since the $c$'s and $d$'s are dummy indices, we can change them to other letters.

*Which may coincide.

So we change both of the present $c$'s to $d$'s and, at the same time, both of the present $d$'s to $c$'s. The result is:

$$\bar{T}^{ba} = A^b_d A^a_c T^{cd},$$

which is identical with

$$\bar{T}^{ba} = A^a_c A^b_d T^{cd}.$$

Comparing this with Equation (7.2) we see that $\bar{T}^{ab} = \bar{T}^{ba}$.

Thus the relation $T^{ab} = T^{ba}$ is an objective one. We say that this $T^{ab}$ is *symmetric* in the indices $a$ and $b$. In a similar way, one finds that the relationship

$$T^{ab} = -T^{ba} \tag{7.4}$$

is also an objective one. Such a tensor is said to be *antisymmetric*, or *skew symmetric.*

**Exercise 7.1** If $T^{ab} = \begin{array}{|c|c|}\hline 1 & 3 \\ \hline 3 & 5 \\ \hline\end{array}$, then $T^{ab} = T^{ba}$. Calculate $\bar{T}^{ab}$ if $\bar{x}^1 = 2x^1 + 4x^2$, $\bar{x}^2 = x^1 + 3x^2$, and verify that $\bar{T}^{ab} = \bar{T}^{ba}$.

**Exercise 7.2** If $T^{ab} = \begin{array}{|c|c|}\hline 0 & 3 \\ \hline -3 & 0 \\ \hline\end{array}$, then $T^{ab} = -T^{ba}$. Calculate $\bar{T}^{ab}$ for the coordinate transformation in Exercise 7.1 and verify that $\bar{T}^{ab} = -\bar{T}^{ba}$.

**Exercise 7.3** Prove that the relationships $T_{ab} = T_{ba}$ and $T_{ab} = -T_{ba}$ are objective.

Given two vectors $U^a$ and $V^a$, we can form the second-order antisymmetric tensor:

$$T^{ab} = U^a V^b - U^b V^a. \tag{7.5}$$

Let us write out its components in the three-dimensional case. We easily see that $T^{11}$, $T^{22}$, and $T^{33}$ are zero. As for $T^{23}$, $T^{31}$, and $T^{12}$, they are respectively $U^2V^3 - U^3V^2$, $U^3V^1 - U^1V^3$, $U^1V^2 - U^2V^1$; and if we look closely we recognize that THEY ARE OF THE SAME FORM AS THE COMPONENTS OF THE CROSS PRODUCT $\mathbf{U} \times \mathbf{V}$ RELATIVE TO A RIGHT-HAND UNIT ORTHOGONAL TRIAD **i, j, k.** This rouses our interest, and we naturally wonder what the remaining components of $T^{ab}$, namely $T^{32}$, $T^{13}$, and $T^{21}$, will be. But, since $T^{ab}$ is antisymmetric, they are obviously the negatives of the above—a fact that reminds us of the trouble we had with the changing sign of $\mathbf{U} \times \mathbf{V}$ when we went from a right-handed to a left-handed reference frame.

It can be shown that *if we confine ourselves to right-handed unit orthogonal reference frames* the components $T^{23}$, $T^{31}$, and $T^{12}$ transform like the components of a vector (as also do the components $T^{32}$, $T^{13}$, and $T^{21}$). It is this fact that caused our luck to hold when we dealt with cross products as though they were vectors. If you look back you will see that we worked with their components only with respect to **ijk** reference frames. Had we used more general frames we would have encountered difficulties. Indeed, even if we had used base vectors that were not of unit magnitude, though still mutually perpendicular, we would have encountered trouble—as we realize from the awkwardness connected with $\mathbf{i} \times \mathbf{j} = \mathbf{k}$.

**Exercise 7.4** Write out the four components of $U^a V^b - U^b V^a$ for the two-dimensional case, and note that two of them are necessarily zero, and the other two, if they are not zero, are equal except for sign. Show that this holds for any second order antisymmetric tensor $T^{ab}$ or $T_{ab}$ in two dimensions.

**Exercise 7.5** Show that in four dimensions a second-order antisymmetric tensor $T^{ab}$ or $T_{ab}$ has four components that are necessarily zero, and twelve components which, if they are not zero, are equal in pairs except for sign.

**Exercise 7.6** In two dimensions, if $T^{ab}$ is an antisymmetric tensor it involves, essentially, the single numerical quantity $T^{12}(= -T^{21})$. Show that $T^{12}$ does not transform like a scalar but that

$$\bar{T}^{12} = (A^1_{\bar{1}} A^2_{\bar{2}} - A^1_{\bar{2}} A^2_{\bar{1}}) T^{12}.$$

[Use the facts that $T^{11} = T^{22} = 0$; $T^{21} = -T^{12}$.]

In Exercise 7.6, if we restrict ourselves to transformations between right-handed unit orthogonal base vectors, it can be shown that $A^1_{\bar{1}} A^2_{\bar{2}} - A^1_{\bar{2}} A^2_{\bar{1}} = 1$. For such reference frames, then, $T^{ab}$ does behave like a scalar. But if we go from a right-handed frame of this sort to a left-handed one, or vice versa, it changes sign. It is called a *pseudoscalar*.

Usually the components of a tensor change when the coordinates are changed. But consider a tensor $T^a_b$ that has components $\delta^a_b$ in the $x^a$ coordinate system. In a new system its components are:

$$\bar{T}^{\bar{a}}_{\bar{b}} = A^{\bar{a}}_c \mathscr{A}^d_{\bar{b}} T^c_d = A^{\bar{a}}_c \mathscr{A}^d_{\bar{b}} \delta^c_d = A^{\bar{a}}_c \mathscr{A}^c_{\bar{b}} = \delta^{\bar{a}}_{\bar{b}},$$

by Equation (2.15). So this particular tensor has the unusual property that its components are the same in every coordinate system. We therefore speak of "the tensor $\delta^a_b$."

Consider a Kronecker delta with two covariant indices:

$$\delta_{ab} = \begin{cases} 1 \text{ if } a = b \\ 0 \text{ if } a \neq b. \end{cases} \tag{7.6}$$

If a tensor $T_{ab}$ has components $\delta_{ab}$ in one coordinate system, does it too have the same components in all systems? We have

$$\bar{T}_{\bar{a}\bar{b}} = \mathscr{A}^c_{\bar{a}} \mathscr{A}^d_{\bar{b}} \delta_{cd},$$

but we cannot continue as we did before. The right side does not, in general, reduce to $\delta_{ab}$. So the answer to our question is *no*. Putting it another way, we can say that "$\delta_{ab}$ is not a tensor."

**Exercise 7.7** Show that "$\delta^{ab}$ is not a tensor."

## 8. MAGNITUDES. THE METRICAL TENSOR

If $(V_x, V_y, V_z)$ are the components of a contravariant vector **V** relative to **i**, **j**, **k**, the square of the magnitude of **V** is given by:

$$(V)^2 = (V_x)^2 + (V_y)^2 + (V_z)^2. \tag{8.1}$$

This fact gave us no qualms when we first encountered it. But now it puzzles us. For we regard $(V)^2$ as a scalar, but the right-hand side of Equation (8.1) is not a scalar. We could express it as $\sum V^a V^a$, but that does not fit the summation convention. So we try $\delta_{ab} V^a V^b$. This *does* fit our notation, and if $\delta_{ab}$ were a tensor, this expression would indeed be a scalar, being the double contraction of a tensor. But as we have just seen $\delta_{ab}$ is not a tensor.

If we insist on $(V)^2$ being a scalar, we have to say that, though $\delta_{ab}$ is not a tensor (in the sense that it does not retain the components $\delta_{ab}$ in all coordinate systems) there nevertheless must exist a tensor, say $g_{ab}$, *that has the components* $\delta_{ab}$ *in an* **ijk** *reference frame*:

$$g_{ab} = \delta_{ab}. \tag{8.2}$$

Then we can write Equation (8.1) in the form

$$(V)^2 = g_{ab} V^a V^b, \tag{8.3}$$

with the right-hand side now manifestly a scalar. Since this tensor equation holds in one coordinate system, it holds in all coordinate systems. The tensor $g_{ab}$ is called the *metrical tensor*. As defined, it is obviously symmetric, since $\delta_{ab} = \delta_{ba}$ and symmetry in one reference frame implies symmetry in all. And we note that we can make Equations (8.2) and (8.3) apply to the $n$-dimensional case by letting $a$, $b$ have the range $1, 2, \ldots, n$, the coordinates being the analogue of those in an **ijk** frame.

Let us get an idea of the role of the metrical tensor by discussing a simple case. Consider a vector **V** with components $V^a$ relative to **ijk**. Suppose we stretch the $x$ coordinate by the transformation

$$\bar{x}^1 = \alpha x^1, \qquad \bar{x}^2 = x^2, \qquad \bar{x}^3 = x^3. \tag{8.4}$$

Then

$$\bar{V}^1 = \alpha V^1, \qquad \bar{V}^2 = V^2, \qquad \bar{V}^3 = V^3. \tag{8.5}$$

Also, since $x^1 = (1/\alpha)\bar{x}^1$, we have $\mathscr{A}^1_1 = 1/\alpha$, and from the law of transformation,

$$\bar{g}_{ab} = \mathscr{A}^c_a \mathscr{A}^d_b g_{cd},$$

we find after a short calculation that

$$\bar{g}_{11} = \frac{1}{\alpha^2}, \qquad \bar{g}_{ab} = g_{ab} \quad \text{for} \quad b \neq 1. \tag{8.6}$$

Had we taken the sum of the squares of the components $\bar{V}^a$ they would not have given the same result as the sum of the squares of the components $V^a$. But the $g_{ab}$ acts as a sort of counterpoise to annul the effect of the stretching of the $x^1$ coordinate, and we have:

$$\bar{g}_{ab}\bar{V}^a\bar{V}^b = \frac{1}{\alpha^2}(\bar{V}^1)^2 + (\bar{V}^2)^2 + (\bar{V}^3)^2$$

$$= \frac{1}{\alpha^2}(\alpha V^1)^2 + (V^2)^2 + (V^3)^2 = (V^1)^2 + (V^2)^2 + (V^3)^2.$$

**Exercise 8.1** Apply the transformation $\bar{x}^1 = \alpha x^1$, $\bar{x}^2 = \beta x^2$, $\bar{x}^3 = \gamma x^3$ to the components of $g_{ab}$ in Equation (8.2) and find $\bar{g}_{ab}$. Also find $\bar{V}^a$ in terms of $V^a$. Then verify that $\bar{g}_{ab}\bar{V}^a\bar{V}^b$ comes to the same thing as $\delta_{ab}V^aV^b$.

**Exercise 8.2** In one dimension, if $V^1 = 2$ and $g_{11} = 9$, what is $V$? [*Ans.* 6.]

**Exercise 8.3** If $V^a = (3, -1)$ and $g_{ab} = \begin{array}{|c|c|}\hline 1 & 2 \\ \hline 2 & 6 \\ \hline\end{array}$, show that $V = \sqrt{3}$.

**Exercise 8.4** If $V^a = (1, 2, 2, 4)$ and $g_{ab} = \begin{array}{|c|c|c|c|}\hline 5 & 0 & 0 & -1 \\ \hline 0 & 2 & 0 & 0 \\ \hline 0 & 0 & 1 & 0 \\ \hline -1 & 0 & 0 & 1 \\ \hline\end{array}$,

show that $V = 5$.

**Exercise 8.5** Given the coordinate transformation $\bar{x}^1 = x^1 + x^2$, $\bar{x}^2 = x^2$, and $\bar{x}^3 = x^3$, solve for the $x$'s in terms of the $\bar{x}$'s and thus obtain $\mathscr{A}^a_b$. Using this, show that the $g_{ab}$ in Equation (8.2) transforms into $\bar{g}_{ab}$ where $\bar{g}_{22} = 2$ and $\bar{g}_{12} = \bar{g}_{21} = -1$, while the other components are unchanged. Then show that $(V)^2 = (\bar{V}^1)^2 + 2(\bar{V}^2)^2 + (\bar{V}^3)^2 - 2\bar{V}^1\bar{V}^2$. [The coefficient $-2$ in the last term arises from the addition of two terms, one involving $\bar{g}_{12}$ and the other, $\bar{g}_{21}$.]

## 9. SCALAR PRODUCTS

Given two contravariant vectors $U^a$ and $V^a$, we can form the scalar $g_{ab}U^aV^b$. What does it represent?

From the analogue of Equation (8.2) in the $n$-dimensional case, we see that in uniform unit rectangular coordinates,

$$g_{ab}U^aV^b = U^1V^1 + U^2V^2 + \cdots + U^nV^n,$$

which, in the three-dimensional case, is just the scalar product $\mathbf{U}\cdot\mathbf{V}$. Since $g_{ab}U^aV^b$ is a scalar, it has the same value in all coordinate systems. Therefore even when the coordinates are such that terms like $5U^1V^1$ and $3U^1V^2$ appear on the right, it still represents the scalar product. Indeed, we call it the scalar product of $\mathbf{U}$ and $\mathbf{V}$ even in the $n$-dimensional case.

If we think of **U** and **V** as arrow-headed line segments starting at $O$, we may write, therefore,

$$g_{ab} U^a V^b = UV \cos\theta, \tag{9.1}$$

where $\theta$ is the angle between them. Therefore,

$$\cos\theta = \frac{(g_{ab} U^a V^b)}{\sqrt{g_{cd} U^c U^d}\sqrt{g_{ef} V^e V^f}}. \tag{9.2}$$

This argument is all very well for three dimensions. But isn't it a little reckless for $n$ dimensions?

Not really. One of the nice things about a tensor equation is that we can check its validity by using any convenient coordinate system. So let us take coordinates such that **U** lies along the $x^1$-axis and has components $U^a = (U^1, 0, 0, \ldots, 0)$, and **V** lies along the $x^2$-axis and has components $V^a = (0, V^2, 0, \ldots, 0)$. Then Equations (9.1) and (9.2), when expanded, yield the same equations we would get if we worked two-dimensionally in the plane containing **U** and **V**.

**Exercise 9.1** Show that $g_{ab} U^a V^b = g_{ab} V^a U^b$. [Use the symmetry of $g_{ab}$.]

**Exercise 9.2** If $U^a = (2, 1)$ and $V^a$, $g_{ab}$ are as in Exercise 8.3, find $g_{ab} U^a V^b$ and $\cos\theta$. [*Ans.* $2, \sqrt{2}/3\sqrt{3}$.]

**Exercise 9.3** If $U^a = (3, 1)$, $V^a = (2, 5)$, and $g_{ab} = \begin{array}{|c|c|} \hline 2 & 1 \\ \hline 1 & 4 \\ \hline \end{array}$, find $U$, $V$, and $\cos\theta$. [*Ans.* $\sqrt{28}, \sqrt{128}, 49/\sqrt{28}\sqrt{128}$.]

**Exercise 9.4** If $U^a = (5, 0, 0, 0)$, and $V^a$ and $g_{ab}$ are as in Exercise 8.4, find $g_{ab} U^a V^b$ and $\cos\theta$. [*Ans.* $5, 1/5\sqrt{5}$.]

**Exercise 9.5** Show that the angle $\theta$ between the $x^1$- and $x^2$-axes is given by $\cos\theta = g_{12}/\sqrt{g_{11}g_{22}}$. [Consider $U^a = (U^1, 0, 0, \ldots, 0)$ and $V^a = (0, V^2, 0, \ldots, 0)$.]

**Exercise 9.6** Show that if $g_{ab} = 0$ when $a \neq b$, the coordinate axes are mutually perpendicular. [Use the preceding exercise.]

**Exercise 9.7** If $U^a = (m, 2)$, and $V^a$ and $g_{ab}$ are as in Exercise 8.3, what must be the value of $m$ if $U^a$ is perpendicular to $V^a$? [*Ans.* 0.]

**Exercise 9.8** Using Exercise 9.5, show that the angle between the $\bar{x}^1$- and $\bar{x}^2$-axes in Exercise 8.5 is 135°. Then see if you can obtain the same result by considering the diagrammatical significance of $\bar{x}^1 = x^1 + x^2$. [The second part is easy once you see it. But there is an enticing false trail that leads to the incorrect answer 45°.]

**Exercise 9.9** Show that in Exercise 8.5 the scale on the $\bar{x}^2$-axis is $\sqrt{2}$ times that along the $\bar{x}^1$- and $\bar{x}^3$-axes. [*Hint*: Consider a vector $\bar{V}^a$ with components $(0, 1, 0)$ which thus lies along the $\bar{x}^2$-axis and stretches from the origin to the "1" mark on that axis. Find its magnitude, using the values of $\bar{g}_{ab}$, and compare it with the magnitudes of vectors with components (1, 0, 0) and (0, 0, 1).]

**Exercise 9.10** If $x^a$ are coordinates in three dimensions associated with an **ijk** frame, and we define new coordinates by $\bar{x}^1 = x^1 + 2x^2 + x^3$, $\bar{x}^2 = x^2$, $\bar{x}^3 = x^3$, find the cosine of the angle between the $\bar{x}^1$- and $\bar{x}^2$-axes, and the relationship of the scales along these axes. [*Ans.* We find that $\bar{g}_{11} = 1$, $\bar{g}_{22} = 5$, and $\bar{g}_{12} = -2$. So $\cos\theta = -2/\sqrt{5}$, and the ratio of the scales is $\sqrt{5}$.]

**Exercise 9.11** What are the components of the metrical tensor in two dimensions when the scale along the $x^1$-axis is such that the "1" mark on it is 2 inches from the origin, the scale along the $x^2$-axis is such that the "1" mark on it is $\sqrt{3}$ inches from the origin, and the cosine of the angle between the axes is $\frac{1}{2}$? [*Ans.* $g_{11} = 4$, $g_{22} = 3$, $g_{12} = \sqrt{3}$.]

When we work with **ijk** frames, the metrical tensor has components $\delta_{ab}$. This makes it practically invisible, because we rarely write coefficients that are 1 or 0. When, for example, in the two-dimensional case we write

$$(V)^2 = (V^1)^2 + (V^2)^2,$$

we are not likely to realize that this is really

$$(V)^2 = 1V^1V^1 + 0V^1V^2 + 0V^2V^1 + 1V^2V^2 = \delta_{ab}V^aV^b = g_{ab}V^aV^b.$$

Since the components of a displacement are pure numbers but its magnitude is a length, we see that the components of the metrical tensor must have the dimensions of the square of a length. And this brings us to something that may have been worrying you for quite a while. When we complained about $\mathbf{i} \times \mathbf{j} = \mathbf{k}$, should we not also have complained about $\mathbf{i}\cdot\mathbf{i} = 1$? That equation, too, is unsatisfactory, since the left-hand side is the square of a length, but the right hand side looks like a pure number.

With the aid of the metrical tensor, we can clarify the situation. The vector **i** has components (1, 0, 0) relative to the **ijk** frame; the 1 is a pure number, not a length. Denote these components of **i** by $i^a$. Then,

$$\mathbf{i}\cdot\mathbf{i} = g_{ab}i^ai^b = g_{11}i^1i^1 = g_{11},$$

and though $g_{11}$ here has the value 1, the right-hand side is now explicitly the square of a length, as it has to be.

**Exercise 9.12** A vector $V^a$ lies along the $x^1$-axis and has components $(V^1, 0, 0, \ldots, 0)$. Show that its magnitude is not $V^1$ but $\sqrt{g_{11}}\, V^1$.

## 10. WHAT THEN IS A VECTOR?

This being a book about vectors, we have presented only the sketchiest account of tensors—barely enough to illustrate the advantages of thinking of vectors in terms of the way their components transform.

We have one final point to make. Notice that we defined contravariant vectors and covariant vectors—indeed, tensors of all ranks—*before we introduced the metrical tensor*. Suppose there were no metrical tensor. What could we then say about the magnitudes of vectors? Or about the cosines of the angles between them?

You may be tempted to argue that such questions prove that there *has* to be a metrical tensor. But actually there does not. Mathematicians often work with spaces that do not possess one; they call them *nonmetrical spaces*.

Thus vectors do not *have* to have magnitudes. And this is as good a place as any to stop.

# INDEX

Renate Klein, Bernard Wallner (Eds.)

# Conflict, Gender, and Violence

Renate Klein, Bernard Wallner (Eds.)

# Conflict, Gender, and Violence

StudienVerlag
Innsbruck
Wien
München
Bozen

by StudienVerlag Ges.m.b.H.
Amraser Straße 118, A-6020 Innsbruck
e-mail: order@studienverlag.at
**www.studienverlag.at**

Gedruckt mit Unterstützung durch das Bundesministerium für Bildung, Wissenschaft und Kultur in Wien.

Buchgestaltung nach Entwürfen von Kurt Höretzeder
Satz: Studienverlag / Helmut Mangott
Umschlag: Studienverlag / Tommi Bergmann

Gedruckt auf umweltfreundlichem,
chlor- und säurefrei gebleichtem Papier.

Bibliografische Information Der Deutschen Bibliothek
Die Deutsche Bibliothek verzeichnet diese Publikation in der Deutschen Nationalbibliografie; detaillierte bibliografische Daten sind im Internet über http://dnb.ddb.de abrufbar.

ISBN 3-7065-1829-5

# Table of Contents

# Foreword

This volume is based on research presented at the 6th Interdisciplinary Conference on Conflict, Gender, and Violence in Vienna 2001. The meeting was organized by the European Research Network on Conflict, Gender, and Violence in collaboration with Women against Violence Europe (WAVE) and local support from the Vienna Project Center of Women's Studies and Gender Research at the University of Vienna.

This broad-based collaboration reflects a promising multidisciplinary approach to the intersections of gender, conflict, violence, and culture. Current knowledge on violent behavior shows how difficult it is to identify its cultural and physiological origins. Many explanations for violent behavior refer to individual crimes or current conflict situations. To find answers and explanations in a global context is one of the challenging tasks for our society. For this, it may be necessary to step back from the premises of any particular discipline and take a wider view that opens up fruitful, cross-disciplinary debate. The contributions to this volume do exactly this and provide a much-needed, wide-angled view of the field.

This volume also contributes to the ongoing publications on Gender Studies at the University of Vienna, many of which the Studienverlag has published. As Vice-Rector, two of my goals have been to establish and strengthen gender research at the University of Vienna and to found an advisory board for sexually harassed university employees. From my perspective, this book contributes to the field of interdisciplinary gender research and provides useful information for those working on sexual harassment. I am pleased that the 6th International Conference of the European Research Network was held in Vienna and I want to thank organizers and participants for a productive conference, and both editors, Renate Klein and Bernard Wallner, for realizing the book project.

*Gabriele Moser*

# Preface

When Carol Hagemann-White approached the Co-ordination Office of WAVE (Women Against Violence Europe) about hosting the 6th Interdisciplinary Conference of the European Research Network on Conflict, Gender and Violence (ENCGV), this request coincided with our own wish to intensify contacts between research and practice in the field of violence against women.

The collaboration between the WAVE European women's shelters and support facilities network and the European Research Network on Conflict, Gender, and Violence began in the year 2000, when the WAVE Co-ordination Office applied to the EU Commission's DAPHNE Programme for funding to set up a European Information Centre Against Violence. The aims of the Information Centre included establishing and intensifying contacts with researchers, reporting on the latest research results in the field of violence, and promoting partnerships between researchers and grassroots women's NGOs *(http://www.wave-network.org)*. WAVE draws on the com-bined experience and expertise of members from forty European countries and – through the European Information Centre Against Violence – can support research and event planning, including access to a pool of specialists working with women and children exposed to violence.

The 6th Interdisciplinary Conference provided an opportunity to continue and expand the collaboration between researchers and practitioners. Ultimately, it will need the concerted efforts of both areas – research and practice – to develop and implement preventive measures to combat violence against women and children. The latest studies on the prevalence of violence against women in intimate relations provide ample evidence underscoring the need for funding for agencies and institutions that work to support victims and end violence. It remains to be stressed that national governments, state and private bodies, and professionals at the local level – most notably the police – have their own roles to play in drawing up a co-ordinated strategy to effectively combat and end violence against women and children.

*Birgit Appelt* and *Verena Kaselitz*
WAVE Network and project co-ordinators

*Renate C.A. Klein, University of Maine*
*Bernard Wallner, University of Vienna*

# Conflict, Gender, and Violence in Europe: Research Advances and Programmatic Challenges

## 1. Background

This book addresses the emerging community of scholars, advocates, and policy makers interested in interdisciplinary, transnational European research on gender and violence. The volume is based on selected contributions to the 6$^{th}$ Interdisciplinary Conference of the European Research Network on Conflict, Gender, and Violence that took place in Vienna, Austria, in October 2001. The volume is not exhaustive of current European research on gender and violence but it highlights key research themes and emerging debates.

This introduction attempts to set a common stage for a collection of highly diverse chapters. We first comment on the use of terminology and then explain why it may be desirable to include in one volume a wide range of topics presented by contributors from Austria, France, Germany, Romania, Spain, and the United Kingdom whose collective disciplinary backgrounds include anthropology, behavioral endocrinology, educational theory, feminist studies, human geography, law, political science, psychology, social policy, sociology, and social work. What unites this group is a willingness to engage, towards the goal of ending gendered violence, in constructive interdisciplinary and international debate. We briefly summarize and discuss each chapter and end the introduction with observations on the current European context for transnational research on gender and violence and some of the challenges and opportunities it may present for the further developments in this field.

## 2. Terminology

In this chapter, but not in reference to the individual chapters nor throughout the book, we use the expressions "gender and violence" or "gendered violence" to refer to violence against women, children, and men and to emphasize that such violence, in both its motivation and primary targets, is woven into cultural constructions of gender identities and practices of "doing gender" (West and Zimmerman 1987). The terms are not entirely satisfying but less bulky than the more accurate "gender and violence against women, children, and men that is committed mostly by men, occasionally by women, rarely by children, most of it in the context of intimate or family relationships".

Naming victims and perpetrators still matters, and is more complicated than it may seem. It matters to name men, and young men in particular, as perpetrators whenever they are perpetrators; the extent of men's violence against other men and against women is well documented in the literature, including young men's share of that violence. It also matters to name women, and young women in particular, as victims of a significant amount of men's violence. The question is if anything else should be named and what the consequences of naming are (e.g., naming men as victims and naming women as perpetrators). Naming occurs in a conceptual and historical context. It was and is important in the violence against women movement to name violence against women as such because it has been (and continues to be) denied or misnamed (e.g., "domestic disturbance"). Naming is important wherever gag orders against naming exist, including the silence surrounding men as victims of violence. Even so, naming can be abused as when naming men's victimization implies victimization by women and is used rhetorically and politically to discount men's violence against women. Finally, as much as it matters to draw attention to the fact that real people with vulnerable bodies engage in or suffer violence, more abstract notions such as "violent masculinity" need to be named as well in order to be able to analyze how violence or victimization is written into gender identities and how stakeholders in cultures around the world promote ideas of desirable maleness or femaleness that prescribe what "real" men or women ought to think, feel, and do.

## 3. Topical diversity

This volume comprises a range of topics that at first glance may not have much to do with each other. Topics include legal reform and policy changes in Austria and Romania, work with violent men in Germany, parenting by violent fathers in the United Kingdom, violence against women in France, the consequences of intimate partner violence for women's health, and an analysis of men's violence against women in terms of brain physiology.

Still, this topical and methodological diversity seems desirable for two reasons. One, inquiry into gender and violence benefits from interdisciplinary approaches, which, with their respective theoretical and empirical tools, can address and illuminate different facets of gendered violence more effectively than any single method or discipline. Second, in order to advance transnational European research in the field it is necessary to acknowledge and take into consideration the differences that exist across Europe with regard to research priorities and infrastructures.

## 4. Contributions to this book

Each chapter is self-contained and represents state-of-the-art developments in its respective area. Throughout the book, terminology and frames of reference vary. Chapters two through eight address different aspects of violence and abuse in the context of intimate and family relationships; Stephanie Condon and colleagues' survey assessed violence against women in public and private settings; Bernard Wallner discusses humans and non-human primates and refers to aggression and dominance within and between the sexes. Some authors use domestic violence and intimate partner violence interchangeably. Others refer to gender violence or gender-based violence. This diversity in terminology reflects the diversity of the research community. In the long-term, a prevailing term may emerge that reflects widespread consensus about usage (Hagemann-White 2002).

Many publications on gender and violence begin with some statement on the scope of the problem and the biological foundations of human behavior. Thus, information on prevalence and "causation" tends to precede everything else. However, as *Hagemann-White and Lenz* (this volume) imply, high-quality prevalence surveys tend to build on, rather than precede, knowledge about gender and violence that has been gained from advocacy and intervention work. Moreover, causation of human behavior is so intimately tied into deliberate cultural activity that we departed from the established custom and ordered the contributions to this volume differently. At the beginning are three chapters on legal reform and policy changes, followed by a chapter on health consequences. Next are two chapters that address issues concerning male perpetrators, followed by a chapter that aims to broaden the debate on men as perpetrators. Then comes the chapter on the prevalence of violence against women in France. We conclude with the chapter on brain physiology that raises interesting conceptual issues for research on gender violence in humans.

### 4.1 Legal Reforms and Policy Changes

*Birgitt Haller* presents initial findings from an evaluation of the Austrian Protection from Violence Act of 1997. Under the Act, new powers were granted to Austrian police to evict domestically violent perpetrators from the residence they share with their victim, and to impose barring orders on perpetrators to stay away from the residence and its immediate vicinity. What is interesting about the Austrian approach is the extent to which the development and implementation of the new offender accountability measures are integrated, through intervention centers, into a system of victim protection and support that operates country-wide since 1999.

The evaluation revealed, among other findings, that 90% of the victims were female and 97% of the perpetrators were male. How victims and perpetrators responded to the new policies varied according to the ways in which police intervention played out in the specific life contexts of those affected. For some, police intervention had successfully disrupted male privilege, causing both victim and perpetrator to rethink their options, while for others police intervention had been unwelcome, or, inadvertently, had reaffirmed existing power abuses. The new policies reflect, and, to some extent, will probably promote, a transformation of traditional dismissive attitudes towards power abuses in intimate relationships. Nevertheless, such cultural change is slow, and in rural areas police still seem to favor the traditional "dispute settlement" approach, that allows violent perpetrator's to continue their power abuses with impunity.

Focusing on the political situation in Austria in the fall of 2001 *Elfriede Fröschl* discusses the implications for women of the conservative shift in the Austrian government following the 2000 elections. While featuring women in influential government positions the coalition between the Christian-conservative Austrian People's Party and the right-wing, conservative Freedom Party made a number of decisions that undermine the status of women. The administration abolished the post of Minister of Women's Affairs and instituted various employment and family policies that impede women's access to the labor market, thus eroding their economic self-reliance and increasing their dependence on male breadwinners.

*Cristina Negutu* speculates that Romania's neglect of violence against women and children may have been partly due to the country's preoccupation with other social problems. While those problems were significant, they alone do not explain the lack of public recognition of violence against women and children. Other countries with much less economic and social hardship have turned a blind eye to gender violence. Nonetheless, for governments and other patriarchal stakeholders alike, the reference to dire economic, social, or political straits has served, and continues to serve, as a convenient excuse for not addressing gender violence.

The recent legal reforms in Romania illustrate the strengths and limitations of criminal justice system responses to gender and violence. On the one hand, the new resolve to "crack down on perpetrators" reflects a welcome intent to redefine violence against women and children as crimes and thus no longer take them for granted as expressions of patriarchal power but consider them intolerable abuses of such power. On the other hand, significant loopholes remain: The legal process often deters victims because it is lengthy, complicated, and may result in secondary victimization. New laws may look good on paper but their reach is often limited by geography and mentality. The implications of the crack down are not always addressed. For example, what happens to victims who depend financially on the perpetrator when he or she sits in prison? The criminal justice system by itself is usually not set up to deal with these consequences but needs to work closely with other state or societal institutions that can provide victim support, or work towards attitude change.

### 4.2 Health Consequences

*Manuela Martinez, M. Isabel García-Linares* and *M. Angeles Pico-Alfonso* give an overview of current, international research on the consequences of domestic violence for women's health and well-being. Mounting evidence worldwide indicates that intimate partner violence is so detrimental to women's physical health and mental health that experts, pushed by battered women's and rape victim's advocates, have begun to view domestic violence as a threat to public health.

The amount of injury, disease, and death directly or indirectly associated with gender violence is staggering. Mental health problems associated with all three forms of gender violence include depression, post-traumatic stress disorder, anxiety, sleeping and eating disorders, suicidal tendencies, and substance abuse. Physical health problems associated with gender violence include illnesses of the neurological, cardiovascular, gastrointestinal, muscular, and reproductive systems.

Gender violence contributes to unwanted pregnancies, miscarriages, stillbirths and sexually transmitted diseases. Beyond the risks associated with physical or psychological attacks, perpetrators of gender violence undermine women's health and well-being by creating a climate of chronic fear and distress, impeding women's access to health care, and isolating women from social support. These factors contribute to a gradual, stress-related deterioration of women's health.

### 4.3 Men and Masculinities

The next three chapters address the roles of men and the development of more informed, more inclusive, and more differentiated ways of thinking and researching the connections between gender and violence. All three chapters comment explicitly on the web of ideological, political, and policy-oriented debates that infuse their respective area of inquiry.

*Hagemann-White* and *Lenz* aim for a more inclusive discussion of gender-based violence that develops ways to address violence against men without minimizing violence against women. Hagemann-White & Lenz argue that the discourse on violence against women, with few exceptions, continues to construct men as the perpetrators of intimate, sexualized violence, while struggling to address men as victims of such violence. If violence against men is noted at all, then primarily in terms of prevalence or incident statistics, while hardly any systematic discourse has developed to address the impact of such violence on men's pain and suffering. There has been some recognition of the victimization of young boys by sexual abuse, while the psychological, interpersonal, and social consequences of the sexual victimization of adolescent boys and adult men continues to be trivialized or ignored.

*Kavemann, Beckmann, Rabe,* and *Leopold* discuss intervention with male perpetrators of domestic violence, the development and current status of this field in Germany, and the associated policy debates and controversies. After laying out the historical and political trajectory of work with perpetrators in Germany, the authors describe different approaches to intervention with male perpetrators and discuss in detail the pros and cons of voluntary versus court-mandated participation in perpetrator programs.

The authors highlight the ambivalence of many feminist projects toward those who work with perpetrators. On the one hand, there is a sense that men should be involved in the work to end gender violence. On the other hand, there is distrust of men who are involved in this work, including concern that work with perpetrators will lull victims into a false sense of security, thus undermining feminist projects' efforts to increase women's safety, and concern that limited funds may be diverted away from women's projects towards programs for perpetrators.

*Harne's* research explores the parenting practices of domestically violent men in the UK. The author positions her study in the contested area of family law policy and associated competing discourses on the alleged benefits of fathering. One such notion presumes that children, in general, will benefit from contact with their natural father post-separation or divorce, even if he has been violent to their mother or to the children themselves. This notion is in contrast to research showing that contact with a domestically abusive father puts children at significant risk of multiple forms of abuse. A related notion presumes that father absence leads to problematic masculinities, whereas father presence prevents various social problems. Finally, there is the idea that fathering benefits violent men, based on the assumption that to engage in nurturing will develop a person's nurturing side.

What the men in Harne's study have to say about their own fathering practices shows that they are more involved in child care than one might think and that the quality of that care casts doubt on the presumed benefits of abusive men's care giving. For example, abusive fathers described how their emotional, verbal, and physical abuse obviously scared and distressed their children. Several fathers blamed their children for the father "losing his temper", much like domestically violent men blame their female partners for the abuse the men inflict on "their" women.

While Harne argues that abusive father's care giving practices reflect and reaffirm aspects of dominant and controlling masculinities, the ideas about the benefits of fathering also reflect and reaffirm notions of psychological functioning based in quasi-automatic cause and effect sequences that underestimate the role of deliberate choice and decision-making. For example, father contact per se is thought to benefit children, regardless of the quality of that contact. And nurturing per se is thought to develop abusive men's gentler side regardless of how abusive father's interpret "nurturing" and how they decide to attend, respond to, or ignore their children's needs.

As political and ideological arguments are wielded there is little explication of which practices make a good parent, or what kind of "enduring relationships" are worth enduring. Instead, political influence is used to bypass an informed discussion of good parenting by decreeing father presence good parenting regardless of actual practice, and declaring extended contact with an abusive father beneficial by virtue of the length of the relationship. Thus, master narratives that employ positively charged terms such as "nurturing" or "best interest of the child" may cloak abusive practices.

### 4.4 Prevalence

*Stephanie Condon, Maryse Jaspard* and the *Enveff research team* report on the recent violence against women survey in France. Based on a sample of almost 7,000 women, the survey was designed to assess the extent, type and contexts of women's victimization experiences in all spheres of life, and to shed light on personal histories of victimization, its consequences, and women's coping strategies.

The French survey illustrates the approach to measuring violence against women that according to Hagemann-White and Lenz underlies the Canadian and many of the European prevalence surveys: Violence against women is considered an expression and means of excercising or regaining power, rather than a conflict strategy, and there is no presumption that physical violence is necessarily worse than psychological violence.

The survey confirms that, across national contexts, only a small fraction of victimization experiences are reported to authorities

(Fisher et al. 2000). Moreover, Condon and colleagues illustrate that surveys are not just acts of measurement; they are, to some extent, interventions in the sense that they not only measure victimization experiences but actively create a social and cultural context that simultaneously acknowledges the occurrence of violence against women, gives permission to address it, and provides opportunities for disclosure.

## 4.5 Physiology of the brain

*Bernard Wallner* approaches gender and violence from the perspective of a biologist. He tackles the evolutionary argument that men's violence against women has served men's reproductive success because such violence is thought to get them access to females able to produce offspring. Evolutionary arguments often raise heckles in the camps of feminists and social constructivists because they seem to ignore the cultural context of gendered interaction and often smack irritatingly of scientific excuses for social injustice. However, Wallner takes a more differentiated approach. For one, he points out that much of the aggression that supposedly serves males' reproductive self-interest is directed against other males rather than females, although he does report evidence of sexualized aggression against females in several non-human primate societies. More importantly, Wallner argues that aggression and propensity for violence in human males may be an evolutionary by-product of a food-shortage alarm system in men's brains. Research on brain physiology suggests that low cholesterol intake reduces serotonin levels in the brain, which in turn seem to promote a range of behavioral manifestations including anxiety, depression, disinhibition, and violence. In terms of food supply, low levels of serotonin would signal low cholesterol levels in the brain, and thus possible food shortage in the environment, and this mechanism would be more pronounced in males than in females because the serotonin-sensitive areas in the brain are bigger in males than in females. Incidentally, low levels of serotonin would also raise males' propensity for violence. In this analysis, males' reproductive success would be enhanced not because but despite of their violence:

Rather than violent males gaining access to females, females would be attracted to resource-sensitive males, despite their occasional violent outbursts.

## 5. Challenges and opportunities for transnational research

### 5.1 The European context

As this book is going to the publisher, the European Union is about to almost double in size and will include even more countries of different historical, cultural, linguistic, political, and economic backgrounds.[1] This diversity of European countries and regions coexists in unprecedented ways with increasing trans-European unity[2]. Both aspects have programmatic and structural implications for research on gender and violence in Europe.

Programmatically, diversity points toward locally specific research needs, whereas unification suggests shared program priorities. Historically, it has been local women's or social change movements that have put gender and violence on the map by naming and challenging abusive practices, and although they share common concerns, their research priorities are likely to differ according to the configuration of local needs, opportunities, and constraints, from locally effective strategies of social change to the strength of local resistance to such change, and the degree to which dominant academic traditions embrace or reject social activism.

Structurally, diversity points toward regional or country-specific forms of organization such as lobbying with state institutions and governments (for example, "state feminism" in the Nordic countries versus "separatism" in the UK; see also Weldon 2002), whereas unification suggests working within transnational networks and supranational organizations such as the Council of Europe and the European Commission.

In response to these programmatic and structural developments several transnational European networks emerged during the 1990 that address gender and violence from pragmatic, political, and scientific perspectives, respectively. They include Women Against Violence Europe (WAVE), the umbrella organization of abused women's projects in Europe, the European Women's Lobby, and the European Research Network on Conflict, Gender, and Violence. Women's Studies International Europe (WISE) concerns women's studies more generally, but its European Journal of Women's Studies occasionally publishes articles that focus on gender and violence.

At the same time, policy recommendations and awareness campaigns at the European level became more prominent. The Council of Europe, through the work of the group of specialists, and a series of international conferences and seminars, produced extensive policy recommendations to its 40plus member states on how to intervene with and end gendered violence. In 1997, the European Parliament passed a resolution calling for a European Union wide zero-tolerance campaign. In response, the European Commission in 1999 and 2000 conducted a campaign to raise awareness of violence against women that included targeted messages via posters, brochures, radio, television, and the internet, and related conferences took place in Germany (1999), Finland (1999), and Portugal (2000). Independent from the work against gendered violence, the European Union recently has been pushing for more integrated European research through the development of a European Research Area, a sort of common market for knowledge, research, and innovation. Although the programmatic priorities of the current European research frameworks have not directly addressed gender and violence, the policy instruments created to advance European research may benefit research on gender and violence in the future.

### 5.2 Challenges and opportunities

The opening of European borders at the end of the Cold War and the spread of market economies across Europe have contributed to the growth of research infrastructure, albeit in disrupted and uneven ways. Formal European integration, economic liberalization, the internet, and the use of English as primary lingua franca have facilitated the development of research networks, summer schools, and international conferences but the available infrastructure still differs greatly across Europe. This is particularly clear for research fields that are not well supported by business and related interests, and who continue to struggle with unreliable power supplies, unaffordable paper and internet access costs, administrative hurdles, and in general the high costs of conducting research and attending conferences.

The bulk of internationally visible research on gender and violence comes from regions that have enjoyed relative wealth and political stability, and have been operating in the English-speaking sphere for some time. For example, virtually all prevalence surveys in Europe have been conducted in prosperous Western European countries that are not currently undergoing drastic political change. In contrast, research produced in other languages or regions still remains largely invisible.

It is difficult to assess the resulting loss of knowledge to the international community. There is increasing recognition that effective work to end gendered violence needs to be grounded locally (Heise 1996), and that research on gender and violence becomes more sophisticated when it is informed by local practice (see Hagemann-White and Lenz, this volume). From this perspective, keeping local experience and research invisible is detrimental to the international community, while making such work more visible would benefit both research and social change.

### 5.3 Culture-presence

Slippery as the delineation of "culture" may be (Goodwin 1999), the European context provides opportunities to develop more "culture-present" analyses of gender and violence, to adapt a term from Hanmer and Hearn (1999). As research collaboration across Europe increases, the need for, and value of culture-present analyses will increase as well. For example, it is difficult to imagine a compelling analysis of the applicability of Austrian police law in Germany that ignored the historical and political differences between the two countries. Cultural factors are likely to figure increasingly in analyses of the transferability of recommended practices across nation states, of the interplay between legal reform, criminal justice and social systems in different countries, and of strategies to increase the health and well-being for resident and migrant populations across Europe.

A comparative approach to prevalence surveys may further our understanding of the cultural context of gendered violence by highlighting which techniques work and which do not work in any given cultural context, and which questions make more or less "cultural" sense. Undoubtedly, there are now more and better surveys on violence against women than there were 20 years ago (Walby 1999). Over the last 15 years at least eleven representative surveys have been conducted in Europe on a regionally or nationally representative level (Hagemann-White 2001). Nationally representative surveys in France (Condon et al., this volume) and Sweden (Lundgren et al 2002) have been completed recently; data of the most recent British crime survey are being analyzed (Walby 2002); and data collection in Germany has just begun (Müller and Schröttle 2002). Yet, trans-national comparisons still are of limited or no use because the existing studies – although each well-crafted in its national context – differ from each other enough in approach and method to make valid comparison ambiguous, if not meaningless (Hagemann-White 2001).

Comparative research in the European context would add significantly to the evidence from comparative anthropological studies on gender and violence. The latter suggest that a low prevalence of violence against women in a society is associated with specific gendered practices such as women's control over wealth and property and their right to divorce (Levinson 1989), the proximity of woman-supportive third parties (Baumgartner 1993), public living arrangements and legal recourse (Counts et al. 1999), and low levels of societal violence in general (Levinson 1989, Ross 1993). While these studies are very informative, their implications for other societies (and eras) are limited. For example, considering the relatively high social status of women in Sweden and the low level of societal violence in that country, shouldn't there be less violence against women than the recent prevalence survey uncovered (Lundgren et al. 2002)?

Finally, culture-present analyses can address how gendered violence may constitute "doing culture" and shed more light on the workings of rigid gender identities and misogynist master narratives in different societies such as definitions of masculinity that deny and silence men's vulnerability, and misogynist master narratives that justify men's ownership over women or children in terms of family values and the best interest of the child.

In any case, transnational European research needs to go beyond rating countries as "better" or "worse" with regard to gendered violence. Even if there were more reliable comparative data, how useful is it to know that there is more violence in country X than in country Y, if we do not understand the reasons for those differences? For those who want to know how to end and prevent gendered violence, the promise of comparative research lies in showing convincingly which factors facilitate or impede gendered violence, and in increasing and expanding what we already know about those factors based on the work of generations of practitioners and activists.

## 6. References


Baumgartner MP (1993). Violent networks: the origins and management of domestic conflict. In Felson RB, Tedeschi JT (eds), Aggression and Violence. American Psychological Association, Washington, DC, pp. 209-231.

Council of Europe (1992). European Charter for Regional or Minority Languages. November 5, Strasbourg, 1992.

Counts DA, Brown JK, Campbell JC. To Have and to Hit: Cultural Perspectives on Wife Beating (2nd. Ed). University of Illinois Press, Urbana/Chicago, IL, 1999.

Fisher BS, Cullen FT, Turner MG. The Sexual Victimization of College Women. National Institute of Justice, 2000.

Goodwin R. Personal Relationships Across Cultures. Routledge, London/New York, 1999.

Hagemann-White C (2001). European research on the prevalence of violence against women. Violence Against Women 7, 732-759.

Hagemann-White C (2002). Violence against women in the European context: Histories, prevalences, theories. In Griffin G, Braidotti R (eds), Thinking differently: A reader in European women's studies. Zed Books, London, pp. 239-251.

Hanmer J, Hearn J (1999). Gendering research on men's violence to known women. In Men and Violence Against Women-Seminar Proceedings. Council of Europe, Strasbourg, pp. 32-40.

Heise L (1996). Violence against women: global organizing for change. In Edleson JL, Eisikovits ZC (eds), Future Interventions with Battered Women and Their Families. Sage, Thousand Oaks, CA, pp. 7-33.

Levinson D. Family Violence in Cross-Cultural Perspective. Sage, Newbury Park, CA, 1989.

Lundgren E, Heimer G, Westerstrand J, Kalliokoski AM. Captured queen. Men's Violence Against Women in 'Equal' Sweden – A Prevalence Study. Fritzes Offentliga Publikationer, Stockholm, 2002.

Müller U, Schröttle M. Violence against women: Research questions and design of the first German survey (starting in 2002). Paper presented at the 7th Interdisciplinary Conference of the European Research Network on Conflict, Gender, and Violence, Valencia, Spain, September 2002.

Ross MH. The Culture of Conflict: Interpretations and Interests in Comparative Perspective. Yale University Press, New Haven, CT, 1993.

Walby S (1999). Comparing methodologies used to study violence against women. In Men and Violence Against Women-Seminar Proceedings. Council of Europe, Strasbourg, pp. 11-22.

Walby S (2002). Comparing innovations in the methodology of large-scale surveys on violence against women: The British crime survey interpersonal violence module in international context. Paper presented at the 7th Interdisciplinary Conference of the European Research Network on Conflict, Gender, and Violence, Valencia, Spain, September 2002.

Weldon SL. Protest, Policy, and the Problem of Violence Against Women. University of Pittsburgh Press, Pittsburgh, PA, 2002.

West C, Zimmerman DH (1987). Doing gender. Gend Soc, 1, 125-151.

## Notes

1 Current member states: Austria, Belgium, Denmark, Finland, France, Germany, Greece, Ireland, Italy, Luxembourg, Netherlands, Portugal, Spain, Sweden, United Kingdom. States currently involved in accession negotiations: Cyprus, Czech Republic, Estonia, Hungary, Latvia, Lithuania, Malta, Poland, Slovakia, Slovenia. Probably next to join: Bulgaria, Romania, and possibly Turkey.

2 As Europe is moving towards integration, pressure is building to recognize and strengthen cultural and linguistic diversity (e.g., protection of European minority languages, Council of Europe, 1992) and to grant regional political autonomy (e.g., autonomous provincial governments within nation states such as Scotland and Catalonia).

*Birgitt Haller, Institute of Conflict Research, Vienna, Austria*

# The Austrian Protection from Violence Act

When the Federal Government launched its action programme against domestic violence in 1994, it committed itself to "state intervention with a view towards protecting the physical safety in particular of socially and physically weaker family members". First initiatives that aimed at having the protection from violence in the private sphere more firmly anchored in law, had been taken in the late 1980s, mainly by feminist circles and in particular by the women's shelters movement, and representatives of the Autonomous Austrian Women's Shelters Network were consistently involved in the drafting of the Protection from Violence Act. The Act eventually entered into force on May 1, 1997. It should be mentioned that it is formulated in gender-neutral language, which, contrary to reality, suggests equal risks for men and women.

The first part of this chapter gives information about the Protection from Violence Act and the second part presents some of the findings resulting from the evaluation of this law by our institute in 1999.[1]

## 1. The statutory provisions

The total sum of provisions designated by the Protection from Violence Act are actually laid down in three different acts: the Security Police Act (Sicherheitspolizeigesetz – SPG), the Enforcement Act (Exekutionsordnung – EO) and the Civil Code (Allgemeines Bürgerliches Gesetzbuch – ABGB). New powers have been granted to the *police*, who are now entitled to impose *eviction* and *barring orders* on perpetrators. Police officers have to make a risk forecast on which they base their decision. Perpetrators can be forcibly removed from the residence. If there is a risk of violence, both measures have to be imposed based on the authority's initiative. They need not be imposed at the same time, however. As a rule, the police are called to the home of both the perpetrator and the person at risk. In such cases, the first step to be taken is to remove the person posing the threat from the home; the barring order will be imposed subsequently. There are cases in which eviction cannot be imposed because the aggressor has left the home before the police arrive.

The barring order imposed by the police is effective for a period of ten days; this period can be extended to 20 days, if the person at risk applies to the Family Court for an interim injunction. Compliance with barring orders has to be checked by the police at least once during the first three days. Violation of a barring order is an offence under administrative law and carries a fine.[2] During the intervention, the police officers are required by law to inform the person subjected to violence about suitable victim protection facilities, in other words about the Intervention Centres.

The second statutory basis for action is the Enforcement Act. Upon the application of the person at risk, the *Family Court* can issue an *interim injunction*, barring the perpetrator from the home of the person at risk (even if it is their common home) and from the immediate vicinity, as well as forbidding the perpetrator to establish any form of contact. This is conditional in a situation in which the person posing the threat makes living with him intolerable by acts of physical violence, threats of violence, or behaviour which seriously impairs the mental health of a close relative. The "next-of-kin" relationship is relatively narrowly defined, presupposing it to be life in a common household at least up until three months before the application has been submitted. This is a problematic rule, since violence frequently occurs or escalates in connection with a separation. An interim injunction can be issued for a maximum of three months. This period can only be extended if divorce proceedings, etc. have been instituted.

Moreover, an amendment of the Civil Code provides that the *youth welfare authorities*, as the guardians of *minors*, can apply for an interim injunction if a child is at risk. This applies both to cases of direct or indirect violence against children, provided that the mother, as the children's statutory representative, has not filed an application herself.

Another vital part of the reform is the setting up of victim protection facilities – the above-mentioned, so-called *Intervention Centres* (Interventionsstellen). These are private facilities, or more precisely incorporated associations, which act at the request of and are funded in equal shares by the Federal Ministries of the Interior and of Social Security and Generations. Their main tasks are, on the one hand, to take care of people subject to violence and on the other hand, to network and to cooperate with all of the authorities and private facilities involved in violence protection. The Intervention Centres have to be notified without delay of any eviction/barring orders imposed by the police. The Centres contact those subject to violence via telephone and invite them to talks in which the threat posed by the perpetrator is assessed and crisis plans as well as safety programmes are put in place together with the victims of violence. Intervention Centres have been operative in every Austrian province since 1999.

## 2. About the evaluation

The evaluation consists of three steps. First, a quantitative evaluation of about 1,000 police files concerning domestic violence was done. These files dated from May 1997 to October 1998 and originated from eight different Austrian regions. Besides the use of the new law in case of domestic violence, there still exists a form of intervention called "dispute settlement". This is a very informal response, sometimes it only means that the police have not judged the situation as being relevant. These 1,000 files documented both forms of reactions: 43 percent of the cases were classified according to the new police law, 52 percent were dispute settlements, and about five percent were charges (mostly involving bodily injury). By analysing these files, data about victims and perpetrators, as well as information on the behaviour of the police and of the courts could be obtained. Second, both persons at risk and perpetrators were interviewed in all of the regions examined. Third, we interviewed representatives of the institutions involved in the intervention process in case of domestic violence: police, Intervention Centres, civil and criminal courts as well as youth welfare authorities.

## 3. Personal data of persons at risk and perpetrators

Nine out of ten victims were female. Nearly two thirds of them were between 25 and 44 years old and were employed. Ninety-seven percent of the perpetrators were male. It was interesting to find that in the initial phase of the law, only one of the examined regions imposed barring orders against women; since then it has also been imposed against women in other regions, but female perpetrators are a small minority. The perpetrators were of the same age as the persons at risk, every second one was a blue collar worker. More than one quarter were unemployed. This concentration of the police intervention on the lower social classes does not indicate that domestic violence is only a problem of the lower classes, but that there are differences in the reporting behaviour to the police, and probably also in the reaction of the police, regarding the social classes. This also has to be considered with regard to the over-representation of migrants. In Austria, about nine percent of the population are migrants, but 20 percent of the victims and 25 percent of the offenders were migrants, and there is no evidence that male migrants are more aggressive than male Austrians.

Only a small number of children witnessed the incident that had caused the police intervention. According to the files, only four out of ten children were witnesses. In one out of ten acts of aggression, children were directly involved as victims.

Three quarters of the victims stated that they had already experienced acts of aggression before the actual intervention, but usually they had not reported them to the police. Nearly ten percent of the offenders recurrently used violence against their partners within the two years examined.

One quarter of the interventions examined led to an interim injunction, sometimes combined with a divorce suit. Among the group of victims that turned to the Family Court for help, migrants were under-represented.

## 4. Interviews with persons at risk and offenders

Twenty-five victims and seven offenders were interviewed, some of the victims twice to get information on the sustainability of the police measures. Among the *persons at risk* was a large range of reactions. Some women opposed barring orders because they wanted to stay together with their partners and thought this measure to be too strict and therefore inadequate. Others judged the barring order as very important for themselves in order to understand that they should separate from their partners. Some women told us that the perpetrators had been shocked by their own behaviour and that their relationship had changed for the better. In some cases, when the offenders did not have to face a barring order but only a dispute settlement, their victims were critical: In their view police intervention had made the perpetrators' actions escalate, when they realized that nobody would stop their aggressive behaviour.

Some of the *offenders* we talked to did not believe that they had done something wrong. Others felt sorry about their behaviour and knew that they had to change if they did not want to lose their partners.

Most of the victims were very satisfied with the behaviour of the police. We found it remarkable that mixed teams were judged as impartial, whereas male officers seemed to be partial to women. Only in the rural areas, women complained about the rural police, the "Gendarmerie", who was often prejudiced against women. They did not believe what the women had told them or blamed the women for their partners' aggressions.

The women appreciated the Intervention Centres' work, as well as the support they received in the women's shelters.[3] It was important for them to get information on the legal situation, as well as emotional support. Youth welfare authorities were not judged as positively as Intervention Centres – some interviewees said that they had not been correctly informed there and others did not feel supported strongly.

## 5. Application of the protection from violence act by the police

The frequency with which the Protection from Violence Act has been applied, since its entry into force, by Austria's two security bodies has varied considerably, the urban police being much more "active" than the rural police. In 2001, nearly six out of ten eviction and barring orders had been imposed by the police – but the urban police are responsible for approximately one third of the Austrian population,

Table 1: *Statistics: Eviction/Barring Orders*

| | **1997** | **1998** | **1999** | **2000** | **2001** |
|---|---|---|---|---|---|
| Federal Police Directorates | 646 (47.3 %) | 1,320 (49.4 %) | 1,654 (53.8 %) | 1,807 (53.9 %) | 1,880 (57.3%) |
| Federal rural police ("Gendarmerie") | 719 (52.7 %) | 1,353 (50.6 %) | 1,422 (46.2 %) | 1,547 (46.1 %) | 1,403 (42.7%) |
| Total | **1,365** (100 %) | **2,673** (100 %) | **3,076** (100 %) | **3,354** (100 %) | **3,283** (100 %) |

Source: Internal statistics of the Ministry of the Interior

Table 2: *Eviction/Barring Orders – annual rates of development*

| | **1998** | **1999** | **2000** | **2001** |
|---|---|---|---|---|
| Federal Police Directorates | + 36.2 % | + 25.3 % | + 9.3 % | + 4.0 % |
| Federal rural police ("Gendarmerie") | + 25.5 % | + 5.1 % | + 8.8 % | – 9.3 % |
| Total | **+ 30.5 %** | **+ 15.1 %** | **+ 9.0 %** | **– 2.2 %** |

Source: In-house calculation based on 1997 figures extrapolated to 12 months

whereas the so-called gendarmerie is in charge of two thirds. From May 1997 until the end of the year 2000, the total number of eviction and barring orders imposed had increased continuously; in 2001 it slightly dropped for the first time.

Before the Protection from Violence Act came into force, the instrument most frequently used to respond to domestic disputes was that of *"dispute settlement"*: The officers spoke with the "conflicting parties", seeking to "calm" them and occasionally suggested the women seek refuge, for instance in a women's shelter. The idea behind the wish to mediate between the victim and perpetrator was that violence in a personal relationship was a "private matter" and that the state and its authorities had no right to interfere. The instrument of dispute settlement is still available to police officers along with eviction and barring orders. Its application is appropriate whenever the cause of police intervention is a loud altercation and if it is not used to trivialise or to ignore violence. In 2001, this instrument was used twice as often throughout Austria as the instruments of eviction and barring. It is striking that dispute settlement is still the method of choice used by the rural police in the event of family conflicts and the reason given is still a reluctance to "interfere" in family affairs.

Such "interference" signals a *change of paradigm* – domestic violence being no longer perceived as a private "conflict", but being associated with the *right to safety.* This is the reason why the decision on imposing eviction and barring orders lies exclusively with the police. It is only in the second stage that those subject to violence are free to decide on applying for an interim injunction and thus for more comprehensive protection. This "division of powers" has its origin in the fact that it is often very hard and even dangerous for victims of violence to leave the relationship on their own initiative. In order to dare to take this step, women have to be supported and empowered and their self-confidence has to be boosted.

Initially, police officers found it hard to meet their obligation of imposing barring orders, especially when the women concerned were reluctant to support them in the performance of their duties. Officers were upset and disappointed when they were asked not to impose a barring order and occasionally complied with such requests. Another obligation police officers were reluctant to meet was that of cooperating with the Intervention Centres. These centres have the legal status of private incorporated associations, which is to say that they are bodies the police were not used to cooperating with. As a result, the obligation to pass on information to the Intervention Centres was occasionally ignored.

## 6. Other government agents: youth welfare authorities and civil and criminal courts

At least during the start-up phase, the attitude of the *youth welfare offices* vis-à-vis the Violence Protection Act was ambivalent. On one hand, the new provisions were welcomed because barring orders prevented children living in the common home of victim and offender of imminent danger and extended the available time frame, thus reducing the pressure on the youth welfare offices. On the other hand, they hardly used the instrument of interim injunction in the interests of children, who were directly or indirectly affected by violence. Such "interference" was occasionally refused with the argument that youth welfare offices are non-partisan institutions, never siding with either parent, but on the contrary, seeking to strengthen and to preserve the family. This attitude has manifestly changed in the meantime: Domestic violence is increasingly also perceived as violence against children, which calls for intervention. Nevertheless, applications for interim injunctions are rarely filed, because this step is conditional on the mother's consent and on her ability to protect the children.

If there are children living with the family, both the youth welfare authorities and the Intervention Centres are immediately notified by the police, once eviction/barring orders have been imposed. However, police reports focus primarily on the situation and condition of the endangered woman, largely ignoring that of the children. So, there is a "blind spot" in regard to the situation of children. Just like other institutions, the youth welfare offices are confronted with the problem of being at times wrongly or incompletely informed by mothers who (justly) dread the reproach of having failed to ensure appropriate protection for the children, as well as the possible consequence of the children being removed from their care.

The initial trend displayed by the *Family Courts* throughout Austria towards issuing interim injunctions was soon curbed by the rulings of the Supreme Court, especially on the issue of unacceptability: Not every family dispute involving violence justified either an eviction order or an interim injunction. However, the fear that these rulings would result in fewer interim injunctions being issued turned out to be unfounded, and protection from violence is being very effectively provided by the civil judiciary.

Concerning the *Criminal Courts*, major shortcomings are still evident. A high percentage of proceedings instituted because of domestic violence are quickly dismissed, not only because victims of violence refuse to give evidence and to authorise criminal prosecution[4], but also because the assault is not deemed punishable. Violence in the private sphere is still perceived as a privileged offence. Another aspect still neglected under formal penal law is the upgrading of victims' rights.

The Protection from Violence Act unequivocally involves *youth welfare authorities and the civil judiciary* into ensuring the safety of persons subject to violence and assigns a clearly defined role to them. The result is (relatively) *effective action* in both of these areas. What makes it particularly difficult for youth welfare offices when dealing with domestic violence, is the fact that they recognise mothers subjected to violence as victims, while, at the same time, having to hold them responsible for their children. This may lead to situations in which women are additionally challenged instead of supported. A case in point would be the threat of the children being removed, should the woman re-admit the evicted partner to the home, which exposes the victim of violence to massive pressure from both the perpetrator and the youth welfare office.

No reference is made in the Protection from Violence Act to the role of the *criminal judiciary*; so there are criminal lawyers who believe that the new legal provisions are none of their concern and they fail to understand that the administration of criminal justice has an important role to play in interpreting and specifying these new legal provisions. A multi-institutional approach involving all the state actors and excluding any controversial messages, is vital in enforcing comprehensive protection from violence.

## 7. Intervention Centres and anti-violence programmes

Apart from having the perpetrator "disciplined" by police and judiciary, one of the Act's fundamental tenets is the setting up of victim protection facilities. Beyond providing safe havens – such as the women's shelters – in situations of acute crisis, such facilities should promote the *empowerment* of the women subject to violence by providing emotional support and legal resources. "Empowerment" is a medium-term strategy designed to establish (restore) women's full capacity to act, thus enabling them to opt for leaving a violent relationship or to adjust the power imbalance within the relationship.

When the Protection from Violence Act entered into force, an important element of victim protection – namely working with the perpetrators – was still lacking. Since 1999, the Federal Ministry of the Interior has promoted the "Training Programme for Men Designed to Put an End to Physical Violence in Relationships between Couples", launched as a model project by the Men's Counselling Centre of Vienna. The programme for perpetrators consists of three parts: the training of perpetrators, which is provided by the Counselling Centre, the support to female partners by the Intervention Centres and the joint coordination of violence-preventing measures as well as the networking with other police, youth welfare and judiciary bodies.

## 8. Summary

A main finding of the evaluation is that the application of the Protection from Violence Act strongly depends on the persons involved in the intervention process, on their commitment and on their attitudes. This can be seen not only in the different reactions of urban police and rural "gendarmerie", but also among the representatives of the courts and of youth welfare authorities. The higher the value of marriage and family is rated, the higher the probability that domestic violence is ignored. As a consequence, trainings for all of the persons who are involved in the prevention of domestic violence were suggested: they have to be more aware of the fact that domestic violence is wrong, and their understanding of women in violent relationships has to be improved.

Responding to violence can never be reduced to a problem that has to be solved by legal experts and law enforcement; there is always a political dimension to it. If measures against violence in the "private" sphere are to have a lasting effect, they must not be restricted to the scope of action of the police and the judiciary, but must ultimately be targeted at changing the gender relationship, taking a broadly-based societal approach.

## Notes

1 The main findings of the evaluation are published in: *Dearing*, Albin/Birgitt *Haller* (Eds.) (2000). Das österreichische Gewaltschutzgesetz, Vienna.
2 Only a few of the perpetrators return within this period, in these cases the police will renew the barring order. If a woman later admits the perpetrator back into the home during the barring order, she can be fined, too.
3 In the regions where no Intervention Centres existed at that time, the women looked for help in women's shelters or in counselling centres.
4 An authorization is essential to the prosecution of threats made against family members and nobody is obliged to testify against family members.

*Elfriede Fröschl, Fachhochschule for Social Work, Vienna, Austria*

# Women's Politics in Austria

In this chapter, I would like to give a brief outline of the present political situation in Austria, with particular relevance for women's affairs. Due to restrictions of space most aspects can only be discussed very briefly.

According to the microcensus 2001 Austria had a population of 8.07 Million (Bundesministerium für soziale Sicherheit und Generationen 2002). In 2000 49% of all Austrians were male and 51% female. For those born in 1999 the average life expectancy for men is 75.1 and 81.2 years for women.

Foreign citizens make up approximately 9% of the resident population in Austria. Amongst them only 46% are women (Bundesministerium für Soziale Sicherheit und Generationen 2002). Two thirds of foreign citizens living in Austria are nationals of ex-Yugoslavia and of Turkey. A significant proportion of them are family members of migrant workers.

The elections in the year 1999 for the National Parliament brought the following results (Figure 1).

Figure 1: *Election results of the Austrian parliament 1999*

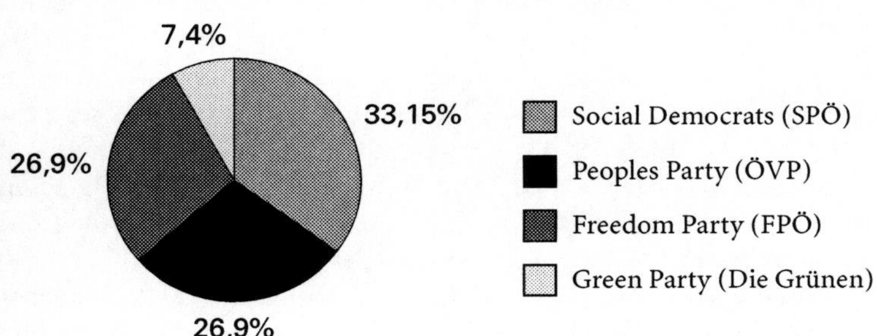

These elections revealed a gender–gap concerning the results of the right-wing "Freedom Party", which is illustrated in Table 1.

Table 1: *Results of the elections in 1999* (Steininger 2002)

| **Elections 1999** | Social Democrats | Peoples Party | Freedom Party | Green Party |
|---|---|---|---|---|
| men | 31 % | 26 % | 32 % | 5 % |
| women | 35 % | 27 % | 21 % | 9 % |
| **Gender gap** | **+4 %** | **+1 %** | **–11 %** | **+4 %** |

After long negotiations, the Austrian Peoples Party (a Christian-conservative party with strong connections to the economic sector) broke a long term taboo and formed a coalition with the so called Freedom Party, which has also strong connections to the economic sector, but at the same time claims to represent the interests of *the man on the street* and has strong racist and xenophobic tendencies. For example, following September 11$^{th}$, one of the remarks from the former head of the Freedom Party, Jörg Haider, was that Austria should only give asylum to refugees coming from Europe! The right wing conservative coalition government began its term in 2000. A legislative period lasts for 4 years. Thus, new elections will probably not be held before 2004.

Beetween the date of the conference and the publication of this article, early General Elections took place on 24 November 2002. The outcome of these elections were the somehow surprising result that the Peoples Party (ÖVP) gained 42.27 percent, the Social Democrats (SPÖ) 36.90 percent, the Freedom Party (FPÖ) 10.16 percent and the Green Party (Die Grünen) 8.96 percent (Figure 2). That would mean 79 seats in parliament for the ÖVP (27 more than in 1999), 69 seats for the SPÖ (plus four), 18 seats for the FPÖ (34 less than in 1999) and 17 for the Greens (plus two). After long negotiations the ÖVP again formed a coalition with their former partner FPÖ. The neoliberal and conservative restructuring process in Austria continues.

Initially, a fairly strong resistance was formed against this government in Austria: there were large demonstrations and the so-called

Figure 2: *Election results of the Austrian parliament 2002*

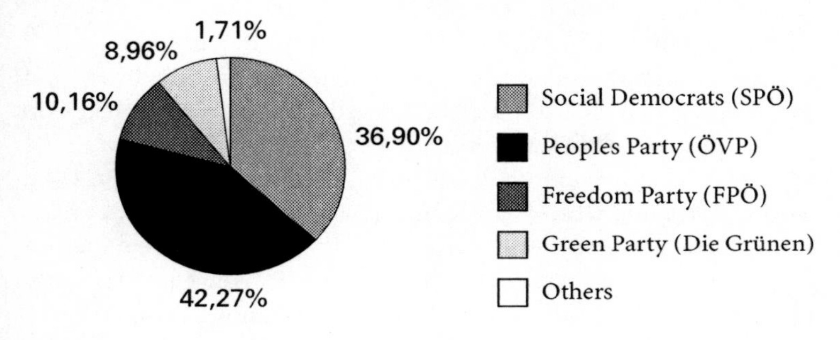


*Thursday demonstrations* have continued up to present. Up to the last elections, the Social Democrats had ruled the country for 30 years and for the last 14 years of their reign they had formed a coalition with the conservative People's Party.

## 1. Contradictions concerning women

Immediately on taking office, the new government in Austria abolished the post of Federal Minister of Women's Affairs. This post had existed for twenty years, first as a Secretary of State, later with the status of Federal Minister. During the negotiations on the formation of the government, the abolition of the post of Women's Affairs Minister was agreed upon without question. Its former agenda was transferred to the Ministry of Social Security and Generations, which is headed by a male minister. One of his first actions was to establish a department for male affairs within his ministry. Thus, a key factor in Austria's active policy on women's affairs and the advancement of gender equality was eliminated with a single stroke. It was an ominous sign for the future of women's policy in Austria. In this connection, it seems doubtful whether the government will be able to meet its obligations as an EU member state concerning gender equality and the advancement of women, when it embarked on the destruction of those structures designed to implement such policies.

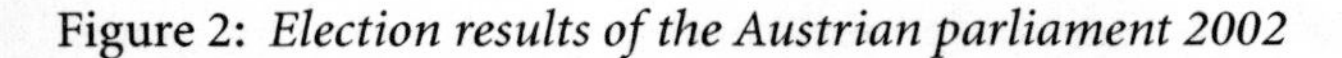

Austria – and hence the government – is committed to fulfil given targets in the advancement of women. Equal opportunity for women in the employment market and guidelines on the advancement of women are featured among the binding agreements within the EU and are enshrined in the Maastricht Accord. Austria also signed and ratified the United Nations *Convention on the Elimination of All Forms of Discrimination against Women* in 1982 (Federal Gazette 443/1982). This convention, which has been elevated to the status of constitutional law, requires that signatory states pursue an active policy on the advancement of women. Furthermore, in 1993 the United Nations General Assembly passed a declaration on the elimination of all forms of violence against women. By the terms of the comprehensive *Platform for Action* drawn up at the *Beijing World Conference on Women* in 1995, states agreed to implement further measures on behalf of the advancement of women.

These documents are just the main international conventions and accords relating to the advancement of women. Compliance is also binding on the Austrian government, which pledged at its swearing-in to respect and abide by the terms of international agreements. But, in practical terms there is nothing concrete going on to bring forth the advancement of women.

The present government propagates a very conservative image of women. However, never before have women held so many key positions in a government despite its conservative and traditional image of women. In former governments under the Social Democrats women traditionally held positions like Minister for Education, for Social and Women Affairs. In the present government, women lead the prestigious Ministries for Foreign Affairs, the Ministry for Research and Education, and even the Vice Chancellor is a woman.

Women in these high ranking and prestigious positions create the illusion of an already established equality between men and women in society, suggesting that under the present government (strong) women have the opportunity to work in high ranking positions and to overcome the so called "victim feminism" of former decades.

Accordingly, the government programme is based on the assumption of an already established equality between men and women.

In practical and actual terms this means that the retirement age was increased by 18 months. In the long run the retirement age for women will be the same as for men. However, there are no programmes to increase women's low employment rate in general. The government argues that we are living in the decade of, from the point of view of government representatives, freedom of choice where every woman can choose her own way of life. Therefore, it seems logical that, for example, not enough money is spent on special programmes for the reintegration of women into the labour market.

According to a recent survey (Bundesministerium für Soziale Sicherheit und Generationen 2002), the mean gross income per year of employed people in 1999 was 16,234 Euro for women and 25,533 Euro for men. In spite of the fact that women earn approximately 30% less than men, the government sees no compelling reason to change anything substantially about this large gender gap. On the contrary, the government promotes an extreme cost cutting policy, which according to Rosenberger (2002) especially affects women in at least three ways, (1) through cutbacks of cash-transfers in connection with social welfare, (2) through cost cutting in the field of social services such as child care facilities and long-term care for older people, and (3) by halting any new employment in the public sector.

## 2. Supporting traditional family models instead of promoting gender equality

Given the extent of insurance-based social expenditures, social care services are relatively under-represented in Austria, especially compared with other European countries such as Sweden or Finland. There are several reasons for these facts. First of all, there is no federal legislation or federal financing for social services for frail elderly people, children and people with disabilities, nor for persons with other social problems.

These services are financed by the nine provinces, which are responsible for almost all care such as residential homes for the elderly and nursing homes, child care, and community care services.

Most of these provinces have a conservative government. For this reason, child care facilities are not at all sufficient and many of them close at noon.

There is still an important emphasis on *family care* (which means unpaid women's work) which is to a great extent expected to compensate for the lack of care facilities for children and people with disabilities, especially in rural areas. For example, the quota of child care facilities for children under 3 years of age is 4% in Austria, one of the lowest in Europe – only Greece has an even lower rate: 3%. (Kommission der Europäischen Gemeinschaften 2001).

Instead of further promoting the improvement of child care facilities, the present government decided to introduce the so called *Child Care Payment*, which parents are entitled to receive for 30 months[1]. It replaces the former payment that employed women received during maternity leave. The new payment was introduced with the official goal to make the decision between child care and career easier for women, because it was promoted to be used also for paying for child care. But in reality it does not change anything in the traditional roles of women and men. For example, while receiving this payment one can only earn up to14,600 Euro. If this sum is exceeded, one loses the payment. Therefore, for well educated and well paid women it is not possible to stay at their job while using the payment for child care as it was promised by the government.

Compared to other countries, Austria has a system of high payments to support traditional families – families according to the model of the traditional breadwinner/head of household-image. The support includes direct payment, tax privileges, and free co-insurance. Despite this massive financial support, Austria has one of the lowest birth rates: 1.3 children per woman.

Furthermore, the right wing conservative government introduced another new law after a divorce, the so called *Joint Custody*, which is another example of how the government reduces the rights of women. Before the introduction of this law, in the case of divorce the person who took care of a child could also decide on all aspects of the child's life. In over 90% of these cases it is the mother who takes care of the child. Now the father has the right to be included in almost every decision concerning the child, without the obligation to take over more responsibilities. This law would be adequate in an utopian society where the average father is not just involved in a few leisure activities of his child. Nowadays, this law is just an opportunity to raise control over women, who take care of their children after divorce.

## 3. Women's projects in Austria

In Austria, the new feminist movement started with the fight for the legalisation of abortion and partially succeeded in 1977 when abortion up to the 12$^{th}$ week of pregnancy was not prosecuted anymore. Afterwards, a period up to the mid 80s followed when numerous women's projects were founded. In 1978, the first women's shelter was opened, followed by counselling centres for women, some of them specialising in fields such as the labour market and others dealing with special problems of girls and young women.

Many of these projects and institutions have celebrated their 20$^{th}$ or 15$^{th}$ anniversary of their foundation over the last years. Most of them still exist because under the social democrat government they could rely on financial support. Unfortunately, these subsidies never became a legal entitlement and consequently it is now easy for the present government to reduce them or stop them all together.

The abolition of the post of Minister of Women's Affairs is having a direct impact on numerous women's facilities and organisations which have hitherto received financing from the Minister and whose financial survival is now in doubt.

## 4. Reduction of women's projects funding

There were so many protests from women's organisations against the government, that the Minister of Social Affairs is very careful about reducing the support for women's projects.

Women's projects that have been especially critical towards the government receive less money, especially feminist magazines. Most

Table 2: *Reduction of funding resources*

| 1999 | 5,959,302 Euro p.a. |
|---|---|
| 2002 | 2,979,651 Euro p.a. |

http://www.ceiberweiber.at/wahl/wien/10dec_01.htm, 06.08.02

of the women's projects concerned with violence against women, like women's shelters and intervention centres, have not experienced any cuts in subsidies and financial support until now.

However, for women working in these organisations, it is extremely frustrating because they have lost almost all of their contacts within the ministries because many restructuring measures are carried out there and civil servants who are known as Social Democrats have lost important positions.

In addition to the negative impact the government had on women's projects and the social situation of women, it needs to be pointed out that Austria is further away than ever from gender-mainstreaming and the implementation of the UN Convention against all forms of discrimination.

## 5. References

Bundesministerium für Soziale Sicherheit und Generationen (2002). Geschlechtsspezifische Disparitäten, Wien (www.bmsg.gv.at).

Kommission der Europäischen Gemeinschaften (2001). Beurteilung der Umsetzung der beschäftigungspolitischen Leitlinien 2001 KOM (2001) 438 endg.

Rosenberger S (2002). Öffentlicher Dienst, Sparpolitik und Frauen. www.ceiberweiber.at/euphorum/bodensee/spar.htm

Steininger B (2002). www.demokratiezentrum.org/download/Steininger.pdf.

www.oestat.at.

www.eswin.net/a/aswfs.htm (2002).

## Notes

1 It can be prolonged up to 36 months if the other parent also takes parental leave. In the year 2000 only 1.7% of the persons who received parental leave were men (www.bmsg.gv.at, 06 08 2002)

*Cristina Alexandra Negutu, Attorney in private practice, Romania*

# Recent Legal Changes in Romania to Protect Women against Domestic and Social Violence

Family violence, too, belongs to the larger sphere of criminality. The extent and seriousness of family violence in Romania demanded legislative and institutional measures to limit it, which have focused primarily on prescribing serious punishments for the perpetrators while providing support for the victims. To a large extent, children and women are the primary targets of domestic aggressors, and for a long time shortcomings in the Romanian legal system have encouraged aggression against those victims.

Before the 1990s, violence in the family in Romania was not visible in public discourse. Perhaps similar to other Eastern European countries such information remained hidden from public view, ignored and dismissed, perhaps, in part, because the country had other economic and social problems.

Romania has changed into a democratic country; yet, it is still marked by cultural differences between one region and the next, between different economic levels and between social categories. The poor are more affected by family violence than the rich; most felonies are committed by members of low-income families.

The country's predominant Christian-Orthodox religion promotes understanding, love, and respect for other members of the family. Yet, a woman, especially in the countryside, is still expected to have more respect for her husband than vice versa, and to accept that he is more important than she or her children. She is expected to bear his physical and psychological violence because – she is taught to think – such submission reflects proper conduct in a Christian-Orthodox family.

The state has its share of guilt as well, to the extent that the waning authority of the state has encouraged, or at least not prevented, an increase in domestic violence accompanied by the use of alcohol and drugs.

A survey conducted in Romania by the United States Centers for Disease Control reported that in 2000, 23.4 percent of women had experienced verbal abuse from their partners, 10 percent had experienced physical abuse, and 1.8 percent sexual abuse. The same survey reported that 22.5 percent of women experienced moderate to severe physical abuse over their lifetimes (http://www.state.gov/g/drl/rls/hrrpt/2001/eur/8327.htm).

## 1. First steps

In 1996, the first center for the assistance and protection of violence victims, located in a Bucharest hospital, was inaugurated. In 1998, new legislation allowed associations and other legal entities that administer social assistance to receive subsidies from local governments. In 1999, The National Fund for Fellowship was set up to contribute to the decrease in poverty among very low-income families that have been particularly plagued by family violence.

The efforts that have been made to reform the domestic legislative framework have aimed at eliminating legislative gaps and discriminatory provisions, and achieving equality between women and men. A department for women's rights was created in the Ministry of Labour and Social Welfare, along with the Office of the Advocate of the People, which assumes ombuds functions and is empowered to protect the human rights of women and children. In December 1999, the law on parental leave entered into force, which is aimed at strengthening the principle of shared responsibility in family and society.

The Romanian government prepared a report to the United Nations' Committee on the Elimination of Discrimination Against Women which details the intention to draft legislation to legalize prostitution, and to promote the presence of women in public affairs and positions. The legislation also addresses family planning and reproductive health as well as measures for the social reintegration of unemployed women.

## 2. Recent legal changes

In November 2000, the Law for the Modification and Completion of the Criminal Code was passed. Many children – too many – are witnesses of family violence. One of the reasons for the number of child witnesses of family violence is that low-income families, including those in rural – and typically poorer – regions, have both higher rates of family violence and more children than more affluent middle class families. In addition, in these areas it is almost considered 'normal' that men batter their women or children. After all, 'they are men' the thinking goes. That said, it is important to note that family violence occurs in more affluent social classes and urban settings as well.

In response to these social issues and the real problems that occurred during the last ten years in Romania, in regard to family violence, government and parliament analyzed the roots of the social phenomenon of violence, which has risen substantially, and decided on measures to curb its continuing climb. These measures include important new pieces of legislation.

For example:

1. Persons who are convicted of assault or other forms of violence that cause physical or psychological harm to family members, and who are sentenced to more than one year in prison can be forced to stay away from the family residence for a maximum period of two years at the judge's discretion.
2. Any sexual act with a family member that involves coercion or otherwise exploits that family member's dependence or vulnerability is punishable by three to ten years in prison. If the victim is a child, and the aggressor has custody of the child or is the child's physician, superintendent, or teacher, the prison term will increase to 15 to 18 years.

3. If a person who has the legal obligation to support another family member abandons this family member or drives him or her away, thus exposing him or her to physical and psychological harm, or if the person does not fulfill his or her obligations of support as established by the judge, the delinquent may be punished by six months to three years in prison.

Assault and other forms of violence against a family member are punishable by six months to one year in prison. If the victim needs medical treatment for up to 20 days, prison time could increase to one to two years. If the victim needs medical treatment for more than 20 days, prison time could increase to one to five years. However, should the parties in the case reconcile, the criminal liability is removed. Moreover, a complicated criminal process discourages domestic violence victims from pressing charges against their perpetrator. The prosecution of rape is particularly difficult because it requires both a medical certificate and a witness.

Members of parliament, alarmed by the increase in family violence, have been analyzing a draft law that would allow official notification of the police and the prosecutor's office about domestic violence cases, and would empower police to use force to enter a home where abuse has been committed. The law would also provide for forensic exams for victims free of charge. The bill also provides special preventive measures, such as restraining orders, so that the aggressor is not allowed to get closer than 200 meters of the victim. Those who disregard the order will be arrested. The law stipulates that domestic violence offenses shall be judged under an emergency procedure. The bill allows investigative entities to help victims bring evidence before the court and testify on their behalf.

Under a new article 202 of the Penal Code that the Romanian government adopted in an emergency decision procedure, the seduction of a minor of the same or opposite sex, with the purpose of committing sexual acts, is punishable by imprisonment from one to five years.

## 3. The Romanian law against family violence (September 2001)

The first big step in the field was made in 2001 with the adoption of the Law Against Family Violence, the first of its kind in Romanian legislation. This legislation defines "violence in the family" as acts by the aggressor against the aggressor's own family that have permanent consequences, whether physical or mental. In the new legislation, victims of family violence can be husbands or wives, lovers, ex-spouses, parents, children born during the marriage, or children born out of wedlock.

The Law Against Family Violence also instituted the Public Service for Social Action, which is overseen by the Ministry of Health and Family and the Ministry of Public Administration. The new program was created to study the extent, gravity, and consequences of family violence, to develop national programs to recognize and combat family violence, and to initiate and develop educational programs. The Public Service for Social Action involves health institutions, local councils, and associations, and consists of a doctor, a psychologist, a law specialist, a police representative, and a social worker.

The main activities of the Service aim to identify cases of family violence, ensure help and therapy for the entire family, start the legal procedures for the defense of the victim and the punishment for the aggressor, organize information, education, prevention and intervention programs, and provide marital reconciliation. Any family member, citizen, or member of health and education institutions can notify the Service of Family Violence Incidents in a confidential manner.

The new legislation provides for 'social shelters' located in health institutions, and akin to orientation centers offering free medical services, therapy, psychological advice, and legal assistance. Victims who are financially dependent on the aggressor may be able to access financial support from the local government.

The new law also introduced two new offenses into the Romanian legal system: psychological harassment and marital abuse, making them punishable offenses for the first time. The punishment is the same for both of them: three months to one year of community service. Although it is a big step forward that these two offenses were introduced into the legal system, the punishment seems not big enough. The law further provides that the aggressors have to attend the national program of education and marital reconciliation.

## 4. Conclusions

Violence in the family has serious consequences for adults, and especially for the children, but also for the entire Romanian society, as it is marked during this transition period by economic poverty and the rise of violence in all societal spheres. The physical and mental health, morality, and well-being of a nation are absolutely necessary and concern all of us. Hence, this present situation demands the intervention of the state and of civil society.

Big steps have been made with regard to the Law Against Family Violence, but we have to change the mentalities on all levels of society that postulate, for religious or other reasons, that women and children ought to accept maltreatment. There still are mentalities and traditions regarding women's status that need to be addressed through the cooperation of central and local authorities, communities, and civil society.

We still are far from an efficient and real protection system for the victims of family violence. The state's efforts to curb family violence are further challenged by difficult circumstances such as the lack of legislation on sexual harassment, the rise in trafficking of women, the high rate of illiteracy among women over 50, the wide gap in literacy between urban and rural women, the relatively high maternal and child mortality rates, and the use of abortion as a means of birth control.

Stiffer penalties and newly established offenses are one step forward in Romanian legislation but it is also necessary to publicize these new measures in the media. Moreover, in order to be effective, cases of family violence ought to be processed faster than other legal cases. Measures are still needed to support and stimulate the political participation of women in order to ensure a greater presence of women in positions of power and in political decision-making so that women can influence and shape the reforms that are designed to benefit them. We also need good practice models and prevention strategies. Cooperation with Western European countries in the field could help to find better solutions that are applicable to the Romanian legal system.

This Law works, primarily, through the power of example. Legislators chose maximum punishments in order to underline how seriously the legislature is taking family violence. Yet, the new law has its drawbacks. For example, it does not address the necessary training of employees or experts in the field, which may compromise its effectiveness, and it does not provide for the collection of data to help decision-makers. Yet, despite the shortcomings, there is progress, so that in years to come we may be able to talk about a decrease in family violence.

*Manuela Martinez, M. Isabel García-Linares, M. Angeles Pico-Alfonso*
*Department of Psychobiology, University of Valencia, Spain*

# Women Victims of Domestic Violence: Consequences for Their Health and the Role of the Health System

## 1. Domestic violence as a public health issue

Considering its great impact on women's health domestic violence (i.e. intimate partner violence) should be considered a public health issue and epidemic problem of human society (Saltzman et al. 2000; Watts and Zimmerman 2002). In October 2002, the World Health Organization published the "World Report on Violence and Health", which indicates that the home is one of the most common settings for violence against women (Krug et al. 2002). In 48 population-based surveys from around the world, between 10% and 69% of women report having been physically assaulted by an intimate male partner at some point in their lives. Studies from Australia, Canada, Israel, South Africa, and the United States show that between 40% and 70% of female murder victims were killed by their husbands or boyfriends, frequently in the context of an ongoing abusive relationship. However, it is important to bear in mind that domestic violence is almost universally under-reported. Thus, the World Health Organization report should be considered as reflecting minimum levels of this type of violence.

Obviously, the worst consequence of domestic violence is the loss of life, and although there are discrepancies between data depending on the source, it is evident in many cases that death is caused by a violent intimate partner. For example, over the last five years in Spain, the number of women reported by the government as having been killed by their partners was 34 in 1998, 45 in 1999, 44 in 2000, 44 in 2001, and 44 as of October 2002. (http://www.mtas.es/mujer/default.htm). However, non-governmental women's organizations argue that the total number is higher because until 2002 only killing by the husband was reported, and, furthermore, women who did not die immediately after the aggression were not included in the official data (http://www.nodo50.org/mujeresred/violencia-red.html).

## 2. Impact on women's health

Although some women lose their lives at the hands of their violent partners, many more suffer terribly by having to live in chronically violent relationships with their aggressors. In these cases, there is little public recognition of the impact of domestic violence on women's health. Yet, the burden of such violence both in terms of societal expenses and the damage to women should be included in the debt that the aggressor has to pay to society.

*Domestic violence as a chronic stressor*

Cross-sectional and longitudinal studies of women from the general population, domestic shelter programs, emergency rooms, and primary health clinic settings consistently demonstrate that intimate partner violence places women at the risk of suffering mental and physical health problems (reviewed by Campbell and Lewandowski 1997). This is not surprising if we consider that women who share their lives with a partner who attacks them physically, sexually, or psychologically are living constantly under threat, and hence stress (Sutherland et al. 2002). It is important to bear in mind that domestic violence is rarely an occasional incident, but may go on over long periods of time, even decades. In general, most of the health consequences for women are not due to the physical attack itself but to the fact of living in constant fear. Furthermore, the abuse may also prevent women from gaining access to information and services, from taking part in public life, and from receiving emotional support from friends and relatives, which altogether may make them unable to look after themselves properly. Thus, the psychological abuse that women experience can be as devastating, if not more so, than the physical violence. Consequently, we should not expect specific illnesses or health problems in abused women but rather a stress-related deterioration of their health. The picture of this deterioration will vary depending on the characteristics of the abuse, the woman herself, and her specific social circumstances. Based on this knowledge, it is necessary to carry out research in order to demonstrate and assess how detrimental it is for women's health to live with a violent partner or know that he is a constant threat.

*Research worldwide*

Relevant research has been carried out during the last two decades, mainly in the United States. In Europe, studies started in the 1980s, mainly in the Scandinavian countries (e.g. Bergman et al. 1987; Alberdi-Olano and Anderson 1983), although in many cases publication in the local language has excluded such research from international recognition (e.g. Bogdanici 2001; Echeburua et al. 1997; Hedin 2002; Kramek et al. 2001; Rasmussen and Knudsen 1996). More recently, research is being carried out in many other countries such as Japan, India, Pakistan, Mexico and South America (e.g. Diaz-Olavarrieta et al. 2002; Ellsberg et al. 1999; Fikree and Bhatti 1999; Martin et al. 1999; Weingourt et al. 2001).

Research has mainly focussed on the impact of physical abuse on mental health and in some cases on the impact of concomitant sexual abuse. Very few studies have paid attention to the accompanying psychological abuse. Even fewer studies have assessed the impact on women's health of psychological abuse alone (Martinez et al. 2002).

It is important to differentiate between those studies carried out with a population-based sample (e.g. Hathaway et al. 2000) and those in which abused women have been recruited from medical health services (e.g. McCauley et al. 1995) or from battered women's shelters (Campbell et al. 1995), which makes them prone to sampling bias.

Moreover, most studies are cross-sectional; very little information is available from longitudinal studies

## 2.1 Consequences for mental health

As previously mentioned, the assessment of the impact of domestic violence on mental health has focussed on physical and sexual abuse, with only very few studies paying attention to psychological abuse alone. Such studies typically compare the incidence of mental disorders between battered women and the general female population or a restricted control group of non-abused women.

*Impact of physical abuse*

In general, among the mental disorders that have a higher incidence in physically abused women are depression, post-traumatic stress disorder, anxiety, sleeping and eating problems, social dysfunction, suicidal tendencies, and increased likelihood of substance abuse (Campbell 2002; Campbell and Lewandowski 1997; Golding 1999; Naumann et al. 1999; Weaver and Clum 1995).

Depression is women's primary mental health response to being battered. Studies indicate that the prevalence of depression in battered women is higher than in the general population, 45-63% and 9.3, respectively, and higher than in non-abused women (Campbell et al. 1997; Gleason 1993; Hathaway et al. 2000; McCauley et al. 1995). Some researchers hypothesize that sex differences (approximately 3 to 1, women/men respectively) in the global incidence of major depression could be attributable to sex differences in intimate partner violence, although this premise has never been specifically tested (Campbell et al. 1996). In one study, 61% of women diagnosed with depression had a lifetime prevalence of domestic violence (Dienemann et al. 2000). In another study with a sample of 394 adult women seeking medical care at a primary care practice medical center, depression was the strongest indicator of intimate partner abuse (Saunders et al. 1993).

Post-traumatic stress disorder (PTSD) is the other most prevalent mental-health sequela of intimate partner violence, although it is not always diagnosed (Astin and Lawrence 1993; Perrin et al. 1997; Saunders 1994; Woods 2000).

Anxiety, sleeping (i.e. insomnia) and eating problems (anorexia, bulimia), and social dysfunction are also higher in physically abused than in non-abused women (Hathaway et al. 2000; Humphreys et al. 1999; McCauley et al. 1995; Ratner 1993). Personality pathologies such as obsessive-compulsive disorder have also been diagnosed in battered women (Gleason 1993).

Suicidal tendencies (ideation/attempts) have been less studied, although they have also been associated with intimate partner violence (Bergman and Brismar 1991b; Gleason 1993; Hathaway et al. 2000; McCauley et al. 1995).

Substance abuse, of alcohol and illicit drugs, is another problem seen in battered women in industrialized countries (Bergman et al. 1987; Brokaw et al. 2002; Diaz-Olavarrieta et al. 2002; McCauley et al. 1995). In one study, physically abused women were eight times more likely to be alcohol dependent than non-abused women (Ratner 1993). Obviously, it is difficult to prove that substance abuse is a consequence of violence and not concomitant to it. One explanation could be that women use alcohol or drugs to calm or cope with the specific symptoms of PTSD, depression, and anxiety (Martin et al. 1998).

*Impact of sexual abuse concomitant to the physical abuse*

It has been estimated that sexual abuse occurs in approximately 40% of all cases of battering (Campbell and Alford 1989). Women who are both physically and sexually abused have a higher incidence of mental problems than those who are only physically abused. For example, in a study on women seeking refuge in battered women's shelters, those who were both physical and sexually abused were more likely to use marijuana and alcohol, to attempt suicide, and to feel a lack of control in their relationships than those with only physical abuse (Wingood et al. 2000).

*Impact of psychological abuse concomitant to the physical/sexual abuse*

Most women who are physically abused are also psychologically abused (Martinez et al. 2002; Ratner 1993), and some studies have been carried out to determine the impact of the concomitant psychological abuse on mental health. For example, it has been reported that psychological abuse, especially the emotional/verbal type (such as verbal attacks and behavior that degrades women), was strong enough to predict PTSD in women who were also physically abused (Street and Arias 2001). In another study, psychological abuse contributed independently to depression and low self-esteem (Sackett and Saunders 1999).

*Impact of psychological abuse alone*

The impact of psychological abuse alone has been relatively understudied. In general, it has been found that psychological abuse can be as devastating, if not more so, than physical violence, increasing symptoms of depression, PTSD, and anxiety, and lowering psychological well-being and self-esteem (Coker et al. 2000b; Follingstad et al. 1990; Martinez et al. 2002; O'Leary 1999). However, policy makers, researchers, and even abused women give psychological abuse considerably less attention than physical abuse (Dutton et al. 1999).

*Recovery from mental disorders and factors that may mediate positive outcome*

One important issue of the impact of domestic violence on women's mental health is the recovery that these women may experience once the violent relationship has ended (women separated from their spouse/partner), although the influence of abuse can persist long after the abuse itself has stopped (Campbell and Soeken 1999). So far, studies have found evidence of depression lessening with decreasing intimate partner violence (Campbell et al. 1995; Sutherland et al. 1998), although the persistence of PTSD symptoms has also been reported in women who were no longer experiencing abuse (Woods 2000).

More specifically, it is important to determine the factors that may act as buffers or as aggravators of health consequences. Studies show that the desistance of abuse, feelings of safety, control over their lives, and the presence of social support are very important for the recovery of battered women, reducing the incidence of PTSD, depression, and anxiety (Campbell et al. 1995, 1997; Coker et al. 2002b; Mertin and Mohr 2001). Thus, longitudinal studies are urgently required to guide the formulation of policy shaping interventions at individual, social, and criminal justice levels.

## 2.2 Consequences for physical health

*Impact of physical abuse*

Physical violence by an intimate male partner has a great impact, over both short- and long-term periods, on women's physical health. The most common immediate and direct consequences are injuries, and the neurological sequelae of violent acts. However, injuries are not as frequent as it is thought. It is the fear and stress associated with violence that result in long-term, chronic stress-related physical health problems. Thus, there is no single physical health problem that is indicative of present or past partner violence. In general, abused women report a lower self-rating of health, and a higher incidence of many signs, somatic symptoms, and illnesses related to the neurological, cardiovascular, gastrointestinal, muscular, urinary, and reproductive systems (reviews in Campbell 2002; Campbell and Lewandowski 1997). Studies show that the most frequent health problems reported by physically abused women are nightmares, lack of energy, constant fatigue, blurred vision, faintness, dizziness, ringing in ears, headache, shortness of breath, back pain, chest pain, stomach pain, muscular and pelvic pain, tachycardia, palpitations, chronic irritable bowel syndrome, hypertension, loss of appetite, and some gynecological problems listed below (Bergman et al. 1987; Campbell et al. 2000, 2002; Eby et al. 1995; Hamberg et al. 1999; Lown and Vega 2001; Martinez et al. 2002; McCauley et al. 1995; Plichta and Abraham 1996).

*Impact of sexual abuse concomitant to the physical abuse*

Although many studies do not differentiate between physical and sexual abuse when studying physical health outcomes, the impact of concomitant sexual abuse is especially high on the reproductive system (Campbell and Alford 1989; Eby et al. 1995). It includes problems related to pregnancy, such as unwanted pregnancies, miscarriages, and stillbirths, and sexually transmitted diseases or HIV infection. Furthermore, gynecological problems such as pelvic inflammatory disease, vaginal and anal tearing and stretching, bladder infections, sexual dysfunction, pelvic pain, dysmenorrhea, anal and vaginal bleeding, leakage of urine, missed menstrual periods, and infertility are frequent in sexually abused women. Moreover, a higher incidence of cervical cancer has been diagnosed in sexually abused women (Coker et al. 2000a). Women who are only sexually abused by their intimate male partner report a greater number of gynecological problems than women who are only physically abused or not abused (Campbell et al. 2002).

*Impact of psychological abuse alone*

Few studies have been conducted with women who have only been psychologically abused. In our study, these women reported a higher incidence of physical health problems than non-abused women (Martinez et al. 2002).

*Impact of abuse during pregnancy*

Batterers may also physically or sexually attack women during pregnancy, risking the death of, or threatening the health of the mother, the fetus, or both (reviewed by Campbell et al. 2000; Murphy et al. 2001; Rasmussen and Knudsen 1996). Prevalence data indicate that abuse during pregnancy ranges from 0.9% to 20.1%, depending on the characteristics of the studies (Gazmararian et al. 1996, Hedin et al. 1999; Moraes and Reichenheim 2002). In general, there is no doubt about the impact of abuse during pregnancy on low infant birth weight (Campbell et al. 1999; McFarlane et al. 1996; Valladares et al. 2002). However, its specific pathway (e.g. low maternal weight gain, anemia, smoking, substance abuse) is not known, although the stress caused by the abuse has been reported to be associated with low birth weight (Altarac and Strobino 2002). Furthermore, severe physical violence is related to spontaneous pre-term delivery (Covington et al. 2001).

*Mechanisms underlying the health impact*

The relationship between abuse and medical morbidity is extraordinarily complex, with health effects occurring along different pathways (Walker and Katon 1996). Although no studies have been carried out so far to determine the exact mechanisms, they could include recurrent injury or stress, and alterations in neurophysiology. Possibly most of the health problems that appear in battered women are stress-related pathologies of psychosomatic origin, although the possible long-term alterations in immune and endocrine functions, which have been associated with persistent stress, have not yet been investigated. In one study, blood samples from abused women compared to non-abused women indicated alterations in the immune system, i.e. a decrease in T-cell function (Constantino et al. 2000). In another study, only slightly lower hemoglobin levels and microcytosis were found in abused women, which the authors did not consider clinically important (Brokaw et al. 2002).

## 2.3 Health care utilization

One important feature of women who are exposed to domestic violence is their higher health care utilization. Studies carried out in the United States estimate that abused women use the emergency department, the primary health care, and the mental health services more frequently than non-abused women, the costs of all these health services being 92% higher in the abused (Kernic et al. 2000; Naumann et al. 1999; Pearlman and Waalen 2000; Wisner et al. 1999). A longitudinal study carried out in Sweden showed that hospital admissions were higher in abused women than in non-abused women (Bergman and Brismar 1991a). No information is available concerning the number of days missed at work.

Domestic violence also has a great impact on children's health (Campbell and Lewandowski 1997; Onyskiw 2002). Thus, if the defense of human rights is not a sufficient reason to reduce domestic violence against women, perhaps the economic impact, which indicates that this violence is very costly, may be effective in establishing efficient interventions to eliminate this violence from our society.

## 3. Role of the health system

One of the issues that arise from the impact of violence on women's health is the role that the health system plays (Garcia-Moreno 2002). As indicated above, women who are victims of domestic violence use the health services more frequently than non-abused women. However, too often neither the health professionals nor the women themselves identify the real reason for those frequent visits. Thus, women are repeatedly treated for the same illnesses, while the real cause is not identified. This is why it is important to clarify the role of the health system.

In the United States there are physician organizations such as the "Physicians for a Violence-Free Society" that have agreed to help end domestic violence. In 1995, the co-founders of this organization, Patricia Salber and Ellen Taliaferro published "The Physician's guide to domestic violence: how to ask the right questions and recognize abuse … another way to save a life". In 1993, the American Medical Association published the "Diagnostic and Treatment Guidelines on Domestic Violence".

In Spain, there is a Protocol for the Intervention of the Health System in Domestic Violence, which is part of the Plan of Action against Domestic Violence, and was published in 1999. Health professionals, both in the private and public practice, should follow such protocols. In general, protocols should address the following actions: identification, intervention, and prevention. Obviously, in order for health personnel to be able to carry out these actions, they need special training and education.

### 3.1 Identification

There are many studies that demonstrate the failure of the health system to identify victims of domestic violence, which has at least one terrible consequence for the victims: the violence continues. Thus, screening for intimate partner violence could provide the opportunity for patients to disclose the violence, and for physicians to compassionately connect patients with appropriate resources (Coker et al. 2000b, 2002a). This screening could even prevent the homicide of women by their intimate partners (Sharps et al. 2001).

*Identification in the emergency department (ED):* Women victims of domestic violence visit the ED more frequently than other women due to traumatisms caused by the partner or by themselves (Dearwater et al. 1998; Ernst and Weiss 2002). Obviously, due to fear of the partner, many women lie when they explain the cause of the traumatism. However, not all abused women visit the ED because of an injury. For example, several studies have found that compared to non-abused women, abused women were more likely to visit the emergency room for psychiatric problems and to be hospitalized for a psychiatric diagnosis or attempted suicide (Kernic et al. 2000; Little 2000). Thus, health professionals should ask these women directly about domestic violence. However, there is no routine screening in the ED. In one study, fewer than 25% of women who visited the ED were asked about domestic violence (Glass et al. 2001).

*Identification in the primary care practice:* Primary health system personnel have a high probability of facing abused women in their practice and so are able to provide a setting in which women can develop an ongoing relationship with them and feel safe to discuss their violence. However, few women are asked directly about domestic violence by their primary care provider (Coker et al. 2002a; Naumann et al. 1999; Richardson et al. 2002). Thus, health personnel should be conversant with those symptoms that abused women suffer most frequently, although as mentioned above, the health problems will vary depending on the woman and the violence (Rodriguez et al. 1999; Waalen et al. 2000). For example, nearly all depressed women experiencing abuse sought general medical rather than mental health care (Scholle et al. 1998).

*Identification in specialty care settings:* Because a large proportion of women receive reproductive health care services each year, these settings also offer an important opportunity to reach women who may be at risk of experiencing intimate partner violence (Parsons et al. 2000; Wiist and McFarlane 1999). As mentioned before, a considerable number of women experience threats, physical, and sexual abuse during pregnancy. Furthermore, domestic violence is a significant problem among women seeking abortion (Leung et al. 2002), which indicates that screening prior to abortion may help health personnel to identify abused women. Thus, in general, there is an obvious need for screening for domestic violence among pregnant women to enhance the safety of women themselves and their unborn babies. Screening should also be carried out in the pediatric emergency department, as child abuse and wife abuse are linked. However, this opportunity is missed by health care providers (Wright et al. 1997).

*Universal screening and available instruments*

Because there is no clinical or demographic profile that would identify women who are at risk of suffering domestic violence and, thus, ought to be recognized by health personnel, universal screening is the best way to identify them (Zachary et al. 2001; Wasson et al. 2000). Routine screening is especially recommended for all women patients in emergency, surgical, primary care, pediatric, prenatal, and mental health settings. Health personnel should directly ask women if they are living in an intimate partner relationship in which there is violence. If they do not ask them directly, women may not talk about it spontaneously (Ernst and Weiss 2002; Gerbert et al. 1999). However, some practitioners do not agree with this universal screening, and prefer selective screening of those women whose symptoms suggest abuse. Furthermore, there are different attitudes and beliefs in health care providers that have long been identified as a barrier to effective clinical response (Maiuro et al. 2000). Here it is important to know that a high percentage of women, abused and non-abused, agree to being routinely screened for domestic violence (Gielsen et al. 2000; Glass et al. 2001; Richardson et al. 2002). Finally, even when physicians are committed to identifying battered women in health care encounters, this is a difficult task (Gerbert et al. 1999).

At present, there are numerous screening tools available that have been validated for use in the detection of domestic violence (Fogarty et al. 2002). These include the Abuse Assessment Screen (McFarlane et al. 1992), the Woman Abuse Screening Tool (Brown et al. 1996), and the Women's Experience with Battering Scale (Smith et al. 1995).

These questions should be asked in a nonjudgmental way and in a confidential setting, in which the woman is alone, without her partner present.

### 3.2 Intervention and prevention

After a woman has been identified as being a victim of domestic violence, a number of interventions are possible that provide help for the woman, putting her in contact with appropriate resources. Reporting of domestic violence to law enforcement officials, which physicians may be obliged to follow, is very common. However, in the opinion of many women, it is important that domestic violence policies and protocols address the safety, autonomy, and confidentiality issues that concern them once the violence has been identified (Gielen et al. 2000). In a population-based survey, most women supported mandatory reporting, although abused women were less supportive, one of the main reasons being the fear of increasing the risk of retaliation by the perpetrator (Sachs et al. 2002). Thus, disclosure of abuse to any third party and reporting to authorities, should be done with the woman's knowledge and consent, and if her safety can be guaranteed completely.

### 3.3 Training of the health personnel

Obviously, in order for all health providers to be able to identify and intervene in cases of domestic violence adequate education about the issue is needed (Wright et al. 1997). In general, health providers are not well trained with respect to domestic violence (Kahan et al. 2000). Thus, information about violence against women or family violence should be integrated into undergraduate curricula for doctors, nurses, and other health care providers.

## Acknowledgements

Special thanks to Miriam Phillips for the revision of the English style. This review was supported by the Institute of the Woman, Ministry of Work and Social Affairs (ref: 53/98; and 102/01) and the Ministry of Science and Technology (ref: BSO2001-3134, and PGC2000-2354-E).

## 4. References

Alberdi-Olano FJ, Aldersen E (1983). Battered women. A prospective study. Ugeskr Laeger 145, 51-53.

Altarac M, Strobino D (2002). Abuse during pregnancy and stress because of abuse during pregnancy and birthweight. J Am Med Womens Assoc 57, 208-214.

Astin MC, Lawrence KJ (1993). Posttraumatic stress disorder among battered women: risk and resilience factors. Violence Vict 8, 17-28.

Bergman B, Brismar B (1991a). A 5-year follow-up study of 117 battered women. Am J. Public Health 81, 1486-1498.

Bergman B, Brismar B (1991b). Suicide attempts by battered wives. Acta Psychiatr Scand 83, 380-384.

Bergman B, Larsson G, Brismar B, Klang M (1987). Psychiatric morbidity and personality characteristics of battered women. Acta Psychiatr Scand 76, 678-683.

Bogdanici C (2001). Incidence of ocular injuries in women. Rev Med Chir Med Nat Iasi 105, 109-110.

Brokaw J, Fullerton-Gleason L, Olson L, Crandall C, McLaughlin S, Sklar D (2002). Health status and intimate partner violence: a cross-sectional study. Ann Emerg Med 39, 31-38.

Brown JB, Lent B, Bratts PJ, Sas G, Pederson LL (1996). Woman Abuse Screening Tool for use in family practice. Fam Med 28, 422-428.

Campbell JC (2002). Health consequences of intimate partner violence. Lancet 359, 1331-1336.

Campbell JC, Alford P (1989). The dark consequence of marital rape. Am J Nurs 89, 946-949.

Campbell JC, Lewandowski LA (1997). Mental and Physical Health effects of intimate partner violence on women and children. Psychiatr Clin North Am 20, 353-373.

Campbell JC, Soeken KL (1999). Women's responses to battering over time: an analysis of change. J Interpers Violence 14, 21-33.

Campbell JC, Kub J, Rose L (1996). Depression in battered women. JAMWA 51, 106-110.

Campbell JC, Kub J, Belknap RA, Templin TN (1997). Predictors of depression in battered women, Violence Against Wom 3, 271-293.

Campbell JC, Woods AB, Chouaf KL, Parker B (2000). Reproductive health consequences of intimate partner violence. A nursing research review. Clin Nurs Res 9, 217-237.

Campbell J, Torres S, Ryan J, Kling C, Campbell DW, Stallings RY, Fuchs SC (1999). Physical and nonphysical partner abuse and other risk factors for low birth weight among full term and preterm babies: a multiethnic case-control study. Am J. Epidemiol 150, 714-726.

Campbell J, Jones AS, Dienemann J, Kub J, Schollenberger J, O'Campo P, Gielen AC, Wynne C (2002). Intimate partner violence and physical health consequences. Arch Intern Med 162, 1157-1163.

Campbell R, Sullivan CM, Davidson WS (1995). Women who use domestic violence shelters: changes in depression over time. Psychol Women Q 19, 237-255.

Coker AL, Sanderson M, Fadden MK, Pirisi L (2000a). Intimate partner violence and cervical neoplasia. Journal of Women's Health and Gender-Based Medicine 9, 1015-1023.

Coker AL, Smith PH, McKeown RE, King MJ (2000b) Frequency and correlates of intimate partner violence by type: physical, sexual, and psychological battering. Am J. Public Health 90, 553-559.

Coker AL, Bethea L, Smith PH, Fadden MK, Branst HM (2002a). Missed opportunities: intimate partner violence in family practice settings. Prev Med 34, 445-454.

Coker AL, Smith PH, Thompson MP, McKeown RE, Bethea L, Davis KE (2002b). Social support protects against the negative effects of partner violence on mental health. J Womens Health Gend Based Med 11, 465-476.

Constantino RE, Sekula LK, Rabin B, Stone C (2000). Negative life experiences, depression, and immune function in abused and nonabused women. Biol Res Nurs 1, 190-198.

Covington DL, Hage M, Hall T, Mathis M (2001). Preterm delivery and the severity of violence during pregnancy. J. Reprod Med 46, 1031-1039.

Dearwater SR, Coben JH, Campbell JC, Nah G, Glass N, McLoughlin E, Bekemeier B (1998). Prevalence of intimate partner abuse in women treated at community hospital emergency departments. JAMA 280, 433-438.

Diagnostic and treatment guidelines on domestic violence. American Medical Association, 1993.

Diaz-Olavarrieta C, Ellertson C, Paz F, de Leon S, Alarcon-Segovia D (2002). Prevalence of battering among 1780 outpatients at an internal medicine institution in Mexico. Soc Sci Med 55, 1589-1602.

Dienemann J, Boyle E, Baker D, Resnick W, Wiederhorn N, Campbell J (2000). Intimate partner abuse among women diagnosed with depression. Issues Ment Health Nurs 21, 499-513.

Dutton MA, Goodman LA, Bennet L (1999). Court-involved battered women's responses to violence: the role of psychological, physical and sexual abuse. Violence Vict 14, 89-104.

Eby KK, Campbell JC, Sullivan CM, Davidson WS (1995). Health effects of experiences of sexual violence for women with abusive partners. Health Care Women Int 16, 563-576.

Echeburua E, Corral P, Amor PJ, Sarasua B, Zubizarreta I (1997). Repercusiones psicopatologicas de la violencia domestica en la mujer: un estudio descriptivo. Revista de Psicopatologia y Psicologia Clinica, 2, 7-19.

Ellsberg M, Caldera T, Herrera A, Winkvist A, Kullgren G (1999). Domestic violence and emotional distress among Nicaraguan women. Results from a population-based study. Am Psychol 54, 30-36.
Ernst AA, Weiss SJ (2002). Intimate partner violence from the emergency medicine perspective. Women Health 35, 71-81.
Fikree FF, Bhatti LI (1999). Domestic violence and health of Pakistani women. Int J Gynaecol Obstet 65, 195-201.
Fogarty CT, Burge S, McCord E (2002). Communicating with patients about intimate partner violence: screening and interviewing approaches. Fam Med 34, 387-393.
Follingstad DR, Rutledge L., Berg BJ, Hause ES, Polek (1990). The role of emotional abuse in physically abusive relationships. J Fam Violence 5, 107-120.
Garcia-Moreno C (2002). Dilemmas and opportunities for an appropriate health-service response to violence against women. Lancet 359, 1509-1514.
Gazmararian JA, Lazorick S, Spitz AM, Ballard TJ, Saltman LE, Marks JS (1996). Prevalence of violence against pregnant women. JAMA 275, 1915-1920.
Gerbert B, Caspers N, Bronstone A, Moe J, Abercrombie P (1999). A qualitative analysis of how physicians with expertise in domestic violence approach the identification of victims. Ann Intern Med 131, 578-584.
Gielen AC, O'Campo PJ, Campbell JC, Schollenberger J, Woods AB, Jones AS, Dienemann JA, Kub J, Wynne EC (2000). Women's opinions about domestic violence screening and mandatory reporting. Am J Prev Med 19, 279-285.
Glass N, Dearwater S, Campbell J (2001). Intimate partner violence screening and intervention: data from eleven Pennsylvania and California community hospital emergency departments. J. Emerg Nurs 27, 141-149.
Gleason WJ (1993). Mental disorders in battered women: an empirical study. Violence Vict 8, 53-68.
Golding JM (1999). Intimate partner violence as a risk factor for mental disorders: a meta-analysis. J Fam Violence 14, 99-132.
Hamberg K, Johansson EE, Lindgren G (1999). "I was always on guard"- an explorarion of woman abuse in a group of women with musculoskeletal pain. Fam Pract 16, 238-244.
Hathaway JF, Mucci LA, Silverman JG, Brooks DR, Mathews R, Pavlos CA (2000). Health status and health care use of Massachusetts women reporting partner abuse. Am J. Prev Med 19, 303-307.
Hedin LW (2002). Abuse of women is a public health problem. All female patients over the age of 14 should be part of a routine screening program. Lakartidningen 99, 2268-9; 2272-4.
Hedin LW, Grimstad H, Moller A, Schei B, Janson PO (1999). Prevalence of physical and sexual abuse before and during pregnancy among Swedish couples. Acta Obstet Gynecol Scand 78, 310-315.
Humphreys JC, Lee K, Neylan T, Marmar CR (1999). Sleep patterns of sheltered battered women. Image J Nurs Sch 3, 139-143.
Kahan E, Rabin S, tzur-Zilberman H, Rabin B, Shofty I, Mehoudar O, Kitai E (2000). Knowledge and attitudes of primary care physicians regarding battered women. Comparison between specialists in family medicine and GPs. Fam Pract 17, 5-9.

Kernic MA, Wolf ME, Holt VL (2000). Rates and relative risk of hospital admission among women in violent intimate partner relationships. Am J Public Health 90, 1416-1420.

Kramek J, Grzymala-Krzyzostaniak A, Celewicz Z, Ronin-Walknowska E (2001). Violence towards pregnant women. Ginekol Pol 72, 1042-1048.

Krug EG, Dahlberg LL, Mercy JA, Zwi AB, Lozano R (eds). World report on violence and health. World Health Organization, Geneva, 2002.

Leung TW, Leung WC, Chan PL, Ho PC (2002). A comparison of the prevalence of domestic violence between patient seeking termination of pregnancy and other general gynecology patients. Int J. Gynaecol Obstect 77, 47-54.

Little KJ (2000). Screening for domestic violence. Identifying, assisting, and empowering adult victims of abuse. Screening for Domestic Violence. Postgrad Med 108, 135-141.

Lown EA, Vega WA. (2001). Intimate partner violence and health: self-assessed health, chronic health, and somatic symptoms among Mexican American women. Psychosom Med 63, 352-360.

Maiuro RD, Vitaliano PP, Sugg NK, Thompson DC, Rivara FP, Thompson RS (2000). Development of a health care provider survey for domestic violence. Psychometric properties. Am J. Prev Med 19, 245-252.

Martin SL, Kilgallen B, Dee DL, Dawson S, Campbell JC (1998). Women in a prenatal care/substance abuse treatment program: links between domestic violence and mental health. Matern Child Health J 2, 85-94.

Martin SL, Kilgallen B, Tsui AO, Maitra K, Singh KK, Kupper LL (1999). Sexual behaviors and reproductive health outcomes. Associations with wife abuse in India. JAMA 282, 1967-1972.

Martinez M, Garcia-Linares MI, Pico-Alfonso MA, Savall-Rodriguez F, Sanchez-Lorente S, Celda-Navarro N, Blasco-Ros C (2002). Impact of physical and psychological domestic violence on women's health: results of a study in the Valencian Community of Spain. 7th Interdisciplinary Conference of the European Research Network on Conflict, Gender and Violence. Valencia, Spain, 22-25 September.

McCauley J, Kern DE, Kolodner K, Schroeder AF, DeChant HK, Ryden J, Bass EB, Derogatis LR (1995). The "Battering Syndrome": prevalence and clinical characteristics of domestic violence in primary care internal medicine practices. Ann Intern Med 123, 717-746.

McFarlane J, Parker B, Soeken K (1996). Abuse during pregnancy: associations with maternal and infant birth weight. Nurs Res 45, 37-42.

McFarlane J, Parker B, Loeken K, Bullock L (1992). Assessing for abuse during pregnancy. JAMA 267, 3176-3178.

Mertin P, Mohr PB (2001). A follow-up study of posttraumatic stress disorder, anxiety, and depression in Australian victims of domestic violence. Violence Vict 16, 645-654.

Moraes CL, Reichenheim ME (2002). Domestic violence during pregnancy in Rio de Janeiro, Brazil. Int J. Gynaecol Obstet 79, 269-277.

Murphy CC, Schei B, Myhr TL, Du Mont J (2001). Abuse: a risk factor for low birth weight? A systematic review and meta-analysis. CMAJ 164, 1567-1572.

Naumann P, Langford D, Torres S, Campbell J, Glass N (1999). Women battering in primary care practice. Fam Pract 16, 343-52.

O'Leary KD (1999). Psychological abuse: a variable deserving critical attention in domestic violence. Violence Vict 14, 3-23.

Onyskiw JE (2002). Health and use of health services of children exposed to violence in their families. Can J Public Health 93, 416-420.

Parsons L, Goodwin MM, Petersen R (2000). Violence against women and reproductive health: toward defining a role for reproductive health care services. Matern Child Health J 4, 135-140.

Pearlman DN, Waalen J. (2000). Violence against women. Charting the impact on health policy, health care delivery, and the law. Am J Prev Med 19, 212-213.

Perrin S, van Hasselt V, Hersen M (1997). Validation of the Keane MMPI-PTSD Scale againsts DSM-III-R criteria in a sample of battered women. Violence Vict 12, 99-104.

Plichta SB, Abraham C (1996). Violence and gynecologic health in women <50 years old. Am J Obstet Gynecol 174, 903-907.

Protocolo sanitario ante los malos tratos domesticos. Consejo Interterritorial. Sistema Nacional de Salud. Edited by Ministry of Labour and Social Affairs, Institute of Women, 1999.

Rasmunsen KL, Knudsen HJ (1996). Violence towards pregnant women. Ugeskr Laeger 158, 2373-2376.

Ratner PA (1993). The incidence of wife abuse and mental health status in abused wives in Edmond, Alberta. Can J. Public Health 84, 246-249.

Richardson J, Coid J, Petruckevitch A, Chung WS, Moorey S, Feder G (2002). Identifying domestic violence: cross sectional study in primary care. BMJ 324, 1-6.

Rodriguez MA, Bauer HM, McLoughlin E, Grumbach K. (1999). Screening and Intervention for Intimate Partner Abuse. Practices and Attitudes of Primary Care Physicians. JAMA 282, 468-474.

Sachs CJ, Koziol-McLain J, Glass N, Webster D, Campbell J (2002). A population-based survey assessing support for mandatory domestic violence reporting by health care personnel. Women Health 35, 121-133.

Sackett LA, Saunders DG (1999). The impact of different forms of psychological abuse on battered women. Violence Vict 14, 105-117.

Salber P, Taliaferro E. The Physician's guide to domestic violence: How to ask the right questions and recognize abuse: Another way to save a life. Volcano Press, 1995.

Saltzman LE, Green YT, Marks JS, Thacker SB (2000). Violence against women as a public health issue: comments from the CDC. Am J Prev Med, 19, 325-329.

Saunders DG (1994). Posttraumatic stress symptom profiles of battered women: a comparison of survivors in two settings. Violence Vict 9, 31-44.

Saunders DG, Hamberger LK, Hovey M (1993). Indicators of woman abuse on a chart review at a family practice center. Arch Fam Med 2, 537-543.

Scholle SH, Rost KM, Golding JM (1998). Physical abuse among depressed women. J Gen Intern Med 13, 607-613.

Sharps PW, Koziol-McLain J, Campbell J, McFarlane J, Sachs C, Xu X (2001). Health care providers' missed opportunities for preventing femicide. Prev Med 33, 373-380.

Smith PH, Earp JA, DeVillis R (1995). Development and validation of the Women's experience with battering (WEB) scale. Womens Health 1, 273-288.
Street AE, Arias I (2001). Psychological abuse and posttraumatic stress disorder in battered women: examining the roles of shame and guilt. Violence Vict 16, 65-78.
Sutherland C, Bybee D, Sullivan C (1998). The long-term effects of battering on women's health. Women's health: Research on Gender, Behavior and Policy 4, 41-70.
Sutherland CA, Bybee DI, Sullivan CM (2002). Beyond bruises and broken bones: the joint effects of stress and injuries on battered women's health. Am J Community Psychol 30, 609-636.
Valladares E, Ellsberg M, Pena R, Hogberg U, Persson LA (2002). Physical partner abuse during pregnancy: a risk factor for low birth weight in Nicaragua. Obstet Gynecol 100, 700-705.
Waalen J, Goodwin MM, Spitz AM, Petersen R, Saltzman LE (2000). Screening for intimate partner violence by health care providers. Barriers and interventions. Am J Prev Med 19, 230-237.
Walker EA, Katon WJ (1996). Researching the health effects of victimization: the next generation. Psychosom Med 58, 16-17.
Wasson JH, Jette AM, Anderson J, Johnson DJ, Nelson EC, Kilo CM (2000). Routine, single-item screening to identify abusive relationships in women. J. Fam. Pract 49, 1017-1022.
Watts C, Zimmerman C (2002). Violence against women: global scope and magnitude. Lancet 359, 1232-1237.
Weaver TL, Clum GA (1995). Psychological distress associated with interpersonal violence: a meta-analysis. Clin Psychol Rev 15, 115-140.
Weingourt R, Maruyama T, Sawada I, Yoshino J (2001). Domestic violence and women's mental health in Japan. Int Nurs Rev 48, 102-108.
Wiist WH, McFarlane J (1999). The effectiveness of an abuse assessment protocol in public health prenatal clinics. Am J Public Health 89, 1217-1221.
Wingood GM, DiClemente RJ, Raj A (2000). Adverse consequences of intimate partner violence among women in non-urban domestic violence shelters. Am J Prev Med 19, 270-275.
Wisner CL, Gilmer TP, Saltzman LE, Zink TM (1999). Intimate partner violence against women: do victims cost health plans more? J Fam Pract 48, 439-443.
Woods SJ (2000). Prevalence and patterns of posttraumatic stress disorder in abused and postabused women. Issues Ment Health Nurs 3, 309-324.
Wright RJ, Wright RO, Issac NE (1997). Response to battered mothers in the pediatric emergency department: a call for interdisciplinary approach to family violence. Pediatrics 99, 186-192.
Zachary MJ, Mulvihill MN, Burton WB, Goldfrank LR (2001). Domestic abuse in the emergency department: can a risk profile be defined? Acad Emerg Med 8, 796-803.

*Carol Hagemann-White, Department of Education and Women's Studies, University of Osnabrück and*
*Hans-Joachim Lenz, Consultant in private practice, Germany*

# Violence Against Women/ Violence Against Men: Comparisons, Differences, Controversies

## 1. Introduction: Beyond the perpetrator-victim dichotomy

Awareness of gender-based violence resulted from feminist struggles to make private abuse an issue of public concern. In the process of recognition, three characteristics became evident: sexualised violence and repeated physical abuse involve acts of men against women; they occur most frequently as violence against known women, in the home, the family, or the environment of everyday life; and such violence is exercised with impunity, socially tolerated and excused, the perpetrators consider it legitimate. Once given the opportunity to share their experiences with others similarly victimized, women described their realizing that the abuse had very little to do with who they were or what they did: Rape or abuse was directed at them as women, and part of the humiliation arose from being made to feel interchangeable, like an object. All of these aspects converged in the naming of "violence against women".

In the years that followed, activists insisted that intervention, services and policies of social change focus clearly on men's violences against women, and in particular towards known women. Men were faced with the classic choice: either be part of the solution, or be justly considered part of the problem. Overcoming gender-based violence seemed to consist in challenging and changing men's role as perpetrators.

Feminist analyses have become both broader and more subtle over time, including much analysis of the ways in which norms of masculinity diminish and harm men while granting them privilege or "patriarchal dividends". Nonetheless, differentiated approaches seem much more difficult when it comes to sexualised and interpersonal violence, where men enter the discourse primarily as (potential) perpetrators. From time to time, data are produced to show that men are targets of violence to a similar extent or even more than women, but even then, the discussion tends to stop with the numbers. Aside from polemical claims and counter-claims, there has been no significant men's movement to challenge the multiple ways in which boys and men are expected to endure violence without taking on the role of a victim. Sexual abuse of young boys has gradually been recognized as a serious problem, but sexual violence against adolescent boys and adult men is still largely denied or trivialized, and there is almost no literature (or practical services) on the numerous types of physical attacks that occur between men from the point of view of the victim suffering pain or injury[1]. Research data that suggest a high prevalence of men-on-men violence rarely focus on the suffering that results due to violence.

There are, however, good reasons for directing serious attention to men's victimization. It seems probable that in any society that denies men visibility and respect as victims, their collective sensibilities for empathy with the pain of another human being will be diminished. They will be trained to meet, or to pre-empt, attacks with counter-attacks, thus learning skills that become part of their repertoire for managing conflict. Both social patterns will increase the probability of men's violence in conditions of domination, where they need not fear counter-attacks.

This said, it must be pointed out that such considerations remain fully within the framework of man-as-perpetrator; it is an indirect recognition of victimization, as when violence against women is addressed out of a primary concern for the ensuing risk to the children. Such considerations are legitimate as long as they do not obscure the fundamental human right to respect, personal safety and freedom from abuse. Addressing violence against men is necessary first and foremost because human rights are indivisible.

The present paper emerges from a tentative discussion seeking ways to make all forms of gender-based violence a serious concern, without obscuring what we have learned over the past thirty years about the specific power dynamics in violence against women and the embeddedness of such violence within patriarchal gender relations (Hagemann-White and Lenz 2002). It is a discussion that lends itself to misunderstandings at every turn, but offers the promise of a new level of insight as well as new alliances in practice.

## 2. The controversy about women and men as victims of domestic violence

Every attempt to open a discussion on men as victims of violence comes up against the great US debate on prevalence data and methodology (see Straus 1991; Gelles and Loseke 1993) and on "battered husbands" (Steinmetz and Lucca 1988; Saunders 1988). This debate needs, first of all, to be put into context. Whereas in Europe violence against women was first brought to public attention by the feminist movement, the US discussion springs from two separate sources. The women's movement turned first to issues of rape and sexual violence, while at the same time a small group of family sociologists began studying the use of physical aggression within the family.

The study of "family violence" was shaped by the desire to measure incidence and prevalence, to be followed by the identification of correlations and further statistical analysis. Thus, the definition of the phenomena to be measured occurred at a very early stage, before extensive qualitative knowledge was available, and in an era in which sexualised and intimate violence was still generally covered by a blanket of silence and shame. Thus, a fairly simple construct of "conflict tactics" was devised, based on the assumption that research was looking for families with inadequate skills in handling disagreements. A list of items, describing acts that might occur in the course of an argument or a fight, was put together and then arranged in a sequence according to the family sociologists' view on what is "more" or "less" severe, and in consideration of the sequential order the instrument was labelled a "conflict tactics scale" (Straus and Gelles 1990).

Over the years, this instrument has been repeatedly tested and validated. Once it had been established as a working tool, further studies faced the expectation to replicate or compare their data, and thus to use the same tool, in a process that soon became self-perpetuating. It is an excellent example of both the strengths and the weaknesses of mainstream US social research. As a pragmatic construction, it generates numbers in multiple settings. This fits well with the dominant model of epidemiology in the US: violence is seen as a public health problem, and the purpose of research is to accumulate data and identify main causes and effects, so that policy can act to eliminate the germ and thus stamp out the disease. The weaknesses of this type of research are also familiar: cultural context is ignored, little attention is given to complex interactions or to exploring the nature of the phenomena; and blatant generalisations enter into all stages of the process. For example, the Conflict Tactics Scale is built on the premise that any form of physical aggression is "more severe" than any form of psychological aggression. In the most influential studies, it is assumed that women and men are equally free to describe to a telephone interviewer how they are treated by their spouses. There is no room for description of abuse unrelated to conflicts, and no way to locate single acts of "hitting" or "kicking" in the situation or sequence of events in which they occur.

In fact, the political agenda driving this research in family sociology is primarily oriented to preventing child abuse, and more broadly, corporal punishment as an accepted means of enforcing norms and/or expressing feelings within the family. The theoretical framework is fairly simple, and Murray Straus, the most influential author of this school of thought, has articulated it repeatedly. It postulates a demarcation line between all kinds of non-physical aggression and the act of hitting. Whoever crosses that line is in danger of using increasingly harmful forms of physical violence, but also of receiving physical violence (Note the complete parallel to the dominant US model of illegal drug use or, in an earlier period, alcohol). Family violence, in this model, is possible because children learn at an early age that hitting people is permitted within the family; if we could teach people never to hit their children for any reason, we could hope for non-violent families. According to this model, women who defend themselves against abuse by hitting, much as we might understand self-defence, are in danger of contributing to this spiral of increasing physical violence. Only prohibition can avoid destructive excesses.

In the US, where an estimated 90% of parents use physical punishment, a research agenda that highlights the damage that can result is certainly addressing an important social problem. At the same time, the theoretical model is far too simple to be generalized, and it lacks sensitivity to gender issues, or, indeed to more complex dynamics of family, social environment, and culture. While the simplicity of the instrument has made it easy to use, it is difficult to know just what the numbers mean that can emerge from using it. It seems clear, by now, that use of the CTS in a representative population gathers very little information on the type of chronic abuse and battering that has appeared in shelters (cf. Straus and Gelles 1990). Numbers sufficient for statistical analysis emerge for families or couples where physical aggression occurs not more than once or twice a year, and the data usually indicate that both women and men report hitting their partners with about equal frequency. Such studies can tell us about habits of family life: In which regions, social classes or life circumstances is causing physical pain to a child, a sibling, or a partner accepted or relatively normal? They are entirely inadequate for gathering information about gender-based violence, showing only the tip of the iceberg. More elaborated instruments have been more successful (Tjaden and Thoennes 2000).

European research and policy have followed a different pattern (Hagemann-White 2002). A social problems approach with a strong current of ethical concerns gave priority to a multifaceted description of the nature of the phenomena, understanding the dynamics of victimization, and gathering data to evaluate what measures and actions might be helpful. These approaches have been closely linked with informing policy on the national and on the European level. In consequence, a quite different model from the epidemiological has emerged, which emphasizes circular connections among gender inequality, discrimination throughout society, and violence against women (see for example Godenzi 1996; Fawcett et al.1996; Hanmer and Itzen 2000). The numerical study of prevalence and incidence has emerged – building on the pioneer work of Römkens (1992; 1997) in 1986 – only since the mid-1990s (Hagemann-White 2001), under the influence of the Canadian national study (Johnson 1996); and this research has drawn on a broad base of qualitative knowledge to develop differentiated instruments as well as responding to ethical concerns, and given great attention to gender sensitivity.

The strength of the European approach rests on a foundation of qualitative exploration and social intervention as the necessary precondition for gathering valid numbers on a large scale. Its corresponding weakness is the dependency of research on prior social movements and social services that create a public discourse, reduce shame and silence, and empower victims to speak out. Since this has not occurred with men who suffer violence, little is known about them from research.

## 3. Denial and silencing of men's victimization

The gender culture of modernity is centred on an "autonomous" masculine subject whose relationship to others and to the world is based on self-assertion, a struggle between competing claims, and undertaking projects to overcome, reshape, and transform what is merely given or natural. In a capitalist economy, competitive achievement is crucial, and in the 19th century, even physical violence between men in the workplace was sometimes encouraged as a sign of the vigorous will to win (Connell 1995). Under the rule of culturally exalted (or hegemonic) forms of masculinity, winning is everything.

In this context, the concept of a masculine victim seems a contradiction in terms: one is either a man or a victim (Lenz 1996). As late as the end of the 1980s, sexual abuse of boys was widely thought to be simply impossible (cf. Enders 1990).

Writings from the men's movement, although widely different in their aims and perspectives, have both described and harshly criticized the initiation practices and forms of cruelty that have been institutionally tolerated or demanded within all-male settings such as boys' boarding schools, sports teams, street gangs, armies, and prisons; much has been written about the expectation that boys and men endure pain without complaint. Yet, very little space has been given to the victims' own perceptions and suffering, to the immediate and long-term damage to their health and their sense of self-esteem. Many progressive male psychotherapists and social workers seem to avoid engaging themselves with male victims of violence by other men (Lenz 1999). Silence on victimization seems to be a requirement of hegemonic masculinity.

Men are socialized to function in a culture of bodily assault as a routine feature of everyday life. They may be subjected to unilateral and practically unlimited violence if they fail to so function, as reports from basic military training in all countries in the past and the present make clear, or if they remind other men of the possibility of failure. Thus, in prisons, the supposed homosexual is subjected to especially brutal gang rape by heterosexually identified prisoners, and the physical weakling will quickly be pressed into sexual and domestic service for a dominant male (Gilligan 1996). Within interactions of civil society, casual physical aggression of men against men is perceived as normal, thus obscuring humiliation and abuse from view even when it is openly visible. Little attention, either in research or in social intervention, has been given to the significant amount of domestic violence within homosexual relationships, due to the implicit attribution of the violations to homosexuality as such (Finke 2000).

Violence is a key instrument of underlining or enforcing social exclusion. Men identified as outsiders – in the EU countries these could be blacks, Arabs, Turks, refugees, or the homeless – may be chased on the streets and beaten or killed. Man-on-man rape and sexual abuse is widespread, profoundly shameful, and routinely denied: by witnesses, by the law, and by the victims (e.g. reinterpreted as sexual initiation) (Hillmann et al. 1990; Mezey and King 1992; Lenz 2000). Little is known about the circumstances and motives: Do the abusers seek to further humiliate a man already perceived as inferior, as is the case during ethnicised wars, or have they extended their concept of the "woman" to include some males, granting themselves a wider range of sadistically tinged sexual satisfaction?

No small amount of violence is exercised by women within the home and the family, particularly when they are in a position to do so without encountering restraint, as when disciplining children (Steinmetz 1980; Elliott 1995; Bange and Enders 1996). Data suggest that domestic violence by women in the couple relationship may increase when men grow older, especially since wives are likely to be younger and may gain more equal physical strength (Wetzels et al. 1995). Within pre-existing power relationships, as between adult and child, or professional and psychiatric patient, women also can and do perpetrate sexual as well as physical abuse, and the victims may be male as well as female. Last year, for example, the sexual attack of a female nurse on a male patient was the trigger for requesting a seminar for the medical and nursing staff in a German psychiatric clinic in order to address sexual violation of patients by clinical personnel. The incident had led to great difference of opinion among the staff, most of whom are women. The question was whether a male patient can actually be raped, and what is the "normal" response of a man to a woman's invasion of his intimate or sexual sphere.

Men who have been victimized, as well as the very few professionals (doctors, psychologists, pedagogues, and social workers) who are receptive to hearing their experiences, report again and again that they encounter considerable resistance to the perception of men's vulnerability. The great majority of doctors and therapists minimize the violation of boys and men or refuse to realise that the events have taken place at all (Lenz 1996).

Counsellors and psychotherapists seem to fear vulnerability in men, because this touches on a dark side of the therapists themselves: their own experience of being at someone's mercy, helpless to defend and protect themselves (Peichl 2000). To recognize the violated man forces the male professional to come to terms with his own "weak", i.e. subordinate feminine side. This painful process can call into question the therapists' own understanding of masculinity as well as his self-image as a competent helper (Lenz 1999).

## 4. Stakeholders and dissenters in the construction of the debates

The issue of violence against men is bound into political struggles in complex ways. Although Western culture and media present physical violence among men and among boys as legitimate, necessary, and normal, when the issue of victimization is raised, the debate often turns on the question of violence by women against men and boys. On the most obvious level, the figure of the "battered husband" may function to challenge recognition of women's victimization, and more fundamentally, all of women's advocacy politics. Gender-neutral surveys, especially those using the CTS, generally find that men report domestic incidents of hitting approximately as often as women do. These data are sometimes used politically – and at first blush, surprisingly – not to call for appropriate (and gender-sensitive) services for the hidden population of battered men, but rather to demand eliminating services or to reject legal intervention altogether. This seems to imply that the women who call a hotline or seek safety in a shelter are the same women whose men report being victimized, thus confirming the old stereotypes of the violence-prone couple with whom all intervention is futile. In fact, however, the argument seldom goes that far, and it seems more likely that the entire debate is occurring on a symbolic level, where the existence of suffering and injured human beings needing support and safety disappears. Public discourse on violence against women is interpreted as a code message about the moral superiority of women over degraded, because violent, men. Proof that women, too, can be violent against men is taken to invalidate the fundamental premise of all equality policy. It sometimes emerges that the entire debate, as it is carried out in the media and on the Internet, cloaks a controversy about child custody and visiting rights when a woman initiates divorce (and perhaps more profoundly about a women's right to initiate separation).

In 2002, after granting money for a prevalence study on violence against women, the German government issued a call for bids on an explorative study on violence against men, with the explicit intention of preparing the ground for a gender-sensitive quantitative study. In the discussions that followed, it was often assumed without question that the aim should be to measure women's violence against men. This exclusive focus on the female perpetrator may distract attention from men's much more frequent and traumatic, real victimization by forms of violence that establish and underpin male hierarchy and dominance, and norms of masculinity. It reasserts silently the premise that men do not suffer when such violence occurs, that the participants in a fist fight, an initiation or punishment ritual, or an abusive attack are warriors, not victims, and that they emerge "bloody but unbowed" having proven themselves "real men". To admit to being a victim thus is equivalent to the public admission of not being a real man.

Research on violence avoids the vulnerability of men. In part, this follows from the construction of the subject of scientific inquiry as a knower simultaneously male and without gender. The gaze of the scientific knower is thus directed outward, seeking objective knowledge of the world as it is. In this world of science, the inner world of thoughts, feelings, and motives is also constructed as a realm of facts. Thus, men have questioned and studied all manner of things, except themselves. As Friebel writes: "The drama of the talented man is that he cannot look at himself. With cool rationality and technical intelligence he has, as it were, displaced himself outward, objectified himself and postulated himself to be 'objective'." (Friebel 1995: 9) This is true for psychology as well, where men only become salient when their behaviour deviates from that masculine norm, for example as criminals (Schmitz 1994: 820).

The past decade has seen the growth of men's studies, both as a critical voice coming from men's projects of change, and within such fields as the sociology of health, psychosomatic medicine, or clinical psychology. It is striking that most of these writings, even those centrally organized around the damaging effects of dominant models of masculinity, such as Hurrelmann and Bründel (1999), fail to make any mention of men's suffering through physical or sexual violence. A familiar theme in masculinity literature is the damage to boys and men by the psychological power of women as mothers or predominant carers, but this discussion also shapes the image of the man as one who must learn to resist, separate himself, (re-)gain control, and overcome susceptibility to feelings that give women too much influence. In critical men's studies, the need to change that very image is a main focus, yet even there, it is the man as (potential) perpetrator due to his lack of empathy and his training in dominance that is seen as the issue. Both in theory and for political action, the connection between men and violence is reduced to men as perpetrators. The violated man is not a subject of political interest. Indeed, we might say that men must act as perpetrators in order to receive attention to their vulnerability. If a man is identified as a perpetrator, he receives the attention of a huge apparatus –from the public prosecutor to the social worker, from the police to the lawyer, all striving to overpower and control him. This extends even to practical projects that do work with men on their experiences of victimization: In the public eye, and when their funding needs justification, they present themselves solely as originators and service providers with the goal of changing the violent behaviour of perpetrators of violence against women.

There are several possible explanations for this avoidance behaviour. Certainly, the claim to work against violence towards women positions male professionals as active protectors and champions of women and children, and defines their work as an active struggle against the forces of evil, which then can be felt – despite all declarations to the contrary – as outside of the self. It should be noted, however, that feminist projects also tend to prefer as allies men who position themselves in these terms – while at the same time distrusting them. The definition of the client as perpetrator also creates an automatic shield of distance, which protects the male speaker (researcher, social worker, therapist, or political activist) from too close an identification; the need for distance may arise from fear of vulnerability, repressed own experiences, or submerged homophobia.

It is notable that controversies on the "correct" methodology of perpetrator programs all seem to focus on issues of control: either it is vital to confront the violent men with effective social control, or it is vital to avoid giving third parties (such as the police, or feminist projects) control over what happens in the men's groups. Much seems to be at stake in avoiding the admission of unmitigated vulnerability and neediness.

## 5. Theoretical considerations: how gender relations shape the patterns and probabilities of victimization

Within the wide range of actions with the potential to be violent, physical assault and injury (including all kinds of coerced sexual penetration) retain a unique and central significance, since there is no real escape from one's own body. The threat to bodily integrity and, in the last resort, to life, lends potency to psychological attacks or social exclusion. Physical violation also mobilizes cultural body practices, especially gender and race; on the interpersonal level, both acts of violence and experiences of violation are probably always gendered, as well as being racially or ethnically shaped. Even when both are men, or women, white or black, violence asserts the proper place of the body in a social order. Thus, we may safely assume, even knowing that some forms of victimization are still invisible, that women and men will have gender-related and different experiences of violence. Gender-neutral research will necessarily gloss over or deny important aspects.

Despite its emphasis on gender, the successful political discourse addressing violence has drawn heavily on a "social problems" approach. Such an approach sees certain groups within society as unfairly disadvantaged, and other groups as unable to function smoothly within institutional and market frameworks (sometimes these are the same groups, and one phenomenon is taken to explain the other). The task of policy, then, is to provide resources, services, or sanctions and thereby reduce the extent of the problem. Within this tradition, particularly strong in the social democratic and corporatist European states, feminist activism was able to divert resources to intervention strategies and create political alliances for change.

At the same time, framing the issue as a social problem may actually tend to reproduce the gender basis of violence. Hegemonic constructions of masculinity and femininity are both supported by, and support the use of violence, but they also define the "legitimate victim". Thus, abuse of the wife within the family could be established as a scandal, underpinned by the norms that women should expect to be protected and provided for when they marry, that the home should be a safe haven complementary to the harsh competitive outside world, a place where needs are met. Attitudes towards sexual violence have remained much more ambivalent, especially when a woman has been willing to strike up a casual acquaintance in a public location such as a bar. Disapproval of trafficking in women is sharp, but it is difficult to mobilize protest against shipping such women back to where they came from, as if it were a sanitary measure cleaning the country of sin. And it remains persistently difficult to get a clear focus on women as actors who sometimes use violence against those dependent on their care.

Hegemonic masculinity defines men as active agents in control of even the most passionate impulses: possessive sexuality and rational violence form a precarious bridge over the contradiction between desires and postulated invulnerability. Men are taught to mask neediness as legitimate demands or rights. Within public discourse on gender-based violence, men appear either as emotional cripples who cannot express their needs any other way than by violence, and thus deserve understanding and help, or as invulnerable actors who must be called to account and punished. These images allow little room for a human both vulnerable and responsible, able to grow and change.

Hegemonic femininity has been challenged by feminism much more thoroughly in its dual definition of women as "strong mothers, weak wives" (so the book title by Miriam Johnson). Nonetheless, speaking of violence seems to lead imperceptibly into recasting the woman in these very terms, as a victim (the term "survivor" is no less dramatic), as one who bravely struggles, for example, to care for her children as best she can, while suffering great wrongs. This image is often reproduced and mirrored in the self-perception of feminist activist projects. As a result, the woman herself may be cast as an object of counselling, services, intervention, and education. Again, vulnerability and responsible decisions seem not to mix; active desire and passive suffering seem unable to exist in the same person.

The feminist movement has used violence against women as emblematic for patriarchy, while at the same time working pragmatically for social change here and now. This, it would seem unavoidably, has partly slipped into reaffirming the very gender relations that were meant to be challenged. At the same time, it must be said that throughout the entire past 30 years of activism in the field, strong – and when necessary unashamedly self-critical – feminist voices have raised all of these issues again and again. Voices of men questioning or rejecting hegemonic masculinity have been fewer and their collective efforts towards change, as Connell (1995) has pointed out, unstable. Our closing question to the discussants of this paper is thus: How can we create a discourse on gender and violence based on mutual respect and empowerment?

## 6. References

Bange D, Enders U. Auch Indianer kennen Schmerz; Sexuelle Gewalt gegen Jungen. Luchterhand, Köln, 1996.

Connell RW. Masculinities. Polity, Cambridge, UK, 1995.

Elliot M (ed). Frauen als Täterinnen: Sexueller Mißbrauch an Mädchen und Jungen. Donna Vita, Ruhnmark, 1995.

Enders U (ed). Zart war ich, bitter war's: Handbuch gegen sexuellen Missbrauch. Kiepenheuer & Witsch, Köln, 1990.

Fawcett B, Featherstone B, Hearn J, Toft C (eds).Violence and Gender Relations: Theories and Interventions. Sage, London, 1996.

Finke B (2000). Schwule als Opfer von ‚häuslicher Gewalt'. In Lenz H-J (ed), Männliche Opfererfahrungen: Problemlagen und Hilfeansätze in der Männerberatung. Juventa, Weinheim/München, pp. 135-148.

Friebel H. Der Mann der Bettler: Risiken im männlichen Lebenszusammenhang. Leske & Budrich, Opladen, 1995.

Gelles RJ, Loseke DR (eds). Current Controversies on Family Violence. Sage, London, 1993.
Gilligan J. Violence: Reflections on a National Epidemic. Vintage, New York, 1996.
Godenzi A. Gewalt im sozialen Nahraum. Helbing & Lichtenhahn, Basel, 1996.
Hagemann-White C (2001). European research on the prevalence of violence against women. Violence Against Women 7, 731-759.
Hagemann-White C (2002). A comparative examination of gender perspectives on violence. In Heitmeyer W, Hagan J (eds), International Handbook of Violence Research. Westview.
Hagemann-White C, Lenz H-J (2002). Gewalterfahrungen von Männern und Frauen. In Hurrelmann K, Kolip P (eds), Geschlecht, Gesundheit und Krankheit:Männer und Frauen im Vergleich. Hans Huber, Bern, pp. 460-487.
Hanmer J, Itzen C. Home Truths About Domestic Violence: Feminist Influences on Policy and Practice: A Reader. Routledge, London, 2000.
Hillmann RJ, Tomlinson D et.al. (1990). Sexual assault of men: a series. Genitourin Med 66, 247-250.
Hurrelmann K, Bründel H. Konkurrenz, Karriere, Kollaps: Männerforschung und der Abschied vom Mythos Mann. Kohlhammer, Stuttgart, 1999.
Johnson H. Dangerous Domains: Violence Against Women in Canada. Nelson Canada, Scarborough, 1996.
Lenz H-J (ed). Männliche Opfererfahrungen: Problemlagen und Hilfeansätze in der Männerberatung. Juventa, Weinheim/München, 2000.
Lenz H-J (1999). Männer als Opfer – ein Paradox? Männliche Gewalterfahrungen und ihre Tabuisierung bei Helfern. Organisationsberatung – Supervision – Clinical Management 6, 117-129.
Lenz H-J. Spirale der Gewalt: Jungen und Männer als Opfer von Gewalt. Morgenbuch, Berlin, 1996.
Mezey GC, King MB (eds). Male Victims of Sexual Assault. Oxford University Press, Oxford/New York/Tokyo, 1992.
Peichl J (2000). Männliche Opfererfahrungen: Rollenklischees und Wahrnehmungsblockaden aus der Sicht eines Psychoanalytikers. Lenz H-J (ed), Männliche Opfererfahrungen: Problemlagen und Hilfeansätze in der Männerberatung. Juventa, Weinheim/München, pp. 307-314.
Römkens R. Gewoon geweld? Omvang, aard, gevolgen en achtergronden van gewald tegen vrouwen in heteroseksueelle relaties. Swets & Zeitlinger, Amsterdam, 1992.
Römkens R (1997). Prevalence of wife abuse in the Netherlands: combining quantitative and qualitative methods in survey research. J Interpers Violence 12, 99-125.
Saunders DG (1988). Wife abuse, husband abuse or mutual combat? In Yllö K, Bograd M (eds), Feminist Perspectives on Wife Abuse. Sage, Beverly Hills, pp. 90-113.
Schmitz J (1994). Psychologie des Mannes. In Grubitzsch S, Rexilius G (eds), Psychologische Grundbegriffe: Mensch und Gesellschaft in der Psychologie. Ein Handbuch. Rowohlt, Reinbek, pp. 820-824.
Steinmetz SK (1980). Women and violence: victims and perpetrators. Am J Psychother 34, 334-350.

Steinmetz SK, Lucca JS (1988). Husband battering. In van Hasselt V, Morrison RL, Belleck AS et al. (eds), Handbook of Family Violence. New York, pp. 233-246.
Straus M, Gelles RJ. Physical Violence in American Families: Risk Factors and Adaptations to Violence in 8.145 Families. Transaction, New Brunswick, NJ, 1990.
Straus MA (1991). New theory and old canards about family violence research. Soc Probl 38, 180-197.
Tjaden P, Thoennes N (2000). Prevalence and consequences of male-to-female and female-to-male intimate partner violence as measured by the National Violence Against Women Survey. Violence Against Women 6, 142-161.
Wetzels P, Greve W, Mecklenburg E, Bilsky W, Pfeiffer C. Kriminalität im Leben alter Menschen: Eine altersvergleichende Untersuchung von Opfererfahrungen, persönlichem Sicherheitsgefühl und Kriminalitätsfurcht. Kohlhammer, Stuttgart, 1995.

## Notes

1 Only since violence in schools has been reframed as "bullying" (especially in Scandinavia and the UK) do we see some studies of the effects on the victim; but this new framework – like the substitution of "mobbing" for harassment at work – tends to suppress considerations of gender.

*Barbara Kavemann, Stefan Beckmann, Heike Rabe, Beate Leopold,*
*Project WiBIG, University of Osnabrück, Germany*

# Work with Perpetrators of Domestic Violence in Germany

## 1. Project WiBIG: "Evaluation of intervention projects against domestic violence"

Even after more than 20 years of public discussion in Germany[1] about the male violence that many women experience in domestic relationships, it continues to be necessary to take initiative to improve women's situation. Although significant progress has been made over the past few years, the problem of violence in gender relations nevertheless has not yet been solved and new strategies are under discussion (Hagemann-White 1992; Hagemann-White et al. 1997).

The intervention projects against domestic violence take this discussion as a starting point for improvements. Intervention projects are institutionalized networks for inter-agency and community cooperation. Since the middle of the nineties, the number of such projects in Germany has been steadily growing. They differ in size, structure, and focus but in the end all pursue the same objectives: to reduce violence against women, prevent its continuation, and ensure its social condemnation. The intervention projects work to hold perpetrators of violence systematically accountable and to optimize intervention and support for women and their children. The projects strive to ensure better access for those seeking help and to reach those groups of women who up to now have not been reached by any support program (Kavemann et al. 2001).

These far-reaching goals are implemented through cooperation forums that aim to include all institutions, agencies, projects, and professions working to overcome domestic violence or that carry a social responsibility for addressing it, such as women's shelters, women's counseling services, the police, professionals in the justice system, men's counseling services, child protection agencies, ministries, and local governments. In addition to a central round table and a coordination office, there are often different working groups of specialists. They coordinate procedures, improve guidelines, and explore the latitude for legal action, in order to make intervention in cases of domestic violence more effective for the victims. The work is interdisciplinary, inter-institutional, binding, and based on the principle of equality. In this context, there is growing interest in programs for the perpetrators of violence.

Since 1998, on behalf of the Federal Ministry for Family Affairs, Senior Citizens, Women and Young People, project WiBIG, the University of Osnabrück-based evaluation of intervention projects against domestic violence, has been evaluating a total of nine intervention projects in Germany (Kavemann et al. 2001).[2] We have taken stock of the existing models of work with violent men as they are practiced within the context of the evaluated intervention projects. This formed the foundation of an evaluation of different programs for perpetrators of violence that we began in late spring of 2001. First results are expected by the end of 2002. This evaluation cannot be more than a snapshot. At the moment, there is a lively policy discussion and practice is constantly evolving. Even in a few months the picture will look very different. In this article we want to give a short – and necessarily incomplete – overview of the development of work with violent men in Germany and present the most important considerations and policy conflicts. Furthermore, we shall very briefly introduce examples from practice.

## 2. Background

Only relatively recently have men in Germany tackled the issue of violence. It was often women who more or less pressured men into dealing with the issue of "men and violence" (Haffner and Spoden

1991, p. 1; Bentheim and Firle 1994, p. 43). Since the beginning of the 70s the second-wave women's movement in particular demanded that men more actively face their own "male" share of violence. In the course of men's engagement with the issue, which since the 70s has slowly become more critical of "masculinities" (Connell 1999), violence has become more and more the focus of counseling work that is offered by men and geared specifically towards men (Bentheim and Firle 1994, p. 50 f.).

Violence was for many years a dominant topic of the second-wave women's movement. Activists from the increasingly specialized areas of the anti-violence work demanded of men a more visible commitment against violence and a shouldering of responsibility for the acts of other men. Dominant issues since the middle of the 70s have been rape (Teubner et al.1983) and battering (Hagemann-White et al. 1981). Since the middle of the 80s, the sexual abuse of children (Kavemann and Lohstöter 1984). More recently, there also has been a focus on sexual harassment at the work place (Holzbecher et al. 1991), sexual trafficking in women, and violence against specific groups of women (e.g. agisra 1990).[3]

Violence was not necessarily a central topic in the counseling of men, but ranged among other issues such as partnership, career, and fatherhood. If there are specific programs for violent men, they usually concentrate on a wide range of forms of violence. Domestic violence is addressed as only one of many forms of violent behavior of men in public and private space. Specific measures for the perpetrators of domestic violence have entered Germany's spectrum of men's counseling only in the last few years, which has been due in part to the creation of the intervention projects against domestic violence.

Even though in Germany the discussion on violence in marriage and partnership was ten years older than that on sexual abuse of children, work with perpetrators of domestic violence was barely developed and discussed. Until recently, work with perpetrators was primarily concerned with the struggle against sexual abuse of children.[4] Here it seemed obvious that the children were developmentally incapable of taking care of their own protection from violence. Because preventive approaches tried to avoid burdening children with this unreasonable responsibility for self-protection intervention strongly favored work with perpetrators in the form of therapy (cf. Kavemann and Bundesverein 1997). In addition, the specific character of child sexual abuse was considered especially outrageous and was often seen as a pathological disorder. This suggested to many of those engaged in social work that imprisonment could not offer lasting protection because of the perpetrators' lack of understanding of their culpability who therefore should be referred to a therapeutic agency. Since the beginning of the 90s experts have discussed policy approaches widely and controversially (cf. Wodke-Werner et al. 1999), even though the actual number of therapies offered is still not very large.[5]

Already at the end of the 70s the discussion on battering in marriage and relationships led rapidly to establishing a range of practical escape and support options for women in the form of women's shelters and women's counseling centers. Today there are more than 300 women's shelters in Germany that house approximately 45,000 women and their children annually and are financed out of public funds. For a long time feminist agencies considered concurrent work with perpetrators not important and regarded it with skepticism or outright rejection. It seemed more pressing to protect women, strengthen them, and encourage a possible separation from the man. The focus was on victim-oriented intervention. Thus, perpetrator-oriented intervention, beyond the demands of the criminal justice system, did not become a significant issue until the 90s, again within the context of the intervention projects.

The Domestic Abuse Intervention Project (DAIP), founded in 1980/81 in Minnesota, USA, became more widely known in Germany in 1991 during the founding phase of the Berlin Intervention Project against Domestic Violence (BIG) (Kavemann et al. 2001, p. 36).[6] The discussion about DAIP led to an intensive debate on the political assessment of the cooperation of feminist institutions and government agencies and as a result also on the cooperation with programs for violent men.

A further important source of inspiration for the programs for perpetrators in Germany was the Domestic Violence Intervention

Project (DVIP) in London, Great Britain. DVIP is based on two primary pillars: the Women's Support Service (WSS) and the Violence Prevention Program (VPP) with the perpetrators (Burton et al. 1998). The cooperation with the women's support agencies and the work in teams of both sexes is now also an issue in Germany.

## 2.1 Conflicts with feminist institutions

There are controversial positions within the feminist women's movement on whether one should work with sexual offenders or batterers at all. The critical discussion continues in parts of the women's movement, and flares up especially strongly within the context of the creation of the intervention projects (Kavemann et al. 2001). The central points of conflict are mostly of a political or ideological nature. Women within the women's movement show a fundamental distrust towards the men in men's counseling services, and question whether the men are capable of avoiding camaraderie with the perpetrators of violence and of really confronting them with the violence and its consequences. Whenever experts working with the perpetrators explain that purely confrontational work is not advisable because a relationship with the client presents the foundation of therapeutic and pedagogical work, then this distrust is reinforced. A similar distrust arises when programs for perpetrators use former batterers as trainers or co-trainers.[7] In addition, many workers at women's shelters and women's counseling services fear that government funding for the work with violent men will be at their expense and will lead to financial cutbacks for their agencies, or that they would be forced against their wishes to collaborate with these new programs.

Not all of the women's projects refuse to cooperate, but even those who consider work with violent men a useful complement, that could very well be in the interest of their clients fight against having programs for violent men financed by the limited funds of the women's departments in state or local governments – thus leading to cutbacks in the support work for women as has happened often enough – but insist that the Department of the Interior or the Justice

Department should finance men's programs. Increasingly, feminist discourse frames work with perpetrators of domestic and sexual violence as a question of inner security for women and children and therefore declare the Department of the Interior to be responsible. An increasing number of women's projects are convinced that work with perpetrators should take place within the context of legal sanctions and therefore that society insist on holding the perpetrators accountable (cf. Kavemann et al. 2001, p. 24 and p. 41). As of now, almost no discussion focuses on the advantages that these programs offer for battered women (cf. Austin and Dankworth 1999).

Other important arguments also play a role in the controversy. Up to now, men's counseling services and other agencies offering programs for violent men have not sufficiently reacted to the concerns of the women's shelters; One of which is that they do not take the protection of the woman sufficiently serious. The experts working in the women's shelters fear that women might be lulled into a false sense of security because the perpetrator participates in a cognitive behavioral training course. Women's advocates insist that women should receive complete information on the possibilities and limits of these programs and that they be urged to continue to look after their own security and to treat with skepticism a man's assertion that he has changed completely. To this end, feminist experts demand a more intensive exchange of information with men's counseling centers, but this is quickly interpreted by the counseling centers as a desire to control them. On the other hand, activists who work with violent men react overly sensitively and defensively to critical inquiries by feminists and the demand to divulge information about their clients. This conflict-ridden discussion will surely be with us for a while.

## 3. Definitions and philosophy of work with violent men in the context of intervention projects in Germany

Work with violent men in the context of the intervention projects against domestic violence concerns only violence in the home. "In the home" is understood to mean violence of adults against their (ex-)partners, regardless of whether or not the partners were married, have shared, or still share a household. The term is distinct from other concepts such as "violence in the family" or "male violence". Domestic violence also includes violence in homosexual relationships. Besides physical violence the term also includes sexual, emotional, and in some cases social and economic violence. The definition is gender neutral and the new intervention instruments and laws also apply in cases in which women are accused of abusing their male partners. However, in the practice of the agencies that cooperate in the intervention projects violence of men against women dominates.

Even if other acts of violence are disclosed in the course of the work with the men, such as sexual offences outside the family or violence against other men, the focus lies on domestic violence. The primary goal of the work with the perpetrators is the improvement of victim protection and the prevention of violence to ensure over both short and long-term the greatest possible protection for the abused women and their children from new violent acts by their partners.

## 4. Differences in intervention philosophy

There are three different models of work with perpetrators in the intervention projects evaluated by WiBIG: therapy, cognitive behavioral training, and counseling. This is not the place to discuss these approaches in detail, approaches that are historically grown, well-discussed methods of working with people's problems. One can say that in our line of work these methods basically never appear in pure form, but always in combinations and hybrid forms that are adapted to the demands of practice. For example, cognitive behavioral training courses that, among other things, are concerned with encouraging self-motivation, work with therapeutic elements as well and also make limited space available for such aspects as clients' socialization experiences and experiences as victims. Similarly, therapeutic groups for perpetrators of violence in some cases use elements of goal-oriented training.

Another difference between existing models in practice is the question if violence and the abuse of power have to be seen in direct connection. Not all of those working against violence stress the connection between violence and the abuse of power equally. Violence as a means of power assertion or violence as a means of averting the feeling of powerlessness often turn up as two differing views in policy discussions, although in reality this frequently polarized debate on which of the two aspects dominates refers to an ambivalence. According to both scenarios, men try to feel in control of the situation experienced, thus to experience their own power consciously. One should ask whether averting the powerlessness, in the end, is nothing but the preservation or restoration of power.

It remains to be seen whether the intervention philosophies that have acquired a sharper profile recently will develop into an independent new approach to work with violent men beyond the classical methods, and if they will gain recognition accordingly. At present these methods are debated intensely and internationally. They have also not yet been evaluated.

## 5. Voluntary participation or court orders

Perpetrators of domestic violence come to programs for violent men in various ways, whether those programs are labeled therapy, counseling, or cognitive behavioral training. A general criterion distinguishing access to the respective programs is whether the men participate voluntarily or have to report by court order.

The men's counseling projects or men's centers that, in Germany, carry out the greater part of the work with perpetrators of domestic violence are divided into two camps. On the one side are those who work exclusively with so-called self-referrals, that is with men who come to the counseling centers out of self-motivation. These agencies strongly reject the work with court ordered offenders. On the other side are those who, while they often work with self-referrals as well, do not reject the work with court ordered offenders, and some of them even favor and promote court orders.

## 5.1 Voluntary versus court ordered participation

Because ideological and conceptual conflicts over the issue of voluntary participation versus court orders persist, the following will present the main points of critique of forced counseling, using the basic principles and assumptions of the Hamburg-based association of "men against male violence" ("Männer gegen Männer-Gewalt®"; in the following abbreviated as MgM). Forced counseling is what MgM calls the work with court mandated violent offenders. In Germany, MgM advocates the most aggressive and most radical espousal of the principle of restricting work with violent men exclusively to so-called self-referrals. MgM is the oldest German anti-violence project in the area of men's work initiated by men, and boasts a complex network of local counseling facilities[8]. Members of MgM argue emphatically against any work in the context of compulsion. In their opinion, compulsory anti-violence counseling cannot be successful. Only if violent men or men inclined towards violence contact anti-violence counseling voluntarily can freedom from violence be achieved permanently. Furthermore, as MgM points out in their criticism of the work with court ordered violent men, the structures of the legal system support the denial strategies of violent men because the legal system endeavors to find exonerating points for acts of violence. This encourages the perpetrator to refuse to accept responsibility for his violence. Hence, MgM concludes, men who have been court mandated to participate in a program for violent men will permanently deny the responsibility for their actions, no matter what happens in counseling (cf. Lempert 2000).

MgM believes that the only adequate treatment for perpetrators of violence who do not have or do not develop any interest in changing their violent behavior is to prosecute them rigorously in the criminal courts with the goal of imprisonment. Since the perpetration of violence is an active deliberate act, and the man has voluntarily decided in favor of that act, one should respect his decision. Anything else amounts to treating the men as minors and not respecting their boundaries. Therefore, work with perpetrators within the context of compulsion violates the boundaries of these men as human beings who, in turn, would see this boundary violation as a reason for further acts of violence in the spirit of self-defense. From this follows the essence of MgM.'s point of view: work with court ordered violent offenders raises their readiness for violence instead of reducing it.

### 5.2 External motivation versus self-referral

Experts who work with perpetrators of violence in the context of compulsion oppose these arguments, among them those experts who cooperate with the police and the legal system in the context of the intervention projects. Those experts believe that a court order, on the one hand, is a necessary external intervention that produces in men an indispensable feeling of suffering, and, on the other hand, is an important preventive signal. Of course, no one would deny that self-motivation is the best prerequisite for a change in behavior. But in the course of group work, court mandated participants of programs for violent men are also very capable of developing an understanding of culpability and accepting responsibility. In fact, court orders and the threat of sanctions are viewed as an especially favorable framework, because they represent a reaction by society that condemns the acts of violence and explicitly makes it clear to the perpetrator that, in the future, the community will no longer accept this behavior. Threatening legal consequences reduces the drop-out rate. Even in the context of compulsory measures it is possible to develop a client's motivation to analyze his own behavior critically. Whether such a self-motivation is achieved probably depends not only on individual prerequisites of the client, but also on the duration and the quality of the program as well as the competency and experience of the instructors. Most of all it depends on the intensity of cooperation and coordination within an intervention system (Gondolf 2002).

However, the threat of criminal sanctions is no guarantee for participation in a program for violent men. In 2000, the HAIP-alliance found that out of the 146 court orders in the context of domestic violence about which the alliance had information, only 68 offenders actually reported to the men's center in Hannover. That is to say more than half of the court ordered men preferred to accept possible criminal sanctions rather than participate in the program for violent men (Hannoversches Interventions Projekt – HAIP – gegen Männer-Gewalt in der Familie, 2001, p. 36). This illustrates that even court ordered offenders make a "voluntary" decision whether or not to participate in programs for violent men. The distinguishing criteria between "voluntary participation" and "court order" are no longer clearly defined.

### 5.3 Limitations of voluntary participation

In the context of domestic violence one usually designates as so-called self-referrals those perpetrators who turn to the counseling centers or other agencies out of a desire to change their behavior but without legal pressure. However, most of the time, all variations of non-judicial external motivation are counted as cases of voluntary participation. For example, one has to count as a non-judicial external motivation the woman's demand on her partner to participate in a program for violent men, or the recommendation of a youth welfare office concerning participation. While men under pressure of such external motivations contact the counseling center or agency on their own, the same is true for court ordered perpetrators of violence.

The court ordered offender also has the freedom to decide whether or not to report to the relevant agency. The consequences of not reporting are different from those for men who are self-referrals but in the end "punishment" always depends on the perception of the offender, and on what he considers to be a punishment. Thus, it can happen that the threat of being deserted by his partner is considered worse than a fine or imprisonment. That is, the necessary feeling of suffering that is required to make a man decide to sign up for such a program can be produced in different ways.

Strictly speaking, one can only classify those men as genuine self-referrals who summon up a motivation for change without any trace of external influences whatsoever, and only on the basis of their own interests and needs, that is to say those who do this out of a feeling of suffering. Self-referrals defined in this way are very difficult to distinguish from others who only appear to report voluntarily to counseling centers or agencies. In the end, the term self-referral distorts the real motives for participation in programs working against domestic violence. In order to be able to make less ambiguous statements on profiles of perpetrators and possible paths of access to programs for violent men, self-referrals have to be differentiated more strongly by motivation and by the local practice of intervention and the structure of available programs. On the basis of a careful evaluation of the programs for violent men such a differentiation could lead to more exact statements about which programs are likely to be more successful for one or the other group of offenders. Thus, the theory that only work with self-referrals leads to the goal of freedom from violence for formerly violent men is only of limited value.

### 5.4 Referral by agencies outside the legal system

In Germany, different individuals and agencies may encourage perpetrators of domestic violence to change their behavior and visit a counseling service. As a rule, these are recommendations so if a man does not follow the recommendation, this does not lead to any further sanctions by the recommending authority or agency.

One of the most common non-judicial motivations for the participation in a program for violent men is the demand of their partner to change something about their violent behavior,[9] often combined with the threat that otherwise she will leave him. The fear of losing their partner often makes men realize for the first time the consequences their own violence can have for themselves.

Up to now, the youth welfare offices or the family courts, in deciding on custody or child contact, have only occasionally suggested to perpetrators of domestic violence that they participate in a program for violent men when it is apparent that domestic violence plays a role in the conflict over child custody. If, after the recommendation by the youth welfare office or the family court, the man does not try to get into a program for violent men, then this could – under some circumstances – lead to disadvantages for him in a custody or child contact decision. In addition, under child and youth welfare law there could be recommendations for perpetrators of domestic violence to participate in such a program. The counseling and support programs by the youth welfare office, with their mandate to guarantee the development of children and adolescents, could be used to make recommendations for participation in a program for violent men, if the youth welfare office knows of violence perpetrated against the mother of the child. In practice, this also happens only rarely.

In situations where violent men are approached directly by police officers or social workers in the context of police intervention, as for example in the case of the Hannover Prevention Program Police and Social Work (PPS), we find a high proportion of perpetrators contacting programs for violent men.

### 5.5 Court orders and stipulations

At present, the strongest sanction available in Germany to try and get perpetrators of domestic violence to participate in a program for violent men is the court order in the context of a suspended prison sentence.[10] If, in the course of the trial, the court pronounces such an order, then this means for the man first of all that he does not have to serve the sentence. If he does not comply with the court order, then he is – among other things – in danger of having to serve the suspended prison sentence. Further legal options are a warning from the court, reserving the right to sentence the perpetrator to a fine until he fulfills the order by, for example, attending a cognitive behavioral training course.[11]

Even before the trial, the public prosecutor has different possibilities[12] of temporarily dropping the proceedings while issuing orders or stipulations.[13] If the man complies with the order or stipulation, then the case is dropped for good. If he does not comply with them, then he has to expect that a charge will be brought against him. Because there is no sentence, this legal procedure considerably reduces the impact of society sanctioning domestic violence. The offender may not have to appear in court or at the public prosecutor's office at all. As a rule, the perpetrator's contact with the legal system is confined to written correspondence, and usually there is no personal confrontation.[14]

On the other hand, the possibility of the public prosecutor dropping the proceedings while issuing an order for cognitive behavioral training allows for reaching a large number of perpetrators. The results of the Berlin intervention project against domestic violence (BIG) show that it is difficult to restrict cognitive behavioral training only to those men convicted by a court to a suspended sentence, because the courts are very reluctant to convict the men of simple bodily harm due to the comparatively high sentences required by the German criminal code.

Experts working with perpetrators in Germany see specific difficulties in the cooperation with the legal system. Some agencies that provide programs for perpetrators of violence are critical of the exchange of information with the legal system. They criticize that they receive only meager feedback from the legal system on what happens to men who stop coming to the program or are excluded from the program, whether or not the threatened sanctions are actually imposed. The importance of this for the credibility of the agencies should not be underestimated. In the opinion of some of the agencies, the courts do not follow up thoroughly enough on the cases in which they imposed orders or stipulations. In cases where men do not follow court orders, there are often no further legal sanctions, even though the court had threatened to impose them. The high number of men not reporting to programs for violent men, even though they have been ordered to do so by the public prosecutors, could be an indication that the men often trust that there will be no further legal consequences if they ignore the order.

## 6. Preliminary standards for cognitive behavioral training

One can assume that a decisive factor in reaching permanent change towards violent-free behavior lies in the duration and quality of the program and the intensity of the accompanying sanctions and controls of society. However, at present there are no obligatory standards in Germany on the number of sessions, the conceptual elements, or the cooperation required of the programs. There is wide agreement on the central goals and basic contents of the programs for violent men, but there are large differences in the specific implementation. The duration of the programs we evaluated ranged from 12 group counseling sessions (intervention groups in KIK-Schleswig-Holstein), to 17 appointments – 4 individual and 13 group sessions – (counseling center in the Packhaus- BiP, Kiel), 24 group sessions (men's center Hannover), up to 20 group sessions (Berlin center for the prevention of violence), and 26 group sessions (Counseling for Men against Violence, Berlin).

In the year 2000 experts from men's counseling centers and other agencies that work with violent men met in a national working group and compiled and published policy considerations that represent a basis for future standards[15]. They serve as a starting point for discussions on further development and improvement of the cognitive behavioral training currently in practice for perpetrators of domestic violence (KIK-Schleswig-Holstein 2001).

A catalog of standards to safeguard quality becomes more and more necessary, as programs for violent men become more popular and state and local governments show a growing readiness to finance them. In the course of this development, institutions that in the past have not worked with perpetrators of violence and have had no practical experience now show interest in this work.

Experts of the working group considered the following elements and methodical aspects as commendable or indispensable. The goals of cognitive behavioral training are violent-free behavior; assumption of responsibility and increase of one's self-control; differentiation of one's self-perception; improvement and training of social skills (ibid. p. 50). The target group are perpetrators of domestic violence, regardless or whether they come voluntarily or are court mandated. However, men who come exclusively through legal pressure with an unshakeable conviction that they are innocent and who show no willingness for a dialogue at all should be excluded. Requirements for participation are the ability to work in groups (recognition of group rules and willingness to participate); written agreement with the rules of participation; sufficient competency in the language in which the group communicates; no cultural or religious reasons for exclusion (ibid. p. 51).

The are several framework conditions of the work with violent perpetrators. If possible, the program for violent men should take place in the form of group work. The staff has to be well qualified and act professionally. They should thoroughly work on individual case histories in a team, should work continuously with an external consultant and should document their work well and, if possible, evaluate it. In order to achieve an external control information on the man's progress, they should be in contact with the (ex-)partner, and should inform her about available support. The staff should have a wide range of different methods at their disposal. This includes e.g. the reconstruction of the act of violence, the cycle of violence and the wheel of violence, the ABC-model, reframing, role playing and psychodrama, the drama-triangle, etc. The staff should develop individual security and emergency plans with the participants. The staff should analyze with the participants their images of men and women (for more details see KIK Schleswig-Holstein 2001, 45-93.)

The participants are confronted with the form and frequency of their violent acts, with their responsibility, and their denial. The staff should question resistance and denial as early as possible, and discuss the consequences of the men's violent acts on their partners, their relationships, their children, their family, their social contacts and their entire lives.

Recommended are follow-up appointments. These serve as a precaution against repeat offenses, since experience has shown that the risk of repeat offenses is highest in the first months after the termination of the program (ibid. p. 89).[16] Additional individual counseling is advisable, for example in crisis situations, in couple counseling or, by request, also for the (ex-) partner (ibid. p. 90). All these factors require a functioning intervention system (cf. Gondolf 2002).

# 7. Work with perpetrators in the context of intervention projects against domestic violence[17]

In the following we present very briefly two different examples of the work with perpetrators in the context of two intervention projects against domestic violence that we are evaluating.

## 7.1 "Männerbüro" (Hannover)

Männerbüro, Hannover, is a men's center within the context of the "Intervention Project against Male Violence in the Family in Hannover" (HAIP). As the name already suggests, HAIP follows the experiences of DAIP. Since 1997, this network of cooperation offers cognitive behavioral training for perpetrators of domestic violence through the men's center Hannover. HAIP is the oldest intervention project in Germany and the Männerbüro has the most experience in the area of cognitive behavioral training for perpetrators of domestic violence. In 2000, 82% of the participants were referred by the police or the public prosecutor, 9% were referred by another counseling agency, and 9% were "self-referrals", mostly due to pressure from the (ex-) partner. In 2000, a total of 144 phone contacts led to 115 men who participated in an initial counseling session. Of these, 56 men decided to participate in a training course; 22 men completed the program in 2000; and 17 still attend the group. Seventeen men were excluded or discontinued the program. This clearly shows the high dropout rate, a problem that is characteristic of the work with perpetrators throughout Germany.

The work of the men's center is financed through modest municipal funds. Only men work here, three on a full-time basis who share two positions between them and three additional men on an hourly-fee basis of just a few hours. The men's center offers about three groups every year. The initial counseling is followed by a training program with 24 weekly sessions. Subsequent follow-up appointments are offered. The men pay for their participation, the fees depend on the individual situation and range between 10 € and 30 € per group session. The programs follow a sequence of modules that build on one another, but respond flexibly to each group and their current need for discussion. The contents follow the goals and contents listed in section 6.

The overall goal of the training courses is the "permanent and complete stop of any acts of violence whatsoever against your partner, and against women and children generally" (Männerbüro 2000, p. 2). However, the course leaders are very well aware that this goal cannot be achieved in the short period of the training courses, even though the number of group sessions is comparatively high, and that therefore the participation does not guarantee a woman freedom from violence. It is seen as an important success of the program if the men agree to a subsequent longer therapy. Contact information: maennerbuero.hannover@t-online.de

7.2 "Beratungsstelle im Packhaus", Kiel

"Beratungsstelle im Packhaus" (BiP),[18] Kiel, is a counseling center in the context of the "Coordination and Intervention Project of Schleswig-Holstein" (KIK). Since 1995, BiP has worked with men who are sexually or physically violent. BiP has many years of experience especially in the area of therapy for sex offenders for which the counseling center is well known throughout Germany. BiP is financed by the Ministry of Justice of the state of Schleswig-Holstein. From 1995 to 1998, BiP worked with perpetrators of violence in the context of the then municipal intervention project Kiel. At the conclusion of this project the state-wide "Coordination and Intervention Project of Schleswig-Holstein (KIK)" was established as a federal model project. BiP is now part of this network of cooperation.

BiP is a specialized therapeutic counseling center. It did not develop as a men's counseling center out of the new men's movement, but rather was designed as a "professional therapeutic expert counseling center for sexual and physical abusers" (Beratungsstelle im Packhaus 2000 b, p. 3). Currently, there are two male and two female employees with permanent positions and one further female employee on an hourly-fee basis. BiP largely works with the same goals and contents as listed in section 6. However, an important difference that corresponds to their therapeutic approach is the strong emphasis on the importance of self-motivation, which has to be developed during therapy with the predominantly externally motivated participants. In addition to the protection of the victims, which the program also especially emphasizes, BiP stresses the idea that anger and dissatisfaction lie at the root of acts of violence, and that men benefit personally from developing violent-free solutions to conflicts. The course of the program is similar to that in Hannover with one procedural difference: The program takes place in open groups, so that new participants can join at any time. A course consists of 13 sessions. These are preceded in each case by four individual counseling sessions.

About 50% of the participants are court ordered, the others are self-referrals who are urged to attend by their (ex-) partners, the youth welfare office, or other counseling centers. Forty-six men participated in the groups in 1998, and 51 in 1999. Five men completed the program in 1998, and six in 1999. BiP also has a very high drop-out rate (Beratungsstelle im Packhaus 1999, p. 7 and Beratungsstelle im Packhaus 2000 a, p. 7). Contact information: profa-packhaus@foni.net

## 8. Reflections on the evaluation of programs for violent men

The crucial question for all programs for violent men is whether they are successful. Which criteria and standards should be used to measure this success remains an open question. Interviews with experts working with violent men and experts working with women need to clarify how to assess the success of such programs. Several issues are currently under discussion. In terms of standards of success, do we expect 100% violent-free behavior? Can we count as success if violent-free behavior is achieved for a limited time or if the frequency and intensity of violence is reduced? Is partial success also laudable or should there be no compromises where the safety of women and their children is concerned? What comparison groups are necessary for the assessment of success of the evaluated programs?

In terms of the range of acts of violence and forms of violence addressed one should take into consideration the fears that women's shelters and women's counseling centers always voice who argue that the reduction of physical violence might lead to an amplification and refinement of psychological repressions.

In terms of desirable protection for women, do training courses sufficiently address the situation of the children and the man's responsibility as a father? In Germany, high expectations are directed towards this issue because men's fight for the right to child contact is often used to control his (ex-) partner. Does a program cooperate with women's protection and counseling services and disclose information to the (ex-) partners of participants? Does the program discuss and condemn violence in general or only violence against women and children? Omitting men's violence against other men can contribute towards strengthening the traditional images of masculinity.

In terms of the methods that should be used to measure success or failure, how can we keep track of possible repeat offenses, for how long should we, and how should we address the legal aspects of data protection? How can we compare our research results with those from other countries, while considering differences in policy and implementation?

In terms of quality criteria and standards, whenever the effectiveness of different training programs is compared, it is important to consider carefully differences in duration, method, didactics, entry requirements, and framework conditions. We need to examine the extent to which a program is integrated into an intervention system, and addresses the security interests of women and children. Likewise, we need to examine underlying images of human beings and the group leaders' understanding of the terms perpetrator and victim. Therapeutic groups and cognitive behavioral training will show differences in implementation. These can only be discovered by participant observation and questions about therapist attitudes.

Since the evaluation has just begun, we cannot yet say much about the effectiveness of the programs for violent men that we presented here. Since the evaluation is designed to be short term, [19] we focus our attention on the number of men in the course, continuity of participation, course progress, development of individual participants during the course of the program, and the assessment of the course leaders. Because the number of cases in the evaluated programs is small, the empirical basis for the forthcoming results will be limited. We may find out more about men who drop out of groups than about those who complete the program positively.

In order to make any reliable statements on the lasting impact of different training programs, they should be evaluated over several years. The question of whether or not men who participated in training programs will permanently refrain from using violence against their partners can only be answered if repeat offenses are documented and become known. So far, official records of renewed use of violence become available only if the public prosecutor's office opens another criminal investigation of domestic violence against a course participant.[20]

However, it is an open secret that in general, violence in close or intimate relationships only rarely becomes public knowledge. In addition, one can assume that a woman who reported the violence to the police, went through preliminary proceedings, and watched her (ex-) partner's mandated participation in a program for violent men, only to realize that all of this was useless, may be less willing to involve the authorities again. Such women might even be subject to repression by their (ex-) partner who may want to prevent new sanctions. Therefore, police and prosecution records provided limited evidence on the actual frequency of repeat offenses. In addition, interviews with the (ex-) partners are necessary to gauge the success of cognitive behavioral training. Finally, the course participants' own views on their development and their learning process as well as possible repeat offenses and their assessment are important sources of information. At present, we can make relatively reliable statements about repeat violence only for the duration of the program. As a recurring issue in the group sessions, repeat violence is part of our evaluation. Sometimes the (ex-) partners also give the agencies corresponding feedback on this issue.

Renewed violence of men against their partners or children during the course of the program can be registered reasonably well if a women's support agency is in touch with the woman at the same time. A pro-active approach of the women's support work therefore seems advisable because it reaches more women. Of special interest is the question to what extent cooperation between agencies that offer programs for violent men and agencies that offer support work for women should occur and what form such cooperation should take.

This touches on questions of confidentiality and the transfer of data. We are investigating these questions further through document analysis and data collection. In particular, we started interviews with experts and participants and since fall 2002 with the (ex-) partners of participants. Only the combination of data from multiple sources will give an approximately accurate image of the effectiveness of the cognitive behavioral training.

## 9. References


agisra. Frauenhandel und Prostitutionstourismus. Trickster-Verlag, München, 1990.

Austin JB, Dankworth J (1999). The impact of a batterers' program on battered women. Violence Against Women 5, 25-42.

Bentheim A, Firle M (1994). Beratungsarbeit für gewalttätige Männer. Ansätze und Projekte. In Diekmann A, Herschelmann M, Pech D, Schmidt K (eds), Gewohnheitstäter – Männer und Gewalt. Papy Rossa Verlag, Köln.

Beratungsstelle im Packhaus. Tätigkeitsbericht 1998. Selbstpublikation, Kiel, 1999.

Beratungsstelle im Packhaus. Tätigkeitsbericht 1999. Selbstpublikation, Kiel, 2000 a.

Beratungsstelle im Packhaus. Konzeptionelle Grundlagen der Arbeit mit körperlich Gewalttätigen im Rahmen des Kieler Interventionskonzeptes gegen häusliche Gewalt. Selbstpublikation, Kiel, 2000 b.

Brückner M (1998). Wege aus der Gewalt gegen Frauen und Mädchen. Fachhochschulverlag, Frankfurt/Main, 1998.

Bullens R (1997). Aufgaben und Möglichkeiten multiprofessioneller Kooperation aus der Sicht der Misshandlertherapie. Informationsdienst Kindesmisshandlung und Vernachlässigung, Sonderband 1.1, 105-114.

Burton S, Regan L, Kelly L. Supporting women and challenging men. Lessons from the Domestic Violence Intervention Project. The Policy Press, Bristol, 1998.

Connell RW. Der gemachte Mann – Krisen und Konstruktionen von Männlichkeiten. Leske & Budrich, Opladen, 1999.

Gondolf E. Batterer Intervention Systems. Sage, London, 2002.

Hagemann-White C, Kavemann B et al. Hilfen für misshandelte Frauen. Schriftenreihe des Bundesministeriums für Jugend, Familie und Gesundheit, Band 124, Kohlhammer, Stuttgart, 1981.
Hagemann-White C. Strategien gegen Gewalt im Geschlechterverhältnis. Centaurus-Verlag, Pfaffenweiler, 1992.
Hagemann-White C, Kavemann B, Ohl D. Parteilichkeit und Solidarität – Praxiserfahrungen und Streitfragen zu Gewalt im Geschlechterverhältnis. Kleine-Verlag, Bielefeld, 1997.
Haffner G, Spoden C. Möglichkeiten zur Veränderung gewalttätiger Männer im Rahmen einer Männerberatungsstelle – Gutachten für die Senatsverwaltung für Jugend und Familie. Berlin, 1991.
Hannoversches Interventionsprojekt HAIP gegen MännerGewalt in der Familie. Selbstverlag, Hannover, 2001.
Holzbecher M et al. Sexuelle Belästigung am Arbeitsplatz. Schriftenreihe des Bundesministeriums für Jugend, Frauen, Familie und Gesundheit. Kohlhammer, Stuttgart, 1991.
Kavemann B, Lohstöter I. Väter als Täter. Rowohlt-Verlag, Reinbek, 1984.
Kavemann B and Bundesverein zur Prävention von sexuellem Missbrauch. Prävention – eine Investition in die Zukunft. Donna Vita, Ruhnmark, 1997.
Kavemann B, Leopold B, Schirrmacher G, Hagemann-White C. Modelle der Kooperation gegen häusliche Gewalt. Ergebnisse der wissenschaftlichen Begleitung des Berliner Interventionsprojekts gegen häusliche Gewalt (BIG), BMFSFJ, Bonn, 2001.
KIK Schleswig-Holstein (ed). Täterarbeit – Programm zur Arbeit mit gewalttätigen Männern. Berlin, 2001.
Lempert J (2000). Therapie als Strafe. Frankfurter Rundschau, 02.12.2000.
Männerbüro Hannover e.V. Konzeption eines mehrwöchigen sozialen Trainingsprogramms für Männer die gegenüber Familienangehörigen, Partnerinnen und ggf. deren Kinder gewalttätig geworden sind. Selbstverlag, Hannover, 2000.
Nini M, Bentheim A, Firle M, Nolte I, Schneble A. Abbau von Beziehungsgewalt als Konfliktlösungsmuster – Abschlußbericht 1994. Kohlhammer, Stuttgart/Berlin/Köln, 1995.
Rösemann U. Untersuchung zur Übertragbarkeit des amerikanischen Modells DAIP: Intervention gegen Gewalt in der Familie, Gladbeck – im Auftrag des Bundesministeriums für Justiz, Familie, Frauen und Gesundheit, 1989.
Schall H, Schirrmacher G. Gewalt gegen Frauen und Möglichkeiten staatlicher Intervention. Boorberg, Stuttgart, 1995.
Schirrmacher G. Neue Reaktionen auf umweltdeliktisches Verhalten – Zugleich ein Beitrag zur Konkretisierung des Anwendungsbereichs des § 153a stopp. Osnabrück, 1999.
Teubner U, Becker I, Steinhage R. Vergewaltigung als soziales Problem – Notruf und Beratung für vergewaltigte Frauen. Bundesministerium für Jugend. Familie und Gesundheit, Kohlhammer-Verlag, Stuttgart, 1983.
Wodke-Werner V et al. (eds). Nicht wegschauen! Vom Umgang mit Sexualstraftätern. Nomos-Verlag, Baden-Baden, 1999.

## Notes

1 In West Germany, the public discussion on violence in gender relations began with the new women's movement at the beginning of the 70s. In East Germany, it developed for the most part after the opening of the wall in 1989.
2 For information on our research project see www.wibig.uni-osnabrueck.de.
3 A good overview of the discussion in Germany can be found in Brückner 1998.
4 In this context policies for juvenile sexual offenders dominate, and here predominantly in-patient therapeutic measures.
5 Some of the agencies that offer programs for perpetrators within the framework of the intervention projects, provide in addition to the groups for perpetrators of domestic violence also groups for sexual offenders.
6 In 1989, the then Federal Ministry of Young people, Family affairs, Women and Health asked the Gladbeck women's counseling center to conduct an "inquiry into the transferability of the American model DAIP" to German conditions (Rösemann 1989).
7 As is the case in Men Against Male Violence (Männer gegen Männergewalt), Hamburg (Nini et al. 1995).
8 Already in the middle of the 80s, MgM began to carry out counseling for violent men in the form of self-help-groups. From 1989 to 1992 the work of MgM within the context of in their cooperation with the program "Victim support Hamburg" was evaluated in the federal model project for work with violent men with the objective of reducing violence in relationships as a pattern of conflict resolution (cf. Nini et al. 1995).
9 The Anglo-American literature often refers to cases where the woman suggests the participation in a program for violent men as "wife-mandated" as opposed to "court-mandated" and "self-mandated". This term seems to gain more and more acceptance in the German-speaking area as well (cf. KIK Schleswig-Holstein 2001, p. 48). A similar discussion exists in the work with sexual offenders in the area of sexual abuse of children, where one also speaks of "court mandated" and "wife mandated" (Bullens 1997).
10 The legal basis for this is §§ 56 I and II as well as 56c I and II of the German penal code (cf. Schall and Schirrmacher 1995, p. 44).
11 §§ 59. of the German penal code.
12 § 153a of the German code of criminal procedure.
13 For more details on § 153a of the German code of criminal procedure and the possibility of its use in cases of domestic violence cf. Schirrmacher 1999; Schall and Schirrmacher 1995, p. 48.
14 In practice, the man in question may very well be summoned to the public prosecutor's office in order to explain to him the consequences of dropping the proceedings against orders or stipulations according to § 153a of the German code of criminal procedure.
15 This included the men's center Hannover, the counseling center in the Packhaus, Kiel (BiP), the Berlin center for the prevention of violence (BZfG), the Munich

information center for men (MIM), and the cooperation and intervention concept Schleswig-Holstein (KIK).

16 All agencies that work with perpetrators find it difficult to maintain contact with the men after the termination of the training. As a rule, the former participants use the follow-up appointments that some agencies offer only very erratically, if at all. In view of an evaluation and the prevention of repeat offenses, one should consider how the follow-up appointments can be designed more attractively, so that the contact to the men is maintained as long as possible after the completion of the program.

17 We present here three examples and give their contact addresses. Further work with perpetrators in intervention projects we evaluated offer for example the "Berlin Center for the Prevention of Violence" www.bzfg.de, as well as "Widerspruch e.V.", Kiel, widerspruch.kiel@t-online.de, ProFa Flensburg +49-461-9092620, "Die Brücke e.V.", Elmshorn +49-4121-840873 and "Neue Wege e.V.", Bremen +49-421-7947118

18 BiP is affiliated with the association Pro Familia Schleswig-Holstein.

19 The evaluation of the men's center Hannover for example began in July 2001 and is expected to end at the end of 2002.

20 We will talk to the public prosecutor's offices on how we can evaluate the repeat offenses on record.

*Lynne Harne, Research Associate at the University of Sunderland*
*Associate member of the Domestic Violence Research Group,*
*University of Bristol*

# Childcare, Violence, and Fathering: Are Violent Fathers Who Look After Their Children Likely to Be Less Abusive?

## 1. Introduction

This article is based on exploratory qualitative research with 20 fathers who had been identified as domestically violent and were separated or divorced from their partners. Most of these fathers had contact with their children in the post-separation context. This is the first UK study with violent men specifically focussing on their fathering practices. The sample was drawn from men who volunteered to be interviewed and were attending perpetrator programmes in different geographical areas of England.[1] The main method used was semi-structured depth interviews and these were supplemented with violence and abuse indexes to assess the extent of the men's violence.[2] Almost half of those interviewed identified themselves as 'serial' abusers in that they had also been violent towards partners and children in second families. All the interviews took place between the years 1998-2000. A separate sample of ten mothers[3] who had experienced domestic violence from partners and were experiencing problems with their children's contact were also interviewed about their ex-partner's fathering practices to provide comparative perspectives.

## 2. Background

### 2.1 The policy context

The research took place in a family law policy context, where there is the general legal presumption that children will benefit from having contact with fathers post-divorce or separation, even where they have been domestically violent and/or have directly abused the children. In these cases the courts have assumed that children will suffer greater harm in the long term if they do not have an "enduring relationship" with both natural parents (Re M (Contact: Welfare Test) [1995] 1 FLR: 274). A defining judgement in the mid-1990s, where there had been severe domestic violence towards the mother, stated that contact is "almost always in the best interests of children," and argued that even though the two year old child in question, was "distressed" by having obligatory contact with the violent father, "the court should take the long term view of the child's development and not accord too much weight to what appears to be short term and transitory problems." (Re O (Contact: Imposition of Conditions) [1995] 2 FLR: 124).

This policy also needs to be seen in the context of general social policy in the UK where, as Williams (1998) has pointed out, lone motherhood and father absence from families has come to be seen as the cause of all problematic masculinities and where it is argued fathers have been deprived of their role as providers for families (see for example, Murray 1990; Dennis and Erdos 1992). However, as Scourfield and Drakeford (2001) have noted there has been a shift in New Labour policy which has stressed the importance of involving fathers in the care of their children and the removal of perceived "institutional barriers to men spending more time with children," as a "solution to social problems" (Scourfield and Drakeford, 2001, p. 4 and p. 5). Moreover, they indicate that there has been an increasing emphasis in policy rhetoric on the notion of the 'unequal' treatment of fathers by mothers who are viewed as blocking fathers' access to participating in children's care (see for example Burgess and Ruxton, 1996; Burgess 1997) and ignoring the "dangers some men can pose in families," (Scourfield and Drakeford 2001, p.5; see also Featherstone 2000). Within such discourses, biological fathers have come to be constructed as the "prime victims of the gender order," and this is a perspective that has also been taken up by some New Labour feminists (Scourfield and Drakeford 2001, p. 4; see also Williams 1998).

Such ideas have also been promoted by the increasingly militant Fathers Rights Movement in family law policy (see for example Guardian 2001) and have influenced practice in this area (Children Act Sub-Committee 2001, 2002). Thus in family law practice, mothers who have blocked a father's contact with his children post-separation in order to protect children's safety in the context of domestic violence, have come to be defined as 'irrationally' hostile (Smart and Neale 1997) and are regarded as being unable to separate their own needs from those of the child (Children Act Sub-Committee 2001, 2002).

In this context, mothers' concerns about child abuse are frequently viewed as false allegations by the family courts and mothers can be viewed as having 'parentally alienated' children who are reluctant to have contact with violent and abusive fathers (Humphreys 1997; Radford et al. 1999; Brown et al. 2000). Moreover, there has been an increasing emphasis on using punitive powers against mothers to enforce such contact, which include the threat of imprisonment to ensure compliance (Radford et al. 1999).

However, by the end of the 1990s English family law policy recognised that very severe domestic violence could be a 'cogent' factor where the courts could deny or limit contact and that some attention needed to be paid to a father's conduct in relation to this (Children Act Sub-Committee 2000; Re L; Re V; Re M; Re H (Contact: Domestic Violence) [2000] 2 FLR: 334). However, the main concern about a father's parenting in the context of domestic violence has been in recognising that such behaviour can have a negative impact on children witnessing it and fears that some children might imitate such behaviour as adults (see for example, Re M (Contact: Violent Parent) [1999] 2 FCR: 56) and there has been little questioning of violent fathers' 'fitness' to look after children, per se (see also Smart and Neale 1997).

Moreover, the limited recognition given to domestic violence has usually meant that the courts may order some form of supported or supervised contact rather than deny such fathers' direct contact altogether.[4] In addition, some research has indicated that such arrangements may progress to 'unsupervised' contact within six months for over half of the children concerned (Radford et al. 1999). Some fathers may also be directed by the courts to attend a perpetrator programme or anger management course to address their domestic violence (Children Act Subcommittee 2000, 2001). However, in general there has been far less attention paid to violent men's fathering practices and how they look after children in their care and in this respect it is assumed that they are 'good enough' fathers, despite their domestic violence.

## 2.2 The aims of the research

One of the aims of this research was to explore the way violent fathers conceptualised their relationships with their children when they were still living with them and their experiences and views about contact in the post separation context. Moreover, since there is now substantial feminist research that indicates inter-connections between physical child abuse and domestic violence from the same perpetrator (Stark and Flitcraft 1988; Bowker et al. 1988; O'Hara 1994; Farmer and Owen 1995), fathers were also questioned about their understanding of child abuse and whether they perceived these relationships as abusive. For example, whilst recognising that mothers may also physically abuse their children, Stark and Flitcraft (1988) have indicated that men are three times more likely to be the prime offenders in the context of domestic violence.

Other research with children and mothers, as well as studies of child protection case records, has suggested that a majority of children living with a domestically violent father experience multiple forms of abuse which includes not only physical abuse, but direct emotional abuse and psychological terrorisation from their fathers, in addition to the negative impact of witnessing violence towards their mothers and that a substantial minority also experience sexual abuse (see for example, Abrahams 1994; Farmer and Owen 1995; Hester and Radford 1996; Hester and Pearson 1998; McGee 2000; Brown et al. 2000). Research with mothers post-separation also indicates that such children continue to be abused in the post-separation context (Abrahams 1994; Forman 1995; Hester and Radford 1996; Hester and Pearson 1998; Radford et al. 1999).

Another salient area in the interviews was to explore men's perceptions of their roles and responsibilities as fathers and to find out how much they were involved in childcare when living with partners. This was significant because of some earlier US research with mothers which had indicated that domestically violent fathers were less involved in child rearing than non-violent men in a comparison group (Holden and Ritchie 1991). Moreover, some later UK survey research on domestic violence and child contact focussing on mothers' perceptions of violent fathers' child caring activities before separation has suggested that, "a low level of involvement in childcare appeared to be matched by a low level of emotional commitment to children and a high level of abuse and neglect which persisted in many cases on contact visits." (Radford et al. 1999, p. 22).

In addition, some psychological perspectives drawing on psychoanalytic arguments (see for example, Chodorow 1978) have argued that men's familial abuse of children is directly related to the sexual division of labour in heterosexual nuclear families and that if fathers are more involved in the care of children, abuse is less likely to occur. Moreover, Burgess, arguing from a New Labour policy perspective of increasing fathers' involvement in the care of their children, has suggested in relation to physical abuse, "it may be that increased time with the children would lead to reduced violence in some cases," (Burgess 1997, p. 213).

Anti-sexist approaches to fatherhood have also suggested that men's greater participation in childcare can serve as a means of recreating more positive and 'nurturant' masculinities (see for example, Segal 1990; Campbell 1993; Pringle 1995; 1998). However, Pringle has also highlighted how the anti-sexist approach has failed to sufficiently take into account the "down-side of men's presence in childcare", (Pringle 1998, p. 319) and suggested that men's involvement can "bring into stark personal relief the destructive potential of dominant forms of masculinity", (Pringle 1998, p. 325).

## 2.3 The findings

The findings in this study, although exploratory, suggest that the relationship between violent fathers' participation in childcare and their direct abuse of children were far more complex and contradictory than indicated in the previous research highlighted above. Moreover, they suggest that there was no simple relation between the level of men's involvement in childcare activities and less abusive practices towards children. There were also indications that such assumptions could be dangerous when considering questions of risk to children in relation to issues of contact post-separation. Alternately, they raise issues about the meaning of men's parenting in the context of domestic violence and the need to recognise that where such parenting is based on assumptions and practices of power and dominance, abuse can continue to occur.

In discussing some of these findings, I have focussed mainly on the fathers' accounts of abusive behaviour when they had specific responsibility for caring for children; although they also described situations where children were clearly being abused when they were 'caught up with' or being deliberately used in the abuse of their mothers, and when children's own safety and the impact of such abuse was completely disregarded by these fathers. Moreover, although I did not expect that fathers would be prepared to talk about any sexual abuse, this did not mean it was not taking place.

## 2.4 Fathers' involvement in childcare when still living with families

Although most of the fathers in this study did not have the main responsibility for looking after the children, when still living with families, their level of involvement was greater than has been indicated from previous research and frequently occurred on a regular basis when mothers were working either full-time or part-time. These circumstances may to some extent, reflect general economic shifts in women's participation in the labour market, even where they have very young children (see for example, Walby 1997).

Two thirds of the fathers from this sample described themselves as having been regularly involved in looking after young children (under six) on their own for a few hours a day, whilst mothers worked part-time and in two cases, fathers described themselves as having been more involved in child caring activities and tasks associated with childcare than partners, because their partners worked. In addition, most fathers described having some regular involvement in childcare at meal-times and bedtimes and only three men in this sample subscribed to the view of the 'traditional' father, who regards childcare as totally the mothers' responsibility. Moreover, whilst some of these fathers may have exaggerated their child caring involvement, almost half of the women from the smaller mothers' sample, described partners undertaking some regular childcare and in three cases this was where mothers worked. In another case, the father had prevented the mother from participating at all in the care of their baby, because according to this mother's account, he wanted total control over the child.

Childcare and household tasks associated with childcare could however be a source of contradiction, conflict, and resentment for some fathers and used as a form of violent and abusive control over the whole family (see also Dobash and Dobash 1998). For example, whilst some fathers stated that they believed in equality in child caring roles they could also be resentful if they felt they were doing too much and this itself could lead to more violence towards mothers. Other fathers stated that they had chosen to do childcare whilst mothers worked, but they could also use violence against mothers if they were not getting what they wanted from the childcare context. For example, one father in 'explaining' some of his controlling violence towards his partner who worked part-time, stated,

"I've always felt that my wife didn't help on the household chores that needed to be done, like dishes, washing – things like that. I did it all – so I didn't have a lot of time to play with my son – I would be cooking tea, bathing X, or getting him ready for bed, so I didn't have a lot of time for the love side of things."

This example highlights the discursive strategies that violent men may employ to account for their domestic violence, which frequently draw on socially acceptable discourses of woman-blame to shift responsibility for their violence onto their victims (see for example Ptacek 1988; Hearn 1998, Cavanagh et al. 2001). What is interesting here is that this father is drawing on acceptable family policy discourses of woman blame to deny his own responsibility for the violence and justify it by arguing that his partner is preventing him from spending more quality time with his son and implying that she was not a good parent.

## 2.5 Abusive child caring practices

Of considerable concern was the kind of care that most fathers described as giving to young children when they were responsible for looking after them on their own and the rationalisations they used to justify any physically and emotionally abusive behaviour towards them. In this context, several violent fathers described children as 'provoking' their 'bad tempers', or 'short fuses' when they were looking after them, including the two fathers who indicated that they were the main carers of children.

For example, one father in describing looking after his two young daughters, aged three and four, whilst his partner worked part-time, argued that these two children could provoke his violent temper by their own behaviour.

"Basically what it is I'm frightened of my temper – the two of them are little sods together and I'm frightened of doing damage to them when I'm on my own with them and they are misbehaving and I can feel my temper and it does frighten me in case I hit them."

Interestingly, there were parallels in this father's justifications for the abuse of his children with his explanations for abuse towards his partner, since he described her as provoking his physical violence towards her because she would not 'shut up' when he was in a bad mood.

This father also described how he frequently smashed objects in the home which included the children's toys and cups. Moreover, although he recognised some of the negative impacts of this violence, he appeared to believe that if he played with them, this would cancel out its effects.

For example he said: "The youngest seems to cry all the time which I think is through the violence and eldest will just cry when I tell her to do anything which I think is due to all the shouting and violence and I'll say I'm not telling you off, I'm not saying you can't do it – wait until tomorrow – we'll play a game of when the police were chasing me – Oh I love them to bits."

Thus, although this father gave indications that sometimes he played with the children and that he had affection for them, the child caring context itself could be a means of providing 'opportunities' for his abuse. This issue was also raised by the mothers who were interviewed about their ex-partners' fathering practices particularly in relation to violent fathers 'play'. For example, one mother described how at times her ex-partner would play with her two young daughters, but then would just stop and lose his temper and shout and swear at them. Another mother described how her ex-partner "would play and cuddle" her four year old son and six year old daughter, "when he felt like it", but at other times would "kick and throw things at them". Both these mothers felt that such behaviour made the children more fearful of their fathers because they could never be sure how they were going to behave towards them. These mothers' views about the negative impact of such behaviour have been confirmed by McGee in her research with 54 children living with domestic violence. For example, McGee has indicated that such unpredictability can result in extreme anxiety in children and can control children's behaviour to such an extent that when they are with their fathers, they are frightened to show any reaction in case it provokes an outburst of violence or hostility.

Fathers also talked about being provoked into abusive behaviour when children failed to conform to their own expectations in the child caring context and this kind of explanation also paralleled some rationalisations for abusive control of mothers when they failed to live up to their partners' expectations of being good wives and mothers (see also Hearn 1998).

For example, one father talked about being regularly involved in looking after his three children after school and 'helping' them with their homework, because he was concerned about their educational achievement. However even though he expressed this concern, at the same time he described how he would shout and hit them if they did not do it properly, as is illustrated in the following extract where he is talking about a five year old girl.

"If they came home late from school ... and I'd get a bit annoyed if they wouldn't sit and concentrate on their homework, because I thought this was important – and if I actually sat say with P and she was trying to read to me and sometimes I would get a bit annoyed if I thought she wasn't trying –"

Q. "So you would hit her?"

A. "Yes, but it was also more shouting."

Q. "So what effect do you think that had on her?"

A. "She would just get terrified and curl up in a corner and wouldn't sit on my knee anymore."

It should be noted that this father had a number of convictions against him for child abuse and was registered as a Schedule One Offender under the Children and Young Persons Act (1933) for cruelty to children. In addition, two other fathers who had convictions for assaulting their children, described these as being related to hitting them "too hard", when they were responsible for putting them to bed, "because they would not go to sleep".

Abuse could also occur when children were perceived as making unreasonable demands and where these conflicted with fathers' own interests. This was illustrated by another father who was responsible for looking after two young daughters (both under five) full-time when his partner had to be hospitalised for two weeks, although the children went to a nursery during the day. He stated, "I would get frustrated with them – they wanted to talk to dad and they were continually badgering me for attention. There was awful shouting and verbal and aggressive abuse to get them to do what I wanted them to do – if they dropped their spoon that would be enough for me to slam my fist down on the table and that would shock them rigid."

## 2.6 Deliberate cruelty

Some fathers talked about being deliberately cruel to children when they failed to meet their own expectations in various childcare contexts, as is illustrated in the following extract, "Just being in the same room was enough in the end – it was mental abuse – they were terrified of me – all I had to do was look – I was quite cruel to be honest with you, for example at meal-times I used to sit there and make them eat things they really did not like and they used to cry. I wanted to make them too perfect – I wanted to make them what I should be like."

This deliberate cruelty to children was also illustrated in the mothers' accounts, although mothers were not always aware that it was occurring at the time. For example, one mother described how her partner was regularly emotionally and physically abusing one of her three children, a boy, when he was looking after the children after school, before she came home from work. She said, "It started when X was about four – his father would get angry at him and call him thick and stupid and hit him around the head and degrade him. When he got glasses he called him 'four-eyes'. When he was about six or seven years old he was still wetting himself at school, so I took him to the doctor and I got some medication to stop it. It was only by sheer chance that X let something slip and I realised why he was wetting himself, because his father was hitting him and shouting at him when he was looking after him."

These accounts contradict previous understandings of domestically violent fathers as having little involvement in childcare and point to the need to recognise that some violent fathers at least can be the prime abusers of children in childcare contexts. Moreover, although several fathers who participated in this study described regularly undertaking childcare, this was not an indication that they were prepared to prioritise the needs of children or understand the child's point of view. For example, only one out of the twenty fathers interviewed, indicated that children's own needs should be addressed, while looking after the children. In addition, most fathers' accounts illustrated that they regarded very young children as being responsible at least in part, for provoking the abuse through their own behaviour. Thus, far from becoming more 'nurturing' or 'caring' these fathers appeared to carry their own assumptions and practices of dominant and violent masculinity which they used to oppress and control women into the parenting context.

Of considerable concern in this research was that fathers were frequently describing ongoing abuse when looking after very young children. Yet, it is these children who are regarded within family law policy as 'needing' fathers most and whose own feelings and views may be disregarded in decisions made by professionals about contact, particularly where such children are fearful of their fathers and reluctant to go on contact visits (see for example Hester et al. 1994; Smart and Neale 1999: Radford et al. 1999).

## 3. Violent fathers' 'emotional commitment' and care of children in the context of child contact

### 3.1 Views on contact – in the name of love

One of the arguments for children's continuing contact with fathers in psychological discourses is the idea that children need a father's love and that this is important to children's emotional development (see for example Sturge and Glaser 2000)[5]. Yet, as has been seen in some of the accounts above, whilst these violent fathers often expressed their love for the children in their accounts, it was apparent that such discourses of 'love' had very different meanings from what might be conventionally understood as an unselfish emotional commitment to children's well-being. For instance, whilst fathers often stated they wanted contact because of their love for children or, more frequently, because of children's love for them, it did not seem to occur to most fathers that children might be reluctant to see them because of their violence and abuse.

Moreover, several fathers stated that they wanted contact because children could provide them with 'unconditional love.' As one father put it, "It's about love – you can't get love like that from anyone else."

Another father stated, "It's because of the unconditional love they give you – its one of the most important things in life". This also meant that within these constructs, children were viewed in a highly romanticised way, where they would always be the providers of unconditional love whatever happened in their lives and whatever fathers did. There was also implicit, within these portrayals, a sense that as permanent providers of love, children would give the men the emotional support which they had expected from women as their partners, but which they could no longer be relied upon to provide. For some fathers it was also because they "loved the children", or as has been seen in one of the accounts of above, because he "loved them to bits". But in both of these kinds of discourse there was a sense that children were perceived as some kind of 'emotional property' or 'investment' where they existed only for the benefit of fathers' own emotional 'needs' and where their love was a form of power over the children. Their perceptions of their child's love could also be used as a form of power against mothers and several fathers suggested in their accounts that their children loved them more than their mothers and implied that they were in competition for children's love.

### 3.2 Fathers' rights and the possession of children

Fathers' love could also be invested with rights of possession over children and in two cases was used to justify the forcible abduction of children from mothers, when they had left the family home because of a partner's violence. Moreover, most men viewed contact as their inviolable right as fathers which could not be interfered with, whatever the children's views and despite having at times acknowledged that their children were often terrified of them and that there had been other harmful impacts as a consequence of their abuse.

This was typically expressed by one father who had been convicted for assaulting his disabled child, when he stated "no one is going to come between me and my children because they are mine". Another father acknowledged that both his children were frightened of him when they came for contact, and that his son, age eight, would "shout and scream", when he was picked up for contact visits. But he explained this in terms of "the insecurity" of his wife and "her jealousy", and drawing on acceptable family law discourses, argued that she had manipulated the children into being fearful of him.

## 4. Fathers' abusive care of children in the contact context

### 4.1 Contact arrangements

At the time of interview, the fathers in this study had various forms of contact. Some fathers stated that because the children were at an age when they could choose whether they wanted to have contact (usually 11 or older), they now saw the children infrequently. For fathers of much younger children, it was clear that initial contact had been limited to a few hours a week, particularly where they had convictions for child abuse, and/ or there had been some involvement from child protection agencies; although only one of these fathers described having contact that was supervised by a social worker. Some of these fathers were still having limited contact, which often took place either in the mother's home or in the presence of grandparents, at the time of interview, although this did not necessarily mean that they were safe from abuse (see Forman 1995). For a few of these fathers however, contact had progressed to having children to visit one day a week or for overnight staying contact and in these situations, contact continued to be in a context where children could be perceived as provoking the father's abuse. For example, one father who was just beginning to have weekly overnight staying contact, and who had earlier stated that he had been abusive towards the children, because they were constantly 'badgering him for attention', when looking after the children full-time, described how he still found himself 'losing patience' and 'the same patterns of abuse coming back'. Especially when the children woke up too early and made demands on him.

This father's ongoing contact was also of particular concern, because he recognised that he had caused some considerable harm to one of his children. He said, "M's four and she's not talking well compared with children half her age. She has these catatonic states – it alarms me – I've seen that when I've been aggressive and smacked her – that's one major effect that I've had on her."

In addition, three other fathers whose contact had been extended to weekend overnight visits, indicated that they continued to use physical abuse and threats in order to control the children.

### 4.2 Emotional abuse

Several fathers indicated that they were emotionally abusing the children during contact visits, through insulting their mothers or through making threats, although there was not necessarily any acknowledgement that they were harming the children in this process. For example, one father who had contact for a few hours a week, described how he had told the children that he would have their mother sent to prison unless she agreed to overnight staying visits. However, in relating this, this father seemed totally unconcerned of how such a threat might be perceived by the children and how anxious they might be for their mother's and their own safety. Another father dismissed any thought that his abusive comments about their mother could have any negative impact on the children, as is illustrated in the following extract.

Q. "Have you said things to the children about their mother?"

A. "Yeh – I've said she was no good – that she's got other boyfriends?"

Q. " How do you think they feel about that?"

A. "They don't know the truth do they? They're just listening to me spouting rubbish from their point of view – they're just thinking I'm not a nice person saying things about their mum – you see I'll say anything because I know I can get back at her."

Numerous examples of this kind of abuse during contact visits were also highlighted in the mothers' accounts (see also Hester and Radford 1996). For example, one mother stated that her five year old son had been diagnosed by a child psychologist as depressed and as having various behavioural and stress related problems including constant bedwetting, attention deficit disorder, and daily kicking and hurting of other children at school, as a direct consequence of what his father was telling him. She said, "He was telling him, mummy's a liar and don't believe a thing she says – she's a thief and she stole your Christmas stocking and daddy's house is better than mummy's and you're going to live at daddy's soon anyway ... because when you lose a mummy it's not so bad – because sometimes mummies die. And you can misbehave here and what you do here you can do at your mum's house and kick your friends."

## 5. Conclusion

This research has indicated that despite being regularly involved in childcare activities, violent fathers can continue to physically and emotionally abuse children. There were also indications that increased child caring responsibilities could provide more 'opportunities' for fathers to abuse and this was indicated both prior to separation and when such fathers gained more contact in the post-separation context. In this context, fathers' deflection of responsibility onto very young children themselves for 'provoking' the abuse and their inconsistent parenting behaviour, was of considerable concern and could have grave implications of harm for the children themselves. In addition, in the post-separation context, their motivations for wanting such contact and their parenting practices suggest that whilst they may express some 'emotional commitment' to children, this is bound up with their own self-interested motivations, and that they are rarely able to prioritise children's interests above their own (see also Sturge and Glaser 2000).

Whilst these findings are exploratory and further research is required in this area, they indicate a need for an assessment of risk of domestically violent fathers' parenting, particularly in relation to considering their contact post-separation, which goes beyond looking at how much involvement they have had in 'caring' for the children or the different activities of childcare, since this is not necessarily an indication that children will not be harmed. They also suggest that the problem is not merely one of poor parenting 'skills', but that violent fathers' practices in relation to children are bound up with their own constructions of dominant masculinity that are integrally connected to their violence and control of women in familial relationships.

## 6. References

Abrahams C. The Hidden Victims: Children and Domestic Violence. NCH Action for Children, London, 1994.
Bowker LH, Arbitell M, McFerron (1988). On the relationship between wife-beating and child abuse. In Yllo K, Bograd M (eds), Feminist Perspectives on Wife Abuse. Sage, London.
Brown T, Frederico M, Hewitt L, Sheehan R (2000). Revealing the existence of child abuse in the context of marital breakdown and custody and access disputes. Child Abuse Negl 24, 849-859.
Burgess A, Ruxton S. Men and Their Children. Proposals for Public Policy. Institute for Public Policy Research, London, 1996.
Burgess A. Fatherhood Reclaimed. The Making of the Modern Father. Vermilion, London, 1997.
Campbell B. Goliath: Britain's Dangerous Places. Methuen, London, 1993.
Cavanagh K, Dobash, RE, Dobash RP, Lewis R (2001). Remedial work: men's strategic responses to their violence against intimate female partners. Sociology 35, 695-714.
Children Act Sub-Committee (Advisory Board on Family Law). A Report to the Lord Chancellor on the Question of Parental Contact in Cases Where There is Domestic Violence Lord Chancellor's Department, London, 2000.
Children Act Sub-Committee (Advisory Board on Family Law). Making Contact Work. A Consultation Paper. Lord Chancellor's Department, London, 2001.
Children Act Sub-Committee (Advisory Board on Family Law). Making Contact Work. A Report. Lord Chancellor's Department, London, 2002.
Chodorow N. The Reproduction of Mothering. University of California Press, Berkeley, CA, 1978.
Dennis N, Erdos G. Families without Fatherhood. Institute of Economic Affairs, London, 1992.
Dobash RE, Dobash RP (1998). Violent men and violent contexts. In Dobash RE, Dobash RP (eds), Rethinking Violence Against Women. Sage, London.
Farmer E, Owen M. Child Protection Practice. Private Risks and Public Remedies. Messages from Research. HMSO, London, 1995.
Featherstone B (2000). Big daddy, big brother: New labour and the politics of fatherhood. Unpublished paper, Social Policy Association Conference, Surrey.
Forman J. Is There a Correlation Between Child Sexual Abuse and Domestic Violence? An Exploratory Study of the Links Between Child Sexual Abuse and Domestic

Violence in a Sample of Interfamilial Child Sexual Abuse Cases. Women's Support Project, Glasgow, 1995.
Guardian Newspaper. Fathers picket judges over child access. October 30, 2001.
Hearn J. The Violences of Men: How Men Talk About and How Agencies Respond to Men's Violence to Women. Sage, London, 1998.
Hester M, Humphries J, Pearson C, Qaiser K, Radford L, Woodfield K (1994). Domestic violence and child contact. In Mullender A, Morley R (eds), Children Living with Domestic Violence: Putting Men's Abuse of Children on the Childcare Agenda. Whiting and Birch, London.
Hester M, Radford L. Domestic Violence and Child Contact Arrangements in England and Denmark. Policy Press, Bristol, 1996.
Hester M, Pearson C. From Periphery to Centre: Domestic Violence in Work with Abused Children, The Policy Press, Bristol, 1998.
Holden G, Ritchie K (1991). Linking extreme marital discord, child rearing and child behaviour problems: evidence from battered women. Child Dev 62, 311-327.
Humphreys C (1997). Child Sexual Abuse Allegations in the context of divorce: issues for mothers. Br J Soc Work 27, 529-44.
O'Hara M (1994). Child deaths in contexts for domestic violence: implications for professional practice. In Mullender A, Morley R (eds), Children Living with Domestic Violence: Putting Men's Abuse of Women on the Child Care Agenda. Whiting and Birch, London.
McGee C. Childhood Experiences of Domestic Violence. Jessica Kingsley Publishers, London, 2000.
Murray C. The Emerging British Underclass. IEA Health and Welfare Unit, London, 1990.
Pringle K. Men, Masculinities and Social Welfare. UCL Press, London, 1995.
Pringle K (1998). Men and childcare: policy and practice. In Popay J, Hearn J, Edwards J (eds), Men, Gender Divisions and Welfare. London, Routledge.
Ptacek J (1988) Why do men batter their wives? In Yllo K, Bograd M (eds), Feminist Perspectives on Wife Abuse. Sage, London.
Radford L, Sayers M, AMICA (1999). Unreasonable Fears? Child Contact in the Context of Domestic Violence: A Survey of Mothers' Perceptions of Harm. Women's Aid Federation of England, Bristol.
Scourfield J, Drakeford M (2001). New labour and the problem of men. Unpublished paper, Belfast, SPA Conference.
Segal L. Slow Motion: Changing Masculinities, Changing Men. Virago, London, 1990.
Smart C, Neale B (1997). Arguments against virtue: must contact be enforced? Fam Law, 332-336.
Smart C, Neale B. Family Fragments? Polity Press, Cambridge, 1999.
Stark E, Flitcraft A (1988). Women and children at risk: a feminist perspective on child abuse. Int J Health Serv 18, 97-118.
Sturge C, Glaser D (2000). Contact and domestic violence – the experts court report. Fam Law, 615-629.
Walby S. Gender transformations. Routledge, London, 1997.

Williams F (1998). Troubled masculinities in social policy discourses: Fatherhood. In Popay J, Hearn J, Edwards J (eds), Men, Gender Divisions and Welfare. Routledge, London.

## Notes

1 These were programmes where fathers either had either been court-mandated to attend as part of probation order, or had volunteered to attend in order to address their domestic violence.

2 Such indexes or checklists are used in perpetrator programmes as means of eliciting violent and abusive behaviours that men are not willing to disclose verbally. They have also been used in previous research with violent men (see for example Dobash et al.; 1996).

3 The mothers' sample was found through two support groups for mothers who had experienced domestic violence and problems over their children's contact.

4 Most contact centres do not offer direct supervision of the father's contact; rather they provide facilities where contact can take place in a neutral environment. The term- supervised contact can cover a range of arrangements, but may mean the contact is overseen by social workers, although this does not mean that children are completely safe from abuse (Hester and Pearson, 1998).

5 Sturge and Glaser (2000) were called upon by the appeal court to provide an expert's opinion on children's contact in the context of domestic violence. Although they outlined a number of benefits to children in having contact, they also outlined several situations where children could be put at risk from a domestically violent father. These included the increased likelihood of direct physical and emotional abuse by a domestically violent perpetrator, and where the father is unable to "consistently sustain the prioritisation of the child's needs". (Sturge and Glaser, 2000, p. 618 to 620).

*Stephanie Condon, Maryse Jaspard and the Enveff*[1] *research team, Paris, France*

# A Quantitative Approach to Understanding Violence Against Women: the French Survey

## 1. Introduction

As in many other European countries, women in France have acquired relative independence, notably regarding birth control and in economic relations. This has been largely thanks to their presence on the labour market and a higher level of education. However, the effects of male domination are still felt. Male violence can be analysed as a fundamental mechanism in the social control of women. The violent acts women are subjected to are an expression of this domination. This does not, however, rule out the fact that they too, practice certain forms of violence. Nor does it mean that male violence is an unavoidable phenomenon.

Gaining knowledge on the subject of violence against women remains problematic in spite of the declared intention, by politicians, social services and other institutions, of changing the image of the victims. The victims of these acts of violence, which are committed on a daily basis and most of the time in the home, still often repress them from their minds to the point of even denying their existence. This invisibility is one of the obstacles that must be overcome in conducting research on violence against women.

## 2. The need for a survey

The Beijing platform drawn up in September 1995, at the Fourth International Conference on Women, encouraged governments to take action to prevent and eliminate violence against women. This action was to include the gathering of reliable data, enabling the provision of statistics relating to the prevalence of the various forms of violence against women and to encourage research into the causes, nature, severity and consequences of such violent acts, as well as an evaluation of the efficiency of preventative measures taken by public authorities.

Quantitative surveys had already taken place in some countries, some on a local scale (for example in London: Hanmer and Saunders 1984), some national (the first, by Statistique Canada 1993). The issue of violence specifically directed towards women had been approached in various ways : as an element of family violence in the USA (Straus and Gelles 1986), as experienced by women in cohabiting couples in Switzerland (Gillioz et al. 1997) and the Netherlands (Römkens 1992), or in all contexts in Finland (Heiskanen and Piispa 1998). Towards the end of the decade, the number of surveys around the world multiplied (Population Reports 1999).

France was still without any means of measuring the scale and nature of the phenomenon within the general population. Existing statistics concerned only acts of violence reported by women to the police or other services. Violence experienced by women contacting crisis centres, refuges, or crisis lines was well understood; since then studies on their activities, numbers and characteristics have been conducted (Cesur-Gers 1998; Collectif féministe contre le viol 1996-97; Cromer 1995). However, a range of other types of violence experienced by other groups of women remained unknown. In such a context, and given the status of statistics – notably national-level statistics – in modern society, surveys concerning the general population become indispensable in order to fully measure the scale of violence individuals are subjected to. It is especially so since debate reveals the existence of a great many preconceived ideas on the issue. The production of reliable data, obtained from a representative sample of the female population residing in France, aimed to supply information useful to the different institutional partners (political, social, the media, etc.) and society at large.

Thus, having found that the pilot survey (November 1998) had been successful in reaching most social groups, a representative survey on a national scale was considered feasible. The data produced by the Enveff survey (2000) will add to existing knowledge of the phenomenon and not compete with previous studies, as well as enabling decision-making, leading to integrated action (Jaspard et al. 2001, 2003).

### 2.1 Methodological issues

Preceding surveys had been based on a variety of protocols and tended to emphasise particular contexts in which violence could occur or specified certain types of violence. Some methodologies are more appropriate than others for studying violence against women. Perusal of the research reports and questionnaires of the Canadian, Finnish and Swiss surveys, as well as those of French crime and security surveys, was particularly useful in the early stage of our research.

The objectives of the survey were to:

1) identify different types of interpersonal violence against women of 20-59 years in all spheres of life, whoever the perpetrator;
2) measure the prevalence of violence against women in the course of the previous 12 months, as well as during their whole life;
3) define the temporal aspects of violence, that is to define whether it is isolated or limited, continuous or repeated, in order to identify situations of harassment and cycles of violence as well as to analyse cumulative mechanisms;
4) analyse the context of violence, using social and demographic characteristics, the living conditions and lifestyle of women and wherever possible, the characteristics of the family and social circles, as well as the characteristics of the perpetrator of the violent act;
5) examine the fear of violence through the use of public space;

6) trace the biographies of women in order to observe the relationship between the family schooling and occupational histories and the emergence of violence;
7) study types of reactions to violence, what kind of help women seek from family and friends and from official or voluntary organisations;
8) identify the consequences of physical and mental health violence on family and social life and on the use of public and private space.

Since we sought to explore the diversity of contexts in which violence occurs, we compiled a questionnaire with separate sections dealing with experiences in different spheres of life (see appendix I for structure of the questionnaire). It was hoped that by such an approach we would obtain a clearer picture of the position of intimate partner violence in relation to other types of violence experienced by women. Six sections refer to violence experienced during the twelve months preceding the interview in different situations: in public places, at work or at school/university, in intimate partner relationships, with a former partner, within the family or circle of contacts. Each module records the occurrence of events or acts[2] described in a more or less identical fashion according to the context; the investigation also concerned the circumstances, as well as the reactions on the part of the women and the help they sought from friends and family or institutions, for the most serious event in the respondent's opinion.

The final section of the questionnaire concerns acts of physical or sexual violence: physical violence experienced since the age of 18 and sexual violence at any moment in the respondent's lifetime. The age, the perpetrators, the circumstances, and legal actions are recorded. In the case of repeated acts, the first and the last act were taken into account.

Our definition of violence was that violence originates within a power struggle or a situation of dominance of one person by another or others and is distinct from situations of conflict, a more normal aspect of a relationship between two persons of fairly equal status. The research design highlighted the fact that violence may be carried out by men or women, that it manifests itself in a multitude of ways and is present in all social contexts but in varying forms. In addition, violent acts are seen to be all the more traumatising when they are out of step with norms and cultural rules in the social group to which the victim belongs, and this fact contributes to subjective definitions of violence. Thus, whilst drawing up the questionnaire, we adhered to a number of precepts:

We have avoided inferring any hierarchy of violence amongst the different types – physical, sexual and psychological – or contexts of violence (see appendix II for types of violence described in questions).

We do not use the word 'violence' in the questionnaire itself referring rather to gestures, acts, words, or incidents without qualifying them as violent.

The gender of the perpetrator of the violent act is not implicit in the questions (no use of 'him', 'his' etc.).

The data was collected using the CATI method (computer-assisted telephone interviewing) and was used to interview 6,970 women. The survey institute worked in close collaboration with the research team, a member, of which was present every day throughout the data gathering period. The training of women interviewers lasted three days including a half day of awareness building provided by members of women's rights associations. Contact with the general population was optimised through the acquisition of an ex-directory list of numbers, the possibility of several calls to each number and of offering an appointment to carry out the interview at a more convenient time.

### 2.2 Qualitative results of the survey

Before going on to look at the statistical results of the survey, three important points must be made with regard to the reliability of the data and to what the survey revealed about the scale of silence surrounding violence against women and the extent to which victims hide the violence.

The survey was well accepted, despite the length of the questionnaire and the intimate nature of many of the questions. We attribute this to the neutral, anonymous interview situation, as well as to a greater awareness of violence against women (rape, domestic violence, sexual harassment) in the general population and a desire amongst women to eradicate such violence.

Over half the women having experienced violence spoke about it for the first time, (particularly so in the case of sexual violence) and thus we feel that the survey helped interviewees describe violence experienced. Some women were stimulated into reassessing situations they had endured and thus mentioned violent acts that they might not immediately have thought of. The existence of violence against women was never denied by interviewees. The women apparently wished to describe all types of violence experienced, including acts committed by other women.

A low level of reported violence was confirmed, whether to official authorities or to intermediate persons such as doctors. Only a small minority of women that experienced violence had reported a violent act, usually in the case of physical violence.

## 3. Findings on violence against women over the twelve-month period

Analysis of the data on violence reported over the twelve months preceding the survey validated the approach by context. The types of violent acts and their meaning to individuals vary and are linked to social relations in each particular context. We will present first of all the overall results by sphere, before examining in more detail violence reported in two contexts.

Before presenting the results, it is necessary to explain how the indicators were constructed. A single instance of physical or sexual assault is sufficient to identify it as an act of violence. However, in other types of abuse, such as insults, denigration, contempt, control and other forms of emotional abuse or psychological pressure, the situation of domination results from an accumulation of events which, if taken separately, might seem quite trivial. Thus in order to establish a scale of these forms of abuse, the indicators had to combine both a measure of frequency and the nature of the events.

Table 1: *Proportion of women declaring acts of verbal, psychological, physical or sexual violence over the preceding 12 months, by sphere and age (in %)*

| Type of violence | 20-24 years | 25-34 years | 35-44 years | 45-59 years | Total |
|---|---|---|---|---|---|
| IN PUBLIC AREAS | n = 717 | n = 1,934 | n = 2,122 | n = 2,197 | n = 6,970 |
| Verbal abuse and threats | 24.6 | 15.2 | 11.7 | 8.6 | 13.2 |
| Physical assault | 2.7 | 1.6 | 1.2 | 1.7 | 1.7 |
| Sexual advances and assault | 6.4 | 2.6 | 0.9 | 0.5 | 1.9 |
| IN THE WORKPLACE | n = 336 | n = 1,410 | n = 1,593 | n = 1,411 | n = 4,756 |
| Verbal abuse and threats | 11.7 | 10.1 | 8.8 | 6.2 | 8.5 |
| Psychological pressure | 20.2 | 18.6 | 15.2 | 15.7 | 16.7 |
| of which emotional harassment | 5.2 | 4.7 | 3.6 | 3.1 | 3.9 |
| Physical assault | 0.6 | 0.6 | 0,7 | 0,5 | 0,6 |
| Sexual advances and assault | 4,3 | 2,8 | 1,9 | 0,8 | 2,0 |
| INTIMATE PARTNER* | n = 464 | n = 1,707 | n = 1,872 | n = 1,865 | n = 5,908 |
| Verbal abuse and threats | 6.1 | 4.1 | 4.3 | 3.9 | 4.2 |
| Emotional blackmail | 2.7 | 1.4 | 2.3 | 1.6 | 1.8 |
| Psychological pressure | 37.6 | 26.1 | 23.0 | 21.0 | 24.3 |
| of which emotional harassment | 12.1 | 8.3 | 7.7 | 6.7 | 7.9 |
| Physical assault | 3.9 | 2.5 | 2.5 | 2.2 | 2.5 |
| Rape and other forced sexual acts | 1.2 | 0.9 | 1.0 | 0.6 | 0.9 |

* violence experienced by women in a couple, living with or apart from partner
Source: National Survey on Violence Against Women in France (Enveff), 2000

The most striking result appearing from Table 1 is that in each life context, whatever the type of violence concerned, young women in the

20-24 age group are those who claim to have been subjected to the most violence (Table 1). It is within intimate partner relationships that most physical violence was found to occur, whereas sexual violence is reported more often in contexts outside the home. Public space is revealed to be a context in which verbal abuse is very frequent.

### 3.1 Intimate partner violence

Violence within couples is not necessarily restricted to the 'domestic' sphere and similarly nor is violence between family members. We studied women living as a couple, whether the couple has legal status or not, whether the partners cohabit or live apart. Of the 6,970 interviewees in the sample, 5,908 women were in such relationships. A minority of these (115) had separated during the previous year and these women have experienced a level of violence three to four times higher than all women in couples.

A global indicator of 'violent situations' was constructed and subdivided into two 'levels' (A and B) so as to show the progression in the seriousness of situations. The 'very serious' level (Level B) refers to an accumulation of acts of violence; in this case, physical and sexual violence are more frequent, are repeated or associated with verbal insults and emotional harassment (2,7% of women).

In addition, a psychological harassment indicator was constructed to explore this type of violence in the context of the couple (see appendix III for list of items). This type of violence is considered to be an insidious form of domination of one gender over the other.

However, in many cases, distinguishing different types of intimate partner violence is not as pertinent as how different forms are intertwined (Brown and Jaspard 2001).

At older ages the proportion of 'serious' violent situations declines, which seems linked more to a decline in emotional harassment. However, 'very serious' situations are encountered at all ages. Married women, and more generally those who live with their partners, report around a third fewer acts of violence than women in a partner relationship who do not share the same dwelling; but the age factor may combine with this characteristic as women in this group are generally younger.

The frequency of such violent situations is not affected by the presence of children or by the length of the partner relationship. There is not necessarily an escalation of violence over time: situations of accumulated acts of violence may be triggered very early and last a long time, in a cycle of violence. On the other hand, situations of emotional harassment may occur during a whole lifetime without evolving towards physical or sexual aggression.

Table 2: *Proportion of women living as a couple and in a situation of domestic violence, according to the socio-occupational category below and the form of violence experienced (in %).*

| Occupational and employment status | Number of women | Global rate | of which | |
|---|---|---|---|---|
| | | | Level A | Level B |
| Farmers, craftswomen, trades women, small business owners | 131 | 7.7 | 6.6 | 1.1 |
| Senior management, higher intellectual professions | 425 | 8.7 | 6.1 | 2.6 |
| Middle management, middle-level professions | 1,189 | 8.3 | 6.8 | 1.5 |
| Clerical, sales or service staff | 1,726 | 8.3 | 6.3 | 2.0 |
| Industrial or other manual workers | 311 | 7.9 | 4.6 | 3.3 |
| Unemployed | 489 | 11.9 | 9.0 | 2.9 |
| Students | 176 | 11.1 | 9.8 | 1.3 |
| Other women not employed | 1,346 | 9.4 | 6.3 | 3.1 |
| Total | 5,793 | 9.0 | 6.7 | 2.3 |

Unemployed women and students report the highest overall level of violence (over 11%). However, there are twice as many unemployed women as there are students reporting a 'very serious' situation of violence (2.9% compared with 1.3%), a proportion similar to women in the blue collar workers group (approximately 3%), and perhaps unexpectedly, also to women in the managerial and senior executive social group (2.6%). Sensitivity to acts of violence differs from one socio-economic group to another, apparently higher amongst more qualified social groups. However, whilst this phenomenon is certainly perceptible in the 'serious' level in which emotional harassment is a relatively important component, it has hardly any impact in situations of accumulated acts of violence. This result confirms that serious domestic violence concerns all social categories.

Although such violence does not increase as one moves 'down' the social scale, high occupational instability and withdrawal from the labour market seem to foster the emergence of situations of accumulated violence (Table 2). Job instability among men has an even greater impact on violence within the partner relationship. This occurs very frequently among the unemployed who do not receive unemployment benefits (16%, of which 8% are 'very serious' violent situations) or others who do not work, no doubt linked to their reaction towards exclusion from the labour market. For either of the partners, having experienced just one period of unemployment increases relatively little the development of situations of violence within the partner relationship. On the other hand, having experienced several periods of unemployment doubles the overall proportion of violent situations and triples that of 'very serious' violence (Table 3).

Although the perpetration of violence within intimate partner relationships seems not to be related to the woman's level of education, when the partner has a lower level of education, this does seem to increase the woman's exposure to the risk of violence (Brown and Jaspard 2001).

Therefore, socio-economic criteria such as socio-economic group, level of education or income, adequate explanatory variables for a number of social phenomena, are not substantially discriminating factors in terms of the underlying mechanisms of violence within couples. Other factors should be looked for that are more closely linked to the social perceptions of male and female roles within couples, and more generally the social representations of women. Such characteristics are of a more socio-cultural order.

Table 3: *Proportion of women living as a couple and in a situation of domestic violence, according to employment status and level of violence (in %).*

| Occupational status | Number of women | Global rate | of which | | Emotional harassment |
|---|---|---|---|---|---|
| | | | Level A | Level B | |
| In employment | 3,794 | 8.2 | 6.2 | 2.0 | 5.6 |
| Unemployed receiving state benefits | 247 | 10.3 | 8.8 | 1.5 | 7.9 |
| Unemployed without state benefits | 182 | 14.0 | 9.7 | 4.3 | 8.6 |
| Prolonged leave from work | 258 | 12.0 | 10.2 | 1.8 | 7.6 |
| Housewife having worked previously | 620 | 9.7 | 5.4 | 4.3 | 6.0 |
| Housewife never having worked | 340 | 8.1 | 6.1 | 2.0 | 6.9 |
| Total | 5,793 | 9.0 | 6.7 | 2.3 | 6.2 |

For example, religion often structures gender relations, setting a moral code by which men and women are expected to live. Not being brought up with religion seems to be linked to less violence in couples than having had a religious education, irrespective of the denomination. In addition a strong correlation is observed between the importance given to religion by women and situations of violence within such relationships, notably very serious situations affecting 5.2% of women who consider religion as important compared with 2% of the others. It is among foreign immigrant women that the highest proportions of women who give importance to religion are to be found. This proportion exceeds 80% among women from the Maghreb. This said, it must be remembered that the survey does not provide information on the religious upbringing or beliefs of the partner. Furthermore, a direct link between religious beliefs and violence is questionable, particularly since patriarchal relations have often existed in societies long before the one or other religion began to predominate in a particular state.

Foreign-born immigrant women more frequently experience situations of partner relationship violence than other women. Women from the Maghreb and Sub-Saharan Africa more often experience partner relationship violence of the least serious type and they report twice as much emotional harassment than other women. Whilst these results are linked to the demographic characteristics of these groups, notably age and marital or family status, they also depend on other criteria such as social isolation, economic hardship and conflict between different sets of cultural norms and values. Whatever level of the indicator is considered, situations of violence are as frequent among mixed couples as among those of the same origin.

For women who are second generation immigrants, with parents from the Maghreb, the global indicator of partner relationship violence is doubled, this high rate is mainly due to situations of 'very serious' violence. No difference was found between women who had partners of the same origin as themselves and the other women in this group. Over half of these women continued education beyond the age of 18, a considerable proportion are full-time students (11%), and one quarter are unemployed, in which each of these characteristics place them in the groups reporting the most violence. Furthermore, many women in this population relate to the image of the young Maghrebin women demanding recognition of their rights as individuals and staking their claim to a place in French society whilst defending their North African roots. In such a social context, women are likely to want to denounce violent situations.

### 3.2 Violence in public areas

The Enveff survey was not a survey on women's safety. Nonetheless, a set of questions based on local surveys conducted in the UK (Valentine 1992; Pain 1997), was included at the beginning of the questionnaire. These sought to measure the extent to which women go out at night alone and to what types of places and then to discover whether they are afraid of walking or being alone in certain areas, depending on the time of day. The responses have been linked to those in reply to the questions on the experience of violence which appear much later on in the questionnaire[3].

Violence in public spaces was the first context explored by the questionnaire. Examples of daily situations were referred to at the outset so that it would be clear that we wanted the interviewee to remember events that had occurred outside the home context or the workplace and that it did not involve their partner or other family members. Situations could include journeys by public transport, walking along a street, on a beach or through a park, an afternoon spent at a sports club, a visit to a shop, a restaurant or a concert hall, and so on. Contrary to preconceived ideas, violence outside the home or the workplace does not necessarily occur in deserted places that women might frequent alone at night. Rather, it takes place in ordinary circumstances in the context of normal and daily use of public places.

Table 4: *Violence experienced in public areas (in %).*

| Type of violence | Global rate | Number of women |
|---|---|---|
| Being followed | 5.2 | 331 |
| Exhibitionism | 2.9 | 190 |
| Fondling, sexual advances | 1.9 | 111 |
| Being robbed | 0.7 | 48 |
| Physical assault | 0.6 | 37 |
| Armed threats or attack | 0.6 | 38 |
| Sexual assault | 0.1 | 8 |

The most frequent type of violent behaviour reported was verbal abuse (Table 4). Thirteen per cent of women said that they had been insulted or verbally abused during the past twelve months and half of these reported having experienced such violence several times. Interestingly, in one quarter of the cases, interviewees were acquainted with the offender. This type of violence is most common in the Paris agglomeration and in other large cities, probably owing partly to the fact that more women travel to work, shops or other amenities by foot or by public transport.

Being followed – a situation usually perceived as having sexual connotations – or being harassed by an exhibitionist is reported mostly by younger women, regardless of their social position or level of education and is more frequent in large cities. Women were most often driving a car when they were followed. Sexual abuse in public places, reported by 2% of interviewees, consists usually of sexual advances or fondling. Sexual assault, attempted rape or rape is much less frequent. A global indicator, constructed to cover all types of 'sexual abuse' reveals the sexist atmosphere prevailing in much of public space, especially in the Paris agglomeration (15% of women as compared to 8% of the total sample) and amongst women aged 20-24 years (22%) (Jaspard et al. 2001).

Acts of physical violence were reported by 2% of women. Although risk of being attacked is not significantly linked to age, young women who were unemployed or living alone were over-represented amongst victims. Whilst assailants were usually male, one in five were female. Women were not always alone and the attacker was sometimes known. Physical assault takes place in social contexts that differ considerably from those in which other types of violence occur. The fact of living alone or of not having a degree in further education (or a primary school certificate), of being unemployed, retired or a student, increases this risk. Acts of physical violence are not only sexist forms of violence, but they are also social forms insofar as they appear to affect to an even greater extent people who are in a situation of relative vulnerability, whether from a social or relational standpoint. This would partly explain why rates of physical violence for women of foreign origin might sometimes be very high: 10% of women from Sub-Saharan Africa and 6% of second generation Algerian or Moroccan women. These women are younger, they have an urban lifestyle; they use public means of transport a great deal or walk and are often alone when travelling. There is also no doubt because they are subjected to a certain amount of discrimination that women of Sub-Saharan African origin are subjected to insults three times more often than in the whole of the surveyed population and Northern or Eastern European women, almost twice as often. Among second generation immigrant women, those whose parents are of Algerian or Moroccan origin – almost one out of five women – reported verbal aggressions.

Physical assault is by far the least common form of violence reported in public space. Rather, it would appear that a combination of different types of harassment (verbal abuse and insults, being insulted, subjected to indecent exposure or receiving unwanted sexual advances) represent a real threat and restrict women's freedom of movement. All these forms of violence are more frequent in large urban areas and cities and the results of the survey demonstrate the sexist atmosphere prevailing in such places.

## 4. Concluding remarks

4.1 What we have attempted to measure:

1) the scale of violence and diversity of social contexts (the fact that most information emanated from studies of women contacting crisis centres, refuges or crisis lines led to the impression of a marginal phenomenon; marginal in the sense that relatively few women are assumed to be affected and also in the sense that women often have a marginal status in society);
2) prevalence rates in a given period, violent situations (repeated nature of acts ...);
3) the relationship between acts of violence and characteristics of women, attempting to go beyond the notion of vulnerability of individual women and to think more in terms of social contexts and socio-demographic factors at risk of being exposed to violence;
4) the extent and forms of psychological violence.

4.2 How can this data assist in the prevention of violence:

1) first and foremost, in raising public awareness;
2) in encouraging women to report violence from the outset.

Great care was taken when the first results were announced to the press to present the guiding methodological principals. We emphasised that the survey covered only women in ordinary households, that it was conducted by telephone, that the word violence was not used in the questionnaire and so forth. The construction of the intimate violence indicator was also explained, stressing the role of emotional abuse and psychological violence. The rate of 9,5% of women living in a relationship having reported being victims of such violence at the hands of their partner became 'one in ten women', a powerful figure for requesting improvement in support systems for women in such situations and changes in legislation. Using extrapolated figures from the general population, we were able to emphasise the scale of certain types of violence reported. Many associations (domestic violence refuges, rape victim support etc) were glad to have issues brought to public attention and some national figures within which they can situate their own experience, whilst some considered the survey results to heavily underestimate the number of women subjected to violence.

In terms of comparability with other surveys, the French survey has contributed to the study of violence against women in the European context. Even though the methodologies and instruments differ from one survey to another, an increasing number of countries now have national or regional level data enabling comparisons of several indicators of violence (Hagemann-White 2001). In terms of intimate partner violence, it would seem that the French study found similar levels of psychological, physical or sexual violence over the previous twelve months to those found in Swiss and Portuguese surveys, again with a substantial increase in rates in the past. However, in contrast to the Swiss survey results, the French study revealed significant differences between age groups and social categories. In cross-referencing one national context to another, we must of course avoid comparing rates of violence. Instead we should examine forms of violence, the relationship between biographical and social characteristics, and obstacles to reporting of violent acts comparing mechanisms and relationships in different contexts.

## 5. References


Brown E, Jaspard M (2001). Violences conjugales, premiers résultats de l'enquête Enveff. Ecole des Parents 2, 36-38.

Cesur-Gers (1998). Les violences conjugales en France: des démarches, des recours, des parcours. IHESI (Institut des hautes études sur la sécurité intérieure), Paris.

Collectif féministe contre le viol (1996-1997). Statistiques générales, 1996-1997. Viols Femmes Informations, 56.

Cromer S (1995). Le harcèlement sexuel en France: la levée d'un tabou, 1985-1990. La Documentation Française, Paris, 228.

Gillioz L, De Puy J, Ducret V (1997). Domination et violence envers la femme dans le couple. Editions Payot Lausanne, Lausanne, 269.

Hagemann-White C (2001). European research on the prevalence of violence against women. Violence against Women 7, 732-759.

Hanmer J, Saunders S. Well-Founded Fear: A Community Study of Violence to Women. Hutchinson, London, 1984.

Heiskanen M, Piispa M (1998). Faith, hope and battering. A survey of mens' violence in Finland. Statistics Finland, Helsinki, 64.

Jaspard M, the Enveff team (2001). Nommer et compter les violences envers les femmes: une première enquête nationale en France. Population et Sociétés 364, INED, Paris, (www.ined.fr/englishversion/publications/pop_et_soc/).

Jaspard M and the Enveff team (2003). Les violences envers les femmes en France: une enquête nationale. La Documentation Française Paris, 310.

Pain R (1997). Social geographies of women's fear of crime. Transactions of the Institute of British Geographers 22, 231-245.

Population Reports (1999). Ending violence against women. Population Reports, vol XXVII (4), December.

Römkens R. Gewoon geweld? Omwang, aard, gevolgen en achtergronden van geweld tegen vrouwen in heteroseksuele relaties. Svets & Zeitlinger, Amsterdam, 1992.

Statistique Canada (1993). L'enquête sur la violence envers les femmes: faits saillants. Le Quotidien, n°11-001F, 10.

Straus MA, Gelles RJ (1986). Societal change and change in family violence from 1975 to 1985 as revealed by two national surveys. Journal of Marriage and the Family 48, 465-479.

Valentine G (1992). Images of danger: women's sources of information about the spatial distribution of male violence. Area 24, 22-29.

# Appendix I

*Structure of the questionnaire*

1. Social and demographic characteristics
- Sexual behaviour, history and fertility history
- Physical and mental health status

2. Measurement of violence experienced during the previous twelve months (in general, number of acts, characteristics of perpetrator
- work, or place of study for full-time students
- public space
- semi-public space (e.g. visit to doctor, private tuition)
- couple
- previous couple
- relationships within the family and close friends

More detailed questions are asked about the act considered most serious by the interviewee, e.g. the duration and location of the violence, reaction of the victim, the presence of witnesses, whether or not help was sought.

3. Measurement of violence experienced prior to the last twelve months
- physical violence since the age of 18 years
- sexual violence since childhood

The questionnaire closes with a couple of questions relating to specific types of violence of a sexist nature. The interviewee then is given the opportunity to comment. A telephone number for information on violence against women is provided.

# Appendix II

*Types of violence described in the questionnaire*

1. Physical violence
- Throwing an object at someone, grabbing hold of, pushing someone.
- Hitting, punching and other physical brutality.
- Attempting to strangle or kill.
- Death threats.
- Threatening with a knife or firearm.
- Preventing someone from leaving a room, family home, building, or abandoning someone on the roadside (from a vehicle)
- Preventing someone from entering the home, a room, etc.

2. Sexual violence
- Forced sex.
- Forced sexual behaviour.
- Unwanted physical, sexual contact.
- Leering, sexual advances.
- Exhibitionism.
- Forced showing of pornographic material, obscene comments.

3. Psychological violence
- Preventing someone from talking to or seeing family and friends.
- Speaking ill of family or friends.
- Criticising, ridiculing what a person does, her work.
- Criticising someone's appearance, her way of dress.
- Ignoring someone, refusing to talk to her.
- Blocking access to money (domestic context), to information (at work).
- Threats, intimidation, emotional blackmail.
- Shouting, insulting or humiliating remarks, giving of orders, coarse remarks.
- Damaging, breaking or throwing of objects belonging to a person.
- Malicious telephone calls with sexual connotations or otherwise.

# Appendix III

Questionnaire items exploring emotional abuse or psychological pressure in the respondent's relationship with her partner

During the last 12 months, did your spouse or partner:
/never/rarely/sometimes/often/regularly

1. prevent you from meeting or talking to friends or family members?
2. prevent you from talking to other men?
3. criticise, ridicule what you were doing?
4. make unpleasant remarks concerning your physical appearance?
5. impose certain clothes, hairstyles, or public behaviour on you?
6. did not take your opinion into account, expressed contempt or tried to tell you what to think? a/ at home b/ in public?
7. insist on knowing where and with whom you had been?
8. refuse to speak to you, refused point-blank to discuss anything?
9. refuse access to household funds for daily expenses?

## Notes

1 Enquête nationale sur les violences envers les femmes en France (National Survey on Violence against Women in France). The members of the research team who conducted the survey in 2000 are Maryse Jaspard, *survey director*, and Elizabeth Brown (Paris University Demography Institute), Stephanie Condon and Jean-Marie Firdion (Ined, National Institute for Demographic Study), Annik Houel (Université de Lyon II) ; Dominique Fougeyrollas-Schwebel, Brigitte Lhomond, Florence Maillochon, and Marie-Ange Schiltz, Cnrs (National Research Institute); Marie-Josèphe Saurel-Cubizolles (Inserm, National Institute for Health and Medical Research).

2 Acts measured are the following:
   - Physical acts of violence: slaps, blows with a blunt instrument, threats made with a weapon, attempts to strangle or kill, other brutal physical acts, locking up or preventing from going out, abandoning on the roadside (from a car), preventing from entering the home.
   - Sexual violence: rape, unwanted sexual acts, unwanted petting, forced sexual acts with other people; at work, sexual harassment, imposing pornographic images.
   - Spiteful acts using an object: throwing, tearing, destroying an object, something made by the other person (a prepared dish, object made by the other person, a piece of artwork …)
   - Obscene and threatening phone calls, with or without sexual connotations.
   - Verbal violence: insults, abusive remarks.

3 Paper by S. Condon forthcoming.

*Bernard Wallner, Office of Evaluation, University of Vienna, Austria*

# Sexual Dimorphism: An Evolutionary View on Violent Behavior in Human Males

This report focuses on differences between the sexes with respect to non-human primates and humans. In the following sections, sexual dimorphic structures of chromosomes and steroid hormones are described, leading up to brain and body morphology. Furthermore, it sheds light on the dominance and aggressive behavior within primate societies and goes into detail concerning the motivational factors of violent male behavior against females and its physiological determinants. In addition, the question is asked, are previously mentioned distinctive sexual features related to male violence against females in primates? As a conclusion, an evolutionary hypothesis of human male violence is proposed. This hypothesis centers on the relationship between cholesterol intake and the activity of the neurotransmitter serotonin in men. Because, low cholesterol dietary intake reduces concentrations of serotonin titers in the brain, resulting in increased anxiety and violent behavior. Due to this men could react more sensitive to cholesterol shortage in the food compared to women, since in men the serotonin-sensitive brain areas are evolutionary forced to be bigger. The approach used for that complex theme is carried out from the field of behavioral endocrinology.

## 1. Dimorphism in chromosomes

In primates, the chromosomal sex corresponds to the typical mammalian expression – homogametic in females (XX) and heterogametic in males (XY). The chromosomal sex is essential to initiate the developmental gonadal sex differentiation in individuals. The major difference between the sexes is located at the functional sex-determining region (SRY) of the Y chromosome. The SRY gene is expressed in the indifferent gonads and it encodes the testis determination factor. This protein binds to a specific hormone response element in the promoter region of other genes. The products of these genes develop the indifferent embryonic gonads into testes. In females, where the SRY region is missing, the gonadal differentiation leads to the formation of the ovaries. Once the gonadal differentiation occurred, the development of sexual dimorphism is not directly driven by the presence or absence of the Y chromosome. It is rather linked to endocrine actions (e.g., the steroid hormone testosterone and the peptide hormone Mullerian Regression Factor). These molecules generate masculine neural structures and the entire body, since the ovaries produce scarce amounts of the steroid hormone estrogen.

## 2. Dimorphism in sex steroids

During adulthood, the sexes secrete different amounts of sex steroids. Females produce higher concentrations of gestagens and estrogens, whereas males produce more androgens. These steroids are responsible for the ontogenetic differentiation of sex characters and can alter the expression of behavior in males and females. However, a closer look at the acting mechanism of the steroids estrogens and androgens gives important insights into their functional effects. Ontogenetically, nature forces the "feminine" on the beginning of each individual development, irrespectively of the genetic sex. Exemplary, the morphological development of the hypothalamic brain structure is discussed.

Normally, pre-natal secreted gonadal androgens have organizational effects in specific hypothalamic areas. Here, the sexual dimorphic preoptic area (SND-POA) and the interstitial nuclei of anterior hypothalamus (INAH) are of major interest. In males, these nuclei increase in size under the influence of androgens. Androgen induced receptor fields are established during specific sensitive periods. In adults, these fields are activated via testosterone originating from mature testes. Functionally, the preoptic area is involved in the generation of sexual behavior. The inhibition of androgens during a critical ontogenetic phase results in a typical morphological female hypothalamus and in female sexual behavior during adulthood. The adult brain of such treated males show organizational effects comparable to females. Even within intact developed males, the adult brain is never completely without control from a "physiological female" component. After the androgen testosterone is bound on specific receptors in the hypothalamus, it must be converted into the estrogen estradiol to maintain characteristic sexual male behavior.

## 3. Dimorphism in brain morphology

In the past, a controversial discussion was carried out on sexual dimorphism in brain structures. The SND-POA area was found to be significantly larger in the male rat brain than in the female (Gorski et al. 1978). Using morphometric methods, Swaab and Hofman (1988) found a sexual dimorphism of the SND-POA between young adult men and women. However, this difference seems to diminish with age fluctuations. In men, the SND-POA cell numbers decline steeply between the ages of 50 and 60 years and in women gradually from the age of 50 years (Hofman and Swaab 1989). Allen et al. (1989) reported significantly larger INAH-2 and INAH-3 areas in men than in women. LeVay's (1991) report on a smaller INAH-3 area in homosexual men compared to heterosexual men generated a vehement debate in the public. However, the most striking dimorphism is observed in the total brain weight between men and women. In samples, encountered total brain size is significantly larger in males than in females. Even the total volumes of the rhombencephalon, the cerebellum and the ventricles are larger in males. Although, when comparing the weight of these structures relative to total brain weight, the difference between males and females becomes insignificant (Holloway et al. 1993).

Other examples of size differences are structures associated with language (planum temporale, dorsolateral prefrontal cortex, superior temporal gyrus) and structures connecting the hemispheres (corpus callosum, anterior commissure, massa intermedia of thalamus). These structures are larger in human females than human males.

## 4. Dimorphism in body morphology

Some species of non-human primates and even humans show sexual dimorphic difference in body size. In non-human primates, these differences are developed in multi-male and multi-female societies and are interpreted as an evolutionary result of sexual selection (Darwin 1871). In non-human primates, both sexes exhibit the same anatomical features, but males are twice as large as females because of more muscle mass and extra bones. The sexes also differ in developed secondary sex characteristics. Prominent examples are the extended canini in males and the conspicuous anogenital swellings in females. A pronounced difference in secondary sex characters is also obvious in humans (indicating a polygyneous evolutionary past of *Homo sapiens*). However, the evolutionary constraints for these traits remain sometimes unclear in humans. Important sexual dimorphic traits in humans are: males are larger and more violent than human females, the latter become sexually mature earlier and differ in their fat distribution; males have higher mortality rates, more rapid senescence and shorter life spans.

Why do these dimorphic characters exist? Darwin's theory of sexual selection predicts that in polygynous societies, males have to compete to get access to females and that females choose males as mating partners. Therefore, this process forces males to be large, strong and aggressive to enhance their reproductive success. This historical view of a role-playing game between males and females has changed in the current field of primatology, because many studies during the last one and half decade showed that both sexes use intra-sexual competition and inter-sexual choice to reproduce optimally.

## 5. Intra-group behavior and dominance

Behaviorally, one has to point out that in non-human primates, both sexes show mating competition among their own sex during reproductive and non-reproductive phases. Females use dominance and apparent aggressive behavior to be successful mothers. Both dominance and aggression are useful to get access to energetic resources to maintain the physiological process of reproduction (cycle, gestation, lactation, and rearing offspring). Males use dominance and high frequencies of aggressive behavior to be successful fertilizers. For this and their body strength, they invest most of their energy budget.

Many studies in primatology refer to inter-sexual aggression as the basic prediction of male dominance over females from the sexual selection theory. Again, according to this Darwinian theory, evolution forces a sexual dimorphism between males and females resulting in larger males, in which males equipped with weapons are more dangerous. These typical traits are a result of male mating competition for females.

To summarize, sexual selection creates morphological features in individuals and aggressive encounters, mainly between members of the same sex to enhance intrasexual competitive mating behavior. Furthermore, it is obvious that evolution acts differently on males and females, resulting in different physiological needs, but forces both sexes to reproduce optimally. However, frequencies of inter-sexual violence are, *per se*, low in non-human primates. Sexually active phases may represent an exception in terms of inter-sexual aggression (see below).

## 6. Motivation of male violent behavior

In many non-human primate species, females receive significantly more aggression from males during estrus periods (Smuts 1992). In other words, primates' male-female relationships are ultimately a result of sexual selection and can include sexual conflict.

Male sexual coercion appears in many primate species. Richard Wrangham and Dale Peterson (1996) suggest in their book *Demonic Males*, that male violence against females, as a behavioral strategy, seems to take place in higher primates, including humans. The authors describe raping behavior in orangutans and battering behavior in chimpanzees. In chimpanzees, males can display very brutal behavior against females when they start to display sexual attractiveness and receptivity. Males force them to copulate via physical attacks. In such encounters, females are often injured. The authors describe in the chapter *Relationship Violence* (p. 127) three parallels in chimpanzee and human battering behavior: (a) it is a predominantly male against female violence, (b) it is a relationship violence and (c) the underlying issue is domination or control. Furthermore, they indicate that a high rank position of females could lower male violence. In fact, bonobo or pygmy chimpanzee (the closest relatives to the common chimpanzee) and some lemur societies are characterized by female dominance accompanied with less male aggression (Smuts 1992; Kappeler 1993). For the bonobo, it is suggested that stronger female-female bonds could prevent them from male violence. Because of these bonds, females can ally themselves against aggressive males.

Generally, male dominance in monkeys and apes along with violence against females represents a by-product of male intra-sexual reproductive competition for females. Even some superficially interpreted unprovoked aggression of male chimpanzees against females during non-reproductive phases seems to be linked to that topic. However, the distinction within humans whether male violence is sexually orientated or not, seems to be sometimes impossible.

David Barash, an evolutionary psychologist, writes in the magazine Chronicle of Higher Education (2002), the maleness (in humans) of violence is so overwhelming that it is rarely even noticed; it is the ocean in which we swim. He continues, whenever seemingly unprovoked and deadly shootings occur in homes and workplaces, men are typically the mass murderers. ... violence may or may not be as American as cherry pie, but it is as male as can be.

In April 1994, the Bureau of Justice Statistics presented selected findings about violent crimes in the USA. They detected a clear sexual difference in who commits violent crimes. 86% of the offenders were males and nearly a quarter of violent crimes occur either in the home or in school. Men are disproportionately the perpetrators and victims of their own violence (40 per 1,000 for men; 25 per 1,000 for women). Further, the crime statistics for England and Wales (1992) found that 81 % of the convictions for indictable crime were for males.

The field of conflict research on humans specifies two major trends in explaining the motivational drive in the application of aggressive behavior between partners. Family scientists argue that conflicts arise from everyday frustrations and stresses of living together and both sexes are approximately equally involved in showing physical aggression. Whereas, feminist and evolutionary researchers view relationship aggression as arising from patriarchy or from proprietary drives of men that underlie the imbalance between male perpetrators and female victims (Archer 2002). In his meta-analyses, Archer found sexual differences in physically aggressive acts between heterosexual partners. Men were more likely to beat up, choke or strangle than were women. Whereas, women used acts such as kicking, biting, punching and attacked using objects. Nevertheless, the explanation of the motivational determinants of partner conflicts in humans is still hard to define. From the evolutionary perspective, different reproductive strategies between men and women seem to be plausible. On the other hand, this view narrows the understanding of such a complex problem. Conflicts in humans can arise from a diverse array of different individual interests, for example, resource and time management and general questions of investment.

An important issue in that context seems to be the general and individual physiological regulation of conflict or aggressive behavior.

## 7. Physiological components of violence

A focus on the physiological origin of dominance and aggressive behavior could give a new explanation for aggressive encounters and their outcome between the sexes in a non-sexual context. Over decades, the steroid testosterone was believed to be the key hormone in modulating male and female aggression. The recent literature reviews indicate that those testosterone concentrations can, but need not be the main influence. There is good evidence that in both humans and non-human primates, low central serotonergic activity is linked to increased aggression or impulsive behavior. Men who carry out impulsive, unplanned violence have decreased 5-HIAA concentrations (5-hydroxyindoleacetic acid is a main metabolite from the neurotransmitter serotonin measured in the cerebrospinal fluid (CSF)) compared to the populations' average (Virkkunen et al.1994a; Virkkunen et al. 1994b). In non-human primates, CSF titers of 5-HIAA were negatively correlated with excessive aggressive behavior (Raleigh and McGuire 1994). Whereas, increased testosterone levels are not automatically related to impulsivity, moreover, it is related to an overall aggressive motivation resulting in social dominance. (Archer 1991; Higley et al. 1996). The latter one can be viewed as one of the main forces for sexual orientated aggression during reproductive phases or as a tool to gain reproductive success.

A collective of serotonin producing neurons are located at the brain stem (raphe nuclei complex at the rhombencephalon; hence, this structure is larger in men than in women). Axons of these serotonergic neurons project in almost all brain regions, communicating with many receptor areas in the brain. One important finding was that cholesterol (a substantial energy component in dietary intake) could affect the binding capacity of serotonin to brain synaptic membranes. Cholesterol affects the fluidity of brain cell membranes, their permeability, and exchange processes. With low cholesterol levels, the cellular membrane fluidity increases and the amount of specific serotonin receptors are decreased in the synaptic cleft (Heron et al 1980; see Almeida-Montes et al. 2000). Because of this information, Jay R. Kaplan and co-workers worked out a series of experiments with crab-eating macaques. In their initial investigation (1991), over a period of two years, they fed 30 males with high or low saturated fat and cholesterol. The main result was that monkeys fed on low-fat diets exhibited more unritualized aggression involving physical contact. An experiment on juvenile females and males of the same species fed with low or high cholesterol diet showed decreased 5-HIAA CSF titers in low cholesterol groups (Kaplan et al. 1996) Because of these results, Kaplan et al. (1997) proposed the cholesterol- serotonin hypothesis of aggression. According to this hypothesis, low cholesterol dietary intake leads to depressed central serotonergic activity resulting in increased aggressive behavior.

## 8. An evolutionary linked hypothesis: cholesterol – serotonin interaction determines violent behavior in human males

About nutrition and the brain, The International College of Applied Kinesiology states in one of its Internet publications: The brain's 200 billion cells have numerous nutritional requirements for good function. These include not only proteins and carbohydrates (amino acids and glucose), but also a number of different vitamins, minerals and water. Fats are also important for good brain function, especially cholesterol (the brain contains more cholesterol than any other area of the body). Any dietary inadequacy can potentially have a dramatic impact on brain function. Numerous symptoms have been associated with a faulty diet's influence on the brain. These include aggression, learning disabilities, crime, depression, hyperactivity and various behavioral problems.

Based on the total brain weight difference between men and women, I hypothesize that:

First, there is a total volume difference of the projection areas of serotonergic axons between the sexes.

Second, therefore, men need more total cholesterol dietary intake to maintain their brain function than women. This could lead to a higher saturation threshold for cholesterol inclusion in the brain membranes of men and make their brains more sensitive for cholesterol fluctuation in the diet than women's.

This suggestion seems to have also evolutionary plausibility. Erickson (1997) points out in his hypotheses entitled: *Lowered Serum Cholesterol, Famine and Aggression: A Darwinian Hypothesis*, that food privation could have lead in primitive man to aggressive behavior resulting in increased risk-taking behavior including hunting. According to that, I suggest:

Third, such a scenario could be selectively advantageous for our male ancestors, because the brain serotonin physiology could act as an alert system for a food shortage. Low central serotonergic activity could represent an intrinsic motivational factor resulting in human male aggressive hunting behavior to require energetic fuels for their mates and offspring.

Forth, human males have a higher tendency to exhibit impulsive and aggressive behavior than human females, because their evolutionarily increased total brain size parallels with increased total cholesterol dietary intake. Compared to other males, these individuals may have represented more reliable reproductive partners for females, because of their earlier (in comparison to females) established cerebral alert system for shortages of essential food items.

## 9. Final Comments

For biologists, the temptation to oversimplify aspects of gender conflicts with a few biological theories or hypotheses is sometimes undoubtedly given. However, the view back to our own history as well as apart to our nearest companions, the non-human primates, provides in some cases a better basic understanding of certain problems. Some hypotheses or theories, irrespective from what scientific field, might either fulfill a lacking part of the evolutionary puzzle or contribute proximate solutions. Complex phenomena, like gender conflict may have multi-dimensional causal origins. These causes can result from both ontogenetic and evolutionary roots, expressed in individual behavioral and physiological strategies exhibited by adults. However, the author suggests that the mentioned hypothesis in this report provides an evolutionary plausible argument of an explanation of why human males are more aggressive than human females. Lenard (2002) writes in his article for the Life Enhancement magazine, ... the 100+ years of research on brain chemistry since Stevenson wrote his Victorian classic – *The Strange Case of Dr. Jekyll and Mr. Hyde* – have revealed that anything that interferes with the actions of serotonin in the brain can bring about a syndrome that resembles Jekyll's transformation to Hyde. While certainly less dramatic than the transformation described by Stevenson, serotonin deficiency bears a striking resemblance in various manifestations, as an increased tendency toward anxiety, depression, out-of-control disinhibition and violence. Conversely, enhancing the activity of the serotonin system may have exactly the opposite effects in many people. Given our current knowledge of neurochemistry, there can be little doubt that if Stevenson were writing today, Jekyll's transforming formula would have been a potent anti-serotonergic agent.

## 10. References


Allen LS, Hines M, Shryne JE, Gorski RA (1989). Two sexually dimorphic cell groups in the human brain. J Neurosci 9, 497-506.

Almeida-Montes LG, Valles-Sanchez V, Moreno-Aguilar J, Chavez-Balderas RA, Garcia-Marin JA, Cortes Sotres JF, Hheinze-Martin G (2000). Relation of serum cholesterol, lipid, serotonin and tryptophan levels to severity of depression and to suicide attempts. J Psychiatry Neurosci 25, 371-377.

Archer J (1991). The influence of testosterone on human aggression. Br J Psychol 82, 1-28.

Archer J (2002). Sex differences in physically aggressive acts between heterosexual partners: a meta-analytic review. Agg Viol Behav 7, 313-351.

Barash DP (2002). Evolution. Males, and Violence. In The Chronicle of Higher Education, issue of date May 24, 2002.

Darwin C. The Descent of Man, and Sexual Selection in Relation to Sex. Princeton University Press, Princeton, NJ, 1981. (Originally published 1871.)

Erickson MT (1997). Lowered serum cholesterol, famine and aggression: a Darwinian hypothesis. Biol Soc Life, 211-222.

Gorski RA, Gordon JH, Shryne JE, Southam AM (1978). Evidence for a morphological sex difference within the medial preoptic area of the rat brain. Brain Res 148, 333-346

Heron DS, Shinitzky M, Hershkowitz M, Samuel D (1980). Lipid fluidity markedly modulates the binding of serotonin to mouse brain membranes. Proc Natl Acad Sci 77, 7463-7467.

Higley JD, Mehlman PT, Poland RE, Taub DM, Vickers J, Suomi SJ, Linnoila M (1996). CSF testosterone and 5-HIAA correlate with different types of aggressive behaviors. Biol Psychiatry 40, 1067-1082.

Hofman MA, Swaab DF (1989). The sexually dimorphic nucleus of the preoptic area in the human brain: a comparative morphometric study. J Anat 164, 55-72.
Holloway RL, Anderson PJ, Defendini R, Harper C (1993). Sexual dimorphism of the human corpus callosum from three independent samples: relative size of the corpus callosum. Am. J. Phys Anthropol 92, 481-498.
Kaplan JR, Fontenot MB, Manuck SB, Muldoon MF (1996). Influence of dietary lipids on agonistic and affiliative behavior in *Macaca fascicularis*. Am J Primatol 38, 333- 347.
Kaplan JR, Potvin Klein K, Manuck SB (1997). Cholesterol meets Darwin: public health and evolutionary implications of the Cholesterol-Serotonin Hypothesis. Evol Anthropol 6, 28-37.
Kaplan JR, Manuck SB, Shively C (1991). The effects of fat and cholesterol on social behavior in monkeys. Psychosom Med 53, 634-642.
Kappeler PM (1993). Female Dominance in primates and other mammals. In Bateson PPG, Klopfer PH, Thomson NS (eds), Perspectives in Ethology, Vol. 10: Behaviour and Evolution, Plenum Press, New York, pp. 143-145.
Lenard L (2002). Reducing aggression and violence. The Serotonin connection. http://www.life-enhancement.com/article_template.asp?ID=208.
LeVay S (1991). A difference in hypothalamic structure between heterosexual and homosexual men. Science 253, 1034-1037.
Raleigh MJ, McGuire MT (1994). Serotonin, aggression, and violence in vervet monkeys. In Masters RD, McGuire MT (eds), The Neurotransmitter Revolution. Southern Illinois University Press, Carbondale, IL, pp. 185-197.
Smuts B (1992). Male Aggression against Women: An Evolutionary Perspective. Human Nature 3, 1-44.
Surveying crime: results from the 1992 British Crime Survey (1992). http://www.homeoffice.gov.uk/rds/pdfs2/r02.pdf
Swaab DF, Hofman MA (1988). Sexual differentiation of the human hypothalamus: ontogeny of the sexually dimorphic nucleus of the preoptic area. Dev. Brain Res. 44, 314-318.
U.S. Department of Justice. Office of Justice Programs. Bureau of Justice Statistics (1994). Violent Crime. http://www.ojp.usdoj.gov/bjs/.
Virkkunen M, Kallio E, Rawlings R, Tokola R, Poland RE, Guidotti A, Nemeroff C, Bissette G, Kalogeras K, Karonen SL, et al. (1994a). Personality profiles and state aggressiveness in Finnish alcoholic, violent offenders, fire setters, and healthy volunteers. Arch Gen Psychiatry 51, 28-33.
Virkkunen M, Rawlings R, Tokola R, Poland RE, Guidotti A, Nemeroff C, Bissette G, Kalogeras K, Karonen SL, Linnoila M (1994b). CSF biochemistries, glucose metabolism, and diurnal activity rhythms in alcoholic, violent offenders, fire setters, and healthy volunteers. Arch Gen Psychiatry 51, 20-27.
Wrangham R, Peterson D. Demonic Males. Houghton Mifflin Company, New York, 1996.

# Contributors

**Stefan Beckmann** holds a Masters Degree in education. For several years he has been working with youth and young boys, and taught at the Institute for Social Pedagogy, at the Technical University in Berlin. Current focus is on gender in work with youth and young boys, gender discourse in relation to socialization and hegemonic power relations.

**Stephanie Condon** trained in England in Languages and Human Geography. Since working on her PhD thesis on Italian immigrants in Lyon (France), her main research interest has been in the migration process, with a focus on gender and the life course, with particular attention to Caribbean migration to France and Britain. She has a permanent research post at the Institut National d'Etudes Démographiques in Paris and contributed to the setting up of the gender and demography unit there. She is a member of the team who conducted the French national survey on violence against women (Enveff, 2000).

**Elfriede Fröschl** is lecturer at the Fachhochschule for social work in Vienna (Sociology, Theory of Social Work, Violence Against Women, Feminist Social Work) and a free-lance trainer for various professional target groups in the field of gender violence.

**Maria I. García-Linares** is a licensed psychologist at the University of Valencia, who works as a researcher in several projects on the impact of domestic violence on women's health in the Valencian Community of Spain.

**Carol Hagemann-White** is Professor of Educational Theory and Feminist Studies at the University of Osnabrück, Germany. In the late 1970s she conducted evaluation research in the first German shelter for battered women in Berlin and has been involved in policy-oriented research close to feminist projects ever since. From 1983 to 1993 she worked as founder and co-director of the Berlin Institute for Social Research and Sociological Practice. From 1985 to 1988 she was Professor of Women's Studies in Political Science at the Free University of Berlin, and from 1992 to 1997, parallel to her chair at the University of Osnabrück, she directed the Research Institute "Women in Society" in Hannover where she conducted state and federally funded research on Women's Organizations, Equality Policy, and Women's Health Care. Since 1996 she has been involved with evaluation research of inter-agency work to address violence against women in community intervention projects across Germany. She is widely published in English and German with over 120 publications to her credit. In 1998 she was awarded the Bank of Sweden Tercentenary Foundation's prize for Internationally Outstanding Research.

**Birgitt Haller** holds a Doctoral Degree in Law and a Masters Degree in Political Science and has been employed as a researcher at the Institute of Conflict Research (Vienna, Austria) since 1994. Main fields of research and publications: research on violence, gender research, Austrian political system.

**Lynne Harne** currently works as a Research Associate at the International Centre for the Study of Violence and Abuse, University of Sunderland. She has recently completed a thesis on Violent Fathering and Child Contact at the University of Bristol, School for Policy Studies. She teaches part-time on professional development courses on domestic violence at the University of Teesside and family and social policy courses at the University of Westminster. She has been active around issues of violence against women and children and family policy for over twenty years.

**Barbara Kavemann**, Ph.D., sociologist. Since 1978 research and teaching on gendered and sexualized violence including the evaluation of the first battered women's shelter in Germany and the first support center for sexually abused girls. Since 1998 Research Associate at the University of Osnabrück working on the evaluation of the intervention projects against domestic violence. Visiting professor at the Catholic University Berlin. Board member of the Association for the Prevention of Sexual Abuse of Girls and Boys.

**Renate Klein** earned a Ph.D. in Psychology from the University of Marburg, Germany. She teaches Family Studies and Women's Studies at the University of Maine, USA and coordinates the European Research Network on Conflict, Gender, and Violence.

**Hans-Joachim Lenz** is a Social Scientist and Gestalt Therapist. He has a private practice for gender research, consultation, and training near Nuremberg/Germany. In the early 1990s he conducted one of the very first adult education and training courses in gender sensitization for men. Numerous publications concerning the gender perspective of men, men's experience of violence, and gender relations. Established the groundwork for a pilot project he is currently working on to assess *Violence Towards Men*, funded by the German Ministry for the Family, Senior Citizens, Women and Youth.

**Beate Leopold**, sociologist. Since 1987 research on women and drug abuse, and on prostitution. Since 1989, evaluation of model projects in the areas of HIV/AIDS prevention, prostitution, and gendered violence.

**Manuela Martinez** is a Titular Professor of Psychobiology at the University of Valencia, Spain. She received her Ph.D. in Medicine in 1987. She has always done research and taught in the field of aggression. Since 1995 she has specialized in the impact that aggression has on the victims both in animal models and human beings. She is a project director of several studies funded by the Ministry of Work and Social Affairs (Institute of Women), and the Ministry of Science and Technology, which are being carried out in the Valencian Community of Spain on the impact of domestic violence on women's health. She is a member of the Nursing Network on Violence Against Women, International (NNAWVI) and a member of the Council of the International Society for Research on Aggression and organized the 14th World Meeting in 2000. In 2001 she edited a book entitled "Prevention and control of aggression and the impact on its victims".

**Cristina Negutu** is a lawyer in a private law office dealing with civil, family and commercial law issues. She graduated from the Faculty of Law in Bucharest in 1999, and attended a one-year post-graduate studies program organized at Bucharest University by Sorbonne University, on Business Law. She got a Master's Degree in 2002 and is currently involved in the Ph.D. stage within the Romanian Academy of Economic Studies. Cristina Negutu has joined the ENCGV since 2000 and presented at its 6th Conference the paper *Latest Regulations about Family Violence in Romania.*

**Maria A. Pico-Alfonso** is a licensed psychologist at the University of Valencia, who works as a psychologist in a shelter for women in Castellón. She collaborates in several projects on the impact of domestic violence on women's health in the Valencian Community of Spain.

**Heike Rabe**, lawyer. Work on family law and criminal law related to youth. Current focus is on the new German protection against violence law and its implications for child protection. She is currently training to become a mediator and family law specialist.

**Bernard Wallner**, studies of Behavioral Endocrinology and Anthropology at the University of Vienna. PhD at the University of Vienna. Research topics are: Quality Assessment in Higher Education; Physiological and Behavioral Aspects in Non-Human Primate Females During Puberty and Adulthood.

# Contact Information

**Stefan Beckmann**
Project WiBIG
University of Osnabrück, Heger-Tor-Wall 9, 49069 Osnabrück
Germany
E *wibig.berlin@web.de*, E *wibigos@uni-osnabrueck.de*

**Stephanie Condon**
Institut National d'Etudes Démographiques
133, Boulevard Davout, 75980 Paris cedex 20
France
E *condon@ined.fr*

**Elfriede Fröschl**
FH-Campus Wien
Grenzackerstrasse 18, 1100 Vienna
Austria
E *elfriede.froeschl@fh-campuswien.ac.at*

**Maria I. García-Linares**
University of Valencia
Avda. Blasco Ibañez, 21, 46010 Valencia
Spain
E *Maria.I.Garcia@uv.es*

**Carol Hagemann-White**
Allgemeine Pädagogik / Frauenforschung
University of Osnabrück, Heger-Tor-Wall 9, 49069 Osnabrück
Germany
E *chageman@uos.de*

**Birgitt Haller**
Institute of Conflict Research
Lisztstraße 3, 1030 Vienna
Austria
E *birgitt.haller@ikf.ac.at*

**Lynne Harne**
Department of Social and International Studies
University of Sunderland
Langham Tower, Ryhope Road, Sunderland SR2 7EE
UK
E *lynne.harne@sunderland.ac.uk*, E *lynneharne@waitrose.com*

**Barbara Kavemann**
Project WiBIG
University of Osnabrück, Heger-Tor-Wall 9, 49069 Osnabrück
Germany
E *wibig.berlin@web.de*, E *wibigos@uni-osnabrueck.de*

**Renate Klein**
College of Education and Human Development
University of Maine, 330A Merrill Hall
Orono, ME 04469
USA
E *rklein@maine.edu*

**Hans-Joachim Lenz**
Burgweg 33, 90542 Eckenhaid
Germany
E *hj-lenz@t-online.de*

**Beate Leopold**
Project WiBIG
University of Osnabrück, Heger-Tor-Wall 9, 49069 Osnabrück
Germany
E *wibig.berlin@web.de*, E *wibigos@uni-osnabrueck.de*

**Manuela Martinez**
University of Valencia
Avda. Blasco Ibañez, 21, 46010 Valencia
Spain
E *Manuela.Martinez@uv.es*

**Gabriele Moser**
Vice Rector
University of Vienna, Dr. Karl-Lueger-Ring 1, 1010 Vienna
Austria
E *gabriele.moser@univie.ac.at*

**Cristina Alexandra Negutu**
Attorney at Law in a Private Law Office
B-dul Ion Mihalache (fost 1 Mai) 42-52
Bl. 35, Sc. G, App. 235, Sector 1, Bucharest
Romania
E *cristina_negutu@yahoo.com*

**Maria A. Pico-Alfonso**
University of Valencia
Avda. Blasco Ibañez, 21, 46010 Valencia
Spain
E *angeles.pico@uv.es*

**Heike Rabe**
Project WiBIG
University of Osnabrück, Heger-Tor-Wall 9, 49069 Osnabrück
Germany
E *wibig.berlin@web.de*, E *wibigos@uni-osnabrueck.de*

**Bernard Wallner**
Office of Evaluation
University of Vienna, Liechtensteinstraße 22, 1090 Vienna
Austria
E *bernard.wallner@univie.ac.at*

**WAVE**
Birgit Appelt/Verena Kaselitz
Bacherplatz 10/4, 1050 Vienna
Austria
E *office@wave-network.org*